ENZYMOLOGY AND MOLECULAR BIOLOGY OF CARBONYL METABOLISM 3

ADVANCES IN EXPERIMENTAL MEDICINE AND BIOLOGY

Recent Volumes in this Series

ENZYMOLOGY AND MOLECULAR BIOLOGY OF CARBONYL METABOLISM 3

Edited by

Henry Weiner
Purdue University
West Lafayette, Indiana

Bendicht Wermuth
Inselspital
Bern, Switzerland

and

David W. Crabb
Indiana University School of Medicine and
    Veterans Administration Medical Center
Indianapolis, Indiana

PLENUM PRESS • NEW YORK AND LONDON

Proceedings of the Fifth International Workshop on Enzymology and
Molecular Biology of Carbonyl Metabolism, held June 12–15, 1990,
in West Lafayette, Indiana

ISBN-13: 978-1-4684-5903-6   e-ISBN-13: 978-1-4684-5901-2
DOI: 10.1007/ 978-1-4684-5901-2

PREFACE

The Fifth International Workshop on the Enzymology and Molecular Biology of Carbonyl Metabolism was held at Purdue University in June, 1990. This represents the fifth time that I had the privilege of organizing the scientific program. It was the first time that I actually hosted the meeting. I wish to salute my four previous co-organizers and the thousands of scientists who have hosted other meetings. It is much easier to arrange the scientific program and edit the proceedings. No local organization could occur without the help of ones research group and, in this case, my wife. I sincerely thank Esther and my research group for their advise and help.

At this Workshop, similar to the preceeding ones, much new information was presented. It was apparent how molecular biological techniques were influencing the direction of the research on the three families of enzymes discussed. It also was apparent that not all biochemical problems could be solved by using these techniques. Many of the presentations showed how important advances still could be made using more traditional biochemical approaches.

I wish to thank my two co-editors for their help in editing the manuscripts. I also wish once again to thank the National Institute on Alcohol Abuse and Alcoholism for providing some financial support. None of us organizing the first Workshop in 1982 thought of having a second, let alone a fifth Workshop. Now there are plans for the sixth to be held in Ireland in 1992. I invite scientists interested in attending this workshop to contact me.

HENRY WEINER
W. Lafayette, Indiana

CONTENTS

ALDEHYDE DEHYDROGENASE

ALDO-KETO REDUCTASE

ALCOHOL DEHYDROGENASE

ALDEHYDE DEHYDROGENASES: WHAT CAN BE LEARNED

FROM A BAKER'S DOZEN SEQUENCES?

Ronald Lindahl[1] and John Hempel[2]

[1]Department of Biochemistry and Molecular Biology
The University of South Dakota School of Medicine
Vermillion, South Dakota 57069

[2]Department of Biochemistry
The University of Pittsburgh School of Medicine
Pittsburgh, Pennsylvania 15261

Aldehyde dehydrogenases (E. C. 1.2.1.3, ALDH) exist in multiple molecular forms which differ in their physical and/or their functional properties (Weiner, 1979). Aldehyde dehydrogenase has been identified in virtually every organism and tissue examined. Distinct ALDHs have been identified in the mitochondrial, microsomal and cytosolic compartments of the cell (Tottmar et al., 1973; Greenfield and Pietruszko, 1977; Lindahl and Evces, 1984). Some forms are constitutive, some inducible (Deitrich, 1971; Deitrich et al., 1977). Tetrameric and dimeric functional forms are known. Some forms display broad substrate specificity, oxidizing a variety of aliphatic and aromatic aldehydes. Other forms possess much narrower substrate preferences, utilizing small aliphatic aldehydes. Kinetic studies have indicated that acetaldehyde derived from ethanol oxidation, medium chain length aliphatic aldehydes derived from membrane lipid peroxidation and perhaps some aldehydes generated from neurotransmitter metabolism are potential physiological substrates for one or more ALDH forms (Tank et al., 1981; Esterbauer, 1982; Weiner, 1982; Mitchell and Petersen, 1989). While all aldehyde dehydrogenases likely use NAD+ as coenzyme in vivo, some forms can utilize NADP+ in vitro.

Until quite recently, it has not been possible to establish the structural and functional relationships between the various aldehyde dehydrogenases. This has been due to the lack of primary structural information on a sufficient number of aldehyde dehydrogenases from different sources. In late 1988, primary sequences were known for five mammalian aldehyde dehydrogenases, the human and horse cytosolic and mitochondrial pairs and the rat tumor/TCDD form (Hempel, et al., 1984, 1985; von Bahr-Lindstrom et al., 1984; Johansson et al., 1988; Jones et al., 1988). Even with this limited data it was apparent that different classes of ALDHs could be identified (Anonymous, 1989). Class 1, a cytosolic form, Class 2, the mitochondrial forms and Class 3, a constitutive/inducible cytosolic form. Comparison of these structures also provided either confirmation or first identification of amino acid residues believed to be critical for proper enzyme function.

Since 1988 the primary structures of several additional aldehyde dehydrogenases from different species have been reported. The spectrum of sequences now ranges from a bacterium (Pseudomonas oleovorans, Kok et al., 1989) through lower eukaryotes (a fungus, Pickett et al., 1987; a yeast, Saigal and Weiner, 1990) to additional mammalian forms (Dunn et al., 1989; Farres et al., 1989; Guan and Weiner, 1990) and, most recently, a higher plant (Weretilnyk and Hanson, 1990). In all 13 complete or partial sequences are available at this writing. Some sequences have been determined by protein sequencing methods, some from their corresponding cDNAs and others determined by a combination of these methods.

Comparison of the available sequences indicate they vary in length from 452 residues for the Class 3 ALDH to 520 for the precursor form of the bovine Class 2 enzyme. Alignment of the sequences for maximal positional identity indicates that, with proper placement of gaps in certain sequences, all enzymes possess a common core region of approximately 430 amino acids extending from residues 57 to 500 (Class 1 and 2 numbering system). The rat Class 3 and microsomal ALDHS as well as the yeast and bacterial forms, have sequences extending 15 to 21 residues carboxyl to the core sequence. The Class 1, 2, phenobarbital and yeast enzymes have 56 residue extensions to the N-terminus of the core and the spinach and Pseudomonas forms similarly extend amino 44 and 48 amino acids. The Class 2 and spinach chloroplast enzymes also possess amino terminal signal sequences.

Based on positional identity, the 9 mammalian ALDH sequences can be placed into one of the three existing classes (Table 1). The human and horse cytosolic and rat phenobarbital-inducible ALDHs share greater that 85% identity and are all Class 1 enzymes. The four class 2 sequences, from human, horse, rat and cow all share greater than 95% identity. Interestingly, the Class 2 signal peptides share approximately only 70% identity. The Class 3 and rat constitutive microsomal ALDH share more than 77% identity and are members of the same class. It is not possible to assign the other sequences to one of the 3 existing classes. However, some clear associations are suggested by both positional identities and analysis of the sequence alignments. The Aspergillus ALDH is most closely related to the mammalian Class 1 and Class 2 forms, with almost 60% positional identity with these ALDHs vs 25% identity to Class 3. The Pseudomonas enzyme is slightly more closely related to the Class 3 forms than to any other group (45% identity vs 28% and 26% for Class 1 and 2 respectively). The spinach chloroplast enzyme is most closely related to the mammalian Class 1 and 2 and Aspergillus enzymes, sharing up to 42% identity. The Saccharomyces enzyme appears to equally distantly related to any of the other ALDHs.

Within the 430 residue common region, approximately 1300 amino acid substitutions have occurred in the 13 proteins. The amino acid changes are relatively uniformly distributed

2

along the molecule, with approximately one-third of the substitutions occurring in each third of the molecule. However, in maximally aligned sequences, the majority of the gaps occur in the N-terminal third of the sequence, making this portion of the molecule the least conserved.

Table I

Percent Positional Identity Among Aldehyde Dehydrogenase Core Regions

	1	2	3	PB	M	A	S	Ps	S
Class 1	--								
Class 2	68	--							
Class 3	30	27	--						
Phenobarbital	85	62	27	--					
Microsomal	31	30	77	30	--				
Aspergillus	56	57	25	57	31	--			
Saccharomyces	30	30	25	29	26	28	--		
Pseudomonas	28	26	42	30	47	27	21	--	
Spinach	42	38	27	36	29	42	21	31	--

When functionally conservative amino acid substitutions or substitutions arising by single base pair changes are included, the relationships between the classes do not change significantly (approximately 10% increase in relatedness between any 2 enzymes), except in the case of the Class 3 --Pseudomonas enzymes, which become 76% identical. Within classes, the similarities increase to greater than 95% for the Class 1--phenobarbital-inducible and 92% for the Class 3--microsomal forms.

Six regions of strong sequence conservation can be identified in the 430 residue long core region. Four regions are conserved among all 13 sequences. The fifth is present in all but the Saccharomyces enzyme. An additional conserved region is present in those enzymes with amino terminal extensions. Within all these regions, there are occasional residues which differ in one or more enzymes, but the majority of the changes are either conservative substitutions or the result of single base pair changes. Included in the conserved regions are many of the residues identified by chemical modification or mutational analysis to be important in enzyme function (Hempel et al., 1982; Hempel et al., 1985; Hempel et al., 1989; Abriola et al., 1987). Glu 268 and Cys 302 both occur within larger conserved regions as does Gly 245.

Several residues previously believed to be important in ALDH function are not conserved. These include Cys 49 (Tu and Weiner, 1988a and b), which is within the 57 residue

3

amino region not present in the Class 3 and Pseudomonas
enzymes, Cys 162 (Tu and Weiner, 1988a and b) which is a Val
in the Class 3 and spinach enzymes and an Ala in
Saccharomyces. Also not conserved is Serine 74 (Loomes et
al., 1990). The corresponding residue in Spinach and
Saccharomyces is Arg. In the Class 3 enzyme it is Lys and
this residue is missing in the Aspergillus enzyme. The
glutamic acid at 487 (Yoshida et al., 1984), the residue
involved in the "flushing reaction" is also not con-served.
This residue is a His in the Class 3 and Pseudomonas ALDHs,
an Asn in spinach and a Gln in Aspergillus.

Analysis of the primary structures of 13 aldehyde
dehydrogenases from diverse phylogenetic sources allows
several observations to be made. First, the adenine
dinucleotide-dependent oxidation of large variety of aldehyde
substrates can be performed by a large number of different
primary sequences. Based on positional identity and residue
substitution patterns these primary structures can be
organized into several groups (Figure 1). The mammalian
Class 1 and 2 ALDHs appear to be the most recently diverged
forms. The Aspergillus ALDH is surprisingly closely related
to the mammalian Class 1/2 group and the Class 1,2 and
Aspergillus forms could be considered as one superclass. The
Spinach ALDH enzyme also is most closely related to the Class
1/2 -Aspergillus group. The Class 3 and Pseudomonas enzymes
may be considered to form a second major superclass. The
yeast enzyme would appear to be a separate, distantly related
form.

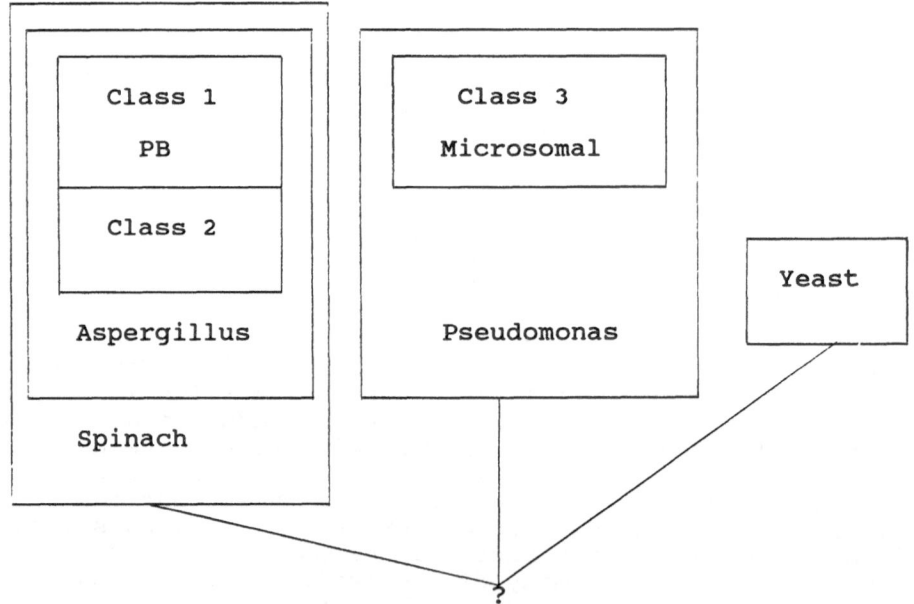

Figure 1 Relationships Between Aldehyde Dehydrogenases

Second, within these 13 structures are certain common
conserved sequences and residues. The strict conservation of
these residues across the broad phylogenetic spectrum
encompassed by the available sequence data indicates that

these regions and residues are critical to enzyme function and/or structure (Bowie et al., 1990). It now appears that six regions of the primary structure are essential for proper enzyme function. These include areas previously suggested to be involved in the active site, the coenzyme binding regions and regions of subunit interaction (Hempel et al., 1987, 1989). For example, it is interesting that the Gly-X-Gly-X-X-Gly motif reportedly characteristic of the coenzyme binding domain of other dehydrogenases is not present in any of the aldehyde dehydrogenases, but that the region most closely approximating this sequence, from residues 242 to 250, is very highly conserved and likely represent the coenzyme binding domain of aldehyde dehydrogenase.

Third, the lack of conservation of certain residues implicated from other studies as being important to enzyme function is also informative. For example, although by mutational analysis the Glu at 487 is critically important for enzyme function (Yoshida et al., 1984), it is not strictly conserved. Those forms which possess a Glu in this position appear to function as tetramers (Hempel et al., 1984; von Bahr-Lindstrom et al., 1984; Hempel et al., 1985; Johansson et al., 1988). Of those that have another residue at position 487 two function as dimers, the Class 3 (Jones et al., 1988), and spinach (Weretilnyk and Hanson, 1990) enzymes. However, molecular weight estimations of the microsomal ALDH and the Pseudomonas enzymes (His at 487) indicate they function as tetramers (Kok et al., 1989). Similarly, Ser 74 has been implicated as a critical residue (Loomes et al., 1990) and is conserved in the Class 1, 2 and Pseudomonas enzymes. However, it is not present in the remaining forms. In three of the enzymes, the corresponding residue is Lys or Arg and a Gly in the fourth. Perhaps Glu 487 and Ser 74 are examples of residues which provide the more subtle functional differences among the various ALDHs.

Analysis of the primary sequences of a large number of related sequences of any protein provides a wealth of information on which to base future work. The aldehyde dehydrogenases are no exception. The cumulative primary structure data will be useful in determining differences in secondary and tertiary structure among the various forms once the 3-dimensional structure of an ALDH has been solved. Work in this area is ongoing in several laboratories. Identification of conserved residues and the patterns of permitted residue substitutions provide much insight into the functional significance of each such residue. Such residues also provide targets for systematic site-directed mutagenesis experiments. Again, such studies are ongoing with the intention of defining the precise role of each critical residue in enzyme function. The rational construction of chimeric proteins using different residue segments from different enzyme forms also promises to provide much information related to functionally critical regions of the protein. In all, deciphering the functional and structural messages in the aldehyde dehydrogenase primary sequences now seems a real possibility.

Acknowledgements

Supported by grants CA21103 to R.L. and AA06985 to J.H.

5

References

Abriola,D. P., Fields, R. Stein, S. MacKerell, A. D. and
    Peitruszko, R. 1987 Active site of human liver aldehyde
    dehydrogenase. Biochemistry 26:5679-5684.
Anonymous 1989 Nomenclature of mammalian aldehyde
    dehydrogenases. In Progress in Clinical and Biological
    Research, vol. 290, H. Weiner and T. G. Flynn, eds.
    Alan. R. Liss, New York, pp. xix-xxi.
Bowie, J. U., Reidhaar-Olson, J. F., Lim, W. A. and Sauer, R.
    T. 1990 Deciphering the message in protein sequences:
    tolerance to amino acid substitutions. Science
    247:1306-1310.
Deitrich, R. A. 1971 Genetic aspects of increase in rat
    liver aldehyde dehydrogenase by phenobarbital. Science
    173:334-336.
Deitrich, R. A., Bludeau, P., Stock, T. and Roper, M. 1977
    Induction of different supernatant aldehyde
    dehydrogenases by phenobarbital and
    tetrachlorodibenzo-p-dioxin. J. Biol. Chem.
    252:6169-6176.
Dunn, T. J., Koleske, A. J., Lindahl, R. and Pitot, H. C.
    1989 Phenobarbital-inducible aldehyde dehydrogenase in
    the rat. cDNA sequence and regulation of the mRNA by
    phenobarbital in responsive rats. J. Biol. Chem. 264:
    13057-13065.
Esterbauer, H. 1982 Aldehydic products of lipid
    peroxidation. In Free Radicals, Lipid Peroxidation and
    Cancer, D. C. H. McBrien and T. F. Slater, eds., Academic
    Press, New york, pp 101-128.
Farres, J., Guan, K.-L. and Weiner, H. 1989 Primary
    structures of rat and bovine liver mitochondrial aldehyde
    dehydrogenases deduced from cDNA sequences. Eur. J.
    Biochem. 180:67-74.
Greenfield, N. J. and Pietruszko, R. 1977 Two aldehyde
    dehydrogenases from human liver: isolation via affinity
    chromatography and characterization of the isozymes.
    Biochim. Biophys. Acta 483:35-45.
Guan, K. and Weiner, H. 1990 Sequence of the precursor of
    bovine liver mitochondrial aldehyde dehydrogenase as
    determined from its cDNA and its functionality. Arch.
    Biochem. Biophys. 277:351-360.
Hempel, J., Harper, K. and Lindahl, R. 1989 Inducible (Class
    3) aldehyde dehydrogenase from rat hepatocellular cinoma
    and 2,3,7,8-tetrachlorodibenzo -p-dioxin-treated liver.
    Distant relationship to the Class 1 and 2 enzymes from
    mammalian liver cytosol/mitochondria. Biochemistry
    28:1160-1167.
Hempel, J., Jornvall, H. and Vallee, B. 1987 Structures of
    human alcohol and aldehyde dehydrogenases. Enzyme
    37:5-18.
Hempel, J., Kaiser, R. and Jornvall, H. 1985 Mitochondrial
    aldehyde dehydrogenase from human liver. Primary
    structure, differences in relation to the cytosolic
    enzyme and functional correlations. Eur. J. Biochem.
    153:13-28.
Hempel, J., Pietruszko, R., Fietzek, P. and Jornvall, H.
    1982 Identification of a segment containing a reactive
    thiol in cytoplasmic human liver aldehyde dehydrogenase
    (isoenzyme E1). Biochemistry 21:6834-6838.

Hempel, J., von Bahr-Lindstrom, H. and Jornvall, H. 1984
    Aldehyde dehydrogenase from human liver. Primary
    structure of the cytoplasmic isoenzyme. Eur. J.
    Biochem. Biochem. 141:21-35.

Johansson, J., von Bahr-Lindstrom, H., Jeck, R., Woenckhaus,
    C. and Jornvall, H. 1988 Mitochondrial aldehyde
    dehydrogenase from horse liver. Correlations of the same
    species variants for both the cytosolic and the
    mitochondrial forms of an enzyme. Eur. J. Biochem.
    172:527-533.

Jones, D. E., Brennan, M. D., Hempel, J. and Lindahl, R.
    1988 Cloning and complete nucleotide sequence of a
    full-length cDNA encoding a catalytically functional
    tumor-associated aldehyde dehydrogenase. Proc. Natl.
    Acad. Sci. 85;1782-1786.

Kok, M., Oldenhuis, R., van der Linden, M. P. G., Meulenberg,
    C. H. C., Kingma, J. and B. Witholt. 1989 The Pseudomonas
    oleovorans alkBAC operon encodes two structurallly
    related rubredoxins and an aldehyde dehydrogenase. J.
    Biol. Chem. 264:5442-5451.

Lindahl, R. and Evces, S. 1984 Comparative subcellular
    distribution of aldehyde dehydrogenase in rat, mouse and
    rabbit liver. Biochem. Pharmacol. 33:3383-3389.

Loomes, K. M., Midwinter, G. G., Blackwell, L. F. and
    Buckley, P. D. 1990 Evidence for reactivity of serine-
    74 with trans-4-(N,N-dimethylamino) cinnamaldehyde during
    oxidation by the cytoplasmic aldehyde dehydrogenase from
    sheep liver. Biochemistry 29:2070-2075.

Mitchell, D. Y. and Petersen, D. R. 1989 Oxidation of
    aldehydic products of lipid peroxidation by rat liver
    microsomal aldehyde dehydrogenase. Arch. Biochem.
    Biophys. 269:11-17.

Pickett, M., Gwynne, D. I., Buxton, F. P., Elliot, R.,
    Davies, R. W., Lockington, R. A., Scazzocchio, C. and
    Sealy-Lewis, H. M. 1987 Cloning and characterization of
    the aldA gene of Aspergillus nidulans. Gene 51:217- 226.

Saigal, D. and Weiner, H., submitted for publication.

Tank, A. W., Weiner, H. and Thurman, J. A. 1981 Enzymology
    and subcellular localization of aldehyde oxidation in rat
    liver. Oxidation of 3,4-dihydroxy-phenylacetaldehyde
    derived from dopamine to 3,4-dihydroxyphenylacetic acid.
    Biochem. Pharmacol. 30:3265-3275.

Tottmar, S. O. C., Pettersson, H. and Kiessling, K. H. 1973
    The subcellular distribution and properties of aldehyde
    dehydrogenases in rat liver. Biochem. J. 135:577-586.

Tu, G.-C. and Weiner, H. 1988a Identification of the
    cysteine residue in the active site of horse liver
    mitochondrial aldehyde dehydrogenase. J. Biol. Chem.
    263:1212-1217.

Tu, G.-C. and Weiner H. 1988b Evidence for two distinct
    active sites on aldehyde dehydrogenase. J. Biol. Chem.
    263:1218-1222.

von Bahr-Lindstrom, H., Hempel, J. and Jornvall, H. 1984 The
    cytoplasmic isoenzyme of horse liver aldehyde
    dehydrogenase. Relationship to the corresponding human
    isoenzyme. Eur. J. Biochem. 141:37-42.

Weiner, H., 1979 Aldehyde dehydrogenase: mechansism of
    action and possible physiological role. In Biochemistry
    and Pharmacology of Ethanol, E. Majchrowicz and E. P.
    Noble eds., Alan R. Liss, New York, pp 107-124.

Weiner, H. 1982 Aldehyde dehydrogenase. In Progress in
    Clinical and Biological Research, vol. 114, H. Weiner and
    B. Wermuth, eds., Alan R. Liss, New York, pp 1-10.
Weretilnyk, E. A. and Hanson, A. D. 1990 Molecular cloning of
    a plant betaine-aldehyde dehydrogenase, an enzyme
    implicated in adaptation to salinity and drought. Proc.
    Natl. Acad. Sci. 87:2745-2749.
Yoshida, A., Huang, I.-Y., Ikawa, M. 1984 Molecular
    abnormality of an inactive aldehyde dehydrogenase variant
    commonly found in orientals. Proc. Natl. Acad. Sci.
    81:258-261.

RAT CLASS 3 ALDEHYDE DEHYDROGENASE: CRYSTALS AND PRELIMINARY ANALYSIS

John Hempel[1], John P. Rose[2], Ingrid Kuo[1], Ronald Lindahl[3] and
Bi-Cheng Wang[2]

[1]Department of Molecular Genetics and Biochemistry, University
of Pittsburgh School of Medicine, Pittsburgh, PA 15261
[2]Departments of Crystallography and Biological Sciences
University of Pittsburgh, Pittsburgh PA 15260, and [3]Department
of Biochemistry and Molecular Biology, University of South
Dakota, Vermillion, SD 57069

INTRODUCTION

Class 3 aldehyde dehydrogenase (AlDH) from rat liver has been expressed
in native form by E. coli from the pTALDH vector (Harper et al., 1988). This
AlDH differs substantially from the class 1 and 2 AlDHs both in catalytic
properties and in primary and quaternary structure. However, secondary
structural predictions suggest that the subunit tertiary structures of all
three AlDH classes are largely similar (Lindahl and Evces, 1984; Hempel et
al., 1989). Although functional residues have been identified from chemical
modification and sequence comparisons, no tertiary structure of an AlDH has
yet been determined.

METHODS

Rat liver class 3 AlDH was isolated in catalytically active form from E.
coli (BH101) transformed with pTALDH. Transformed E. coli were grown in a 10-
liter fermentor and harvested by centrifugation. This represents a ten-fold
increase in starting material over previous purifications (Harper et al.,
1988), and it was not possible to obtain pure enzyme from a single 5' AMP
chromatography. Thus, a CM-Sepharose step has been included for greater
purity. The pelleted cells were stored at -70°C, thawed, suspended in 10mM
sodium acetate, 2mM EDTA, 0.1% mercaptoethanol, pH 5.5, and sonicated (4 x 15
sec bursts). Cellular debris was removed at 48,000 g for 30 min in a Ti 50.2
rotor (33,000 rpm) and the dialyzed supernatant (same buffer) was applied to a
4 x 30 cm column of CM Sepharose 6B. After washing off unbound material the
enzyme was eluted with a linear gradient of NaCl (0-1.0 M). Activity was
measured using NADP and benzaldehyde as substrates (Lindahl and Evces, 1984),
and active fractions were loaded onto a 4 x 12 cm column of 5' AMP Sepharose
(Pharmacia) equilibrated with 25mM potassium phosphate, 1mM EDTA, 0.1% β-
mercaptoethanol, pH 7.5. After washing with 400mM potassium phosphate, 1mM
EDTA, 0.1% β-metcaptoethanol, pH 7.5, the enzyme was eluted with this same
buffer containing 0.5 mg/ml NAD. The purity of the resulting material, as
indicated after SDS PAGE is shown in Figure 1.

Crystals of AlDH were grown by vapor diffusion from 50 μl droplets of
the protein (0.7 mg/ml), in 12.5mM PIPES/NaOH (pH 6.2), 0.05% (v/v)
mercaptoethanol, 0.5mM EDTA, 0.25 mg/ml NAD, with 1.5% (w/v) PEG 8000, and

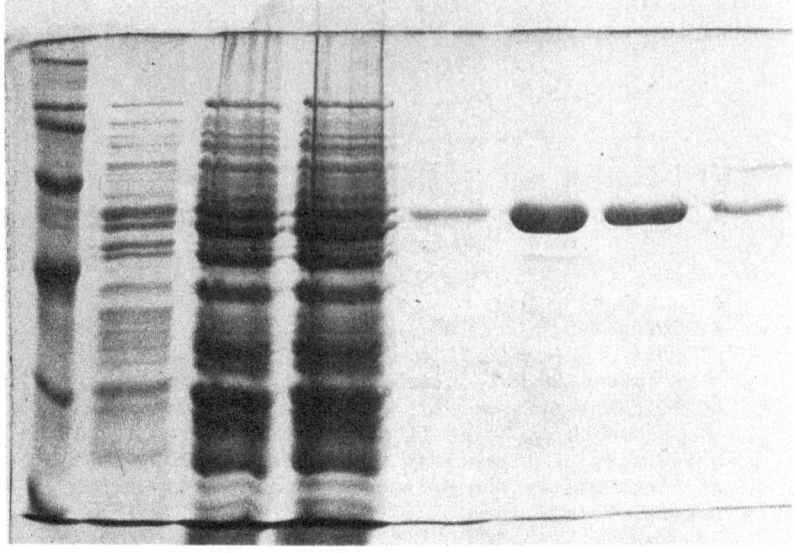

Figure 1. SDS polyacrylamide gel of fractions during preparation of rat class 3 AlDH as expressed by <u>E. coli</u> HB101 transformed with pTALDH. Lane 1: marker proteins. Lane 2: after cell sonication and CM-Sepharose chromatography. Lanes 3 and 4: inactive pass-through from 5'AMP Sepharose column. Lanes 5-8: fractions across peak of activity eluted with NAD.

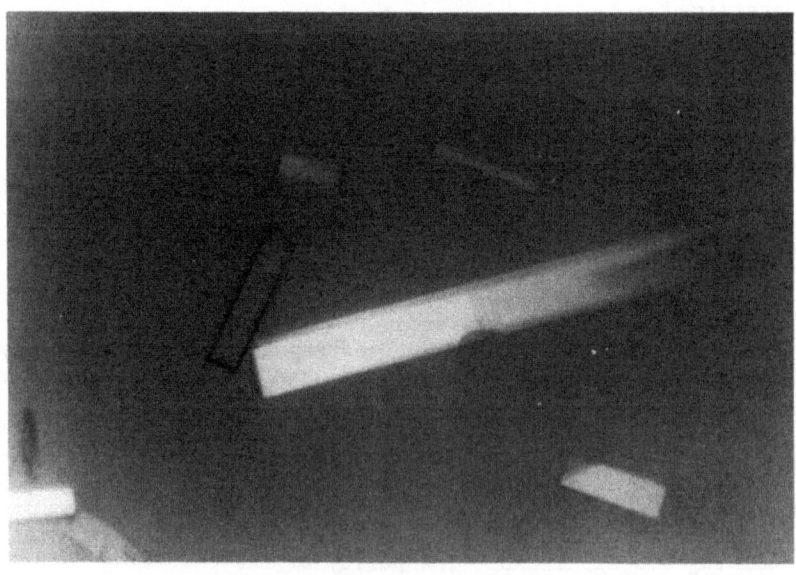

1 mm

Figure 2. Crystals of rat class 3 AlDH under polarized light. The crystals diffract to about 2.5Å resolution.

placed in wells of a 9-well spot plate. The plate was then placed in a sandwich box containing this buffer with 2.5% PEG.

RESULTS AND DISCUSSION

Small rectangular crystals formed from the sitting drops in a few days; some grew as large as 0.8 x 0.3 x 0.2 mm in two weeks (Fig. 2). For X-ray analysis, a crystal was mounted in a thin-walled glass capillary containing a small amount of mother liquor and sealed with diffusion pump oil. Oscillation photographs show diffraction to 2.8 β, with come reflections observed to a resolution of 2.5 β.

Analysis of the three-dimensional data collected using a Siemens X100 area detector system indicates a monoclinic space group with cell parameters a = 65.11 β, b = 170.67 β, c = 47.15 β, and β = 110.5°. Systematic absences in 0k0 for k $\neq$ 2 suggest that the space group is $P2_1$. There is a dimer in the asymmetric unit. Rotation function study indicates that the dimer has a pseudo-dimer axis (Rose et al., 1990). A heavy atom search is underway to conduct a full crystallographic study of this AlDH. Meanwhile, in the absence of a tertiary structure for any AlDH, a suggested distant relationship to the active site of papain, revealed at the primary structural level (Hempel et al., 1990), may provide a useful model.

ACKNOWLEDGEMENTS

This work was supported by AA-06985 (to JH), RR-02877 and GM-17528 (to BCW).

REFERENCES

Harper, K., Jones, D. E., Brennan, M. D., Lindahl, R., 1988. Characterization of a functional recombinant rat liver aldehyde dehydrogenase: Expression as a non-fusion protein in *E. coli*. Biochem. Biophys. Res. Comm. 152:940.
Hempel, J., Harper, K., Lindahl, R., 1989. Inducible class 3 aldehyde dehydrogenase from rat hepatocellular carcinoma and 2,3,7,8-tetrachlorodibenzo-p-dioxin-treated liver: Distant relationship to the Class 1 and 2 enzymes from mammalian liver cytosol/mitochondria. Biochemistry 28:1160.
Hempel, J., Nicholas, H., Jörnvall, H., 1990. Thiol proteases and aldehyde dehydrogenases: Evolution from a common thiolesterase precursor? Proteins: Structure Function and Genetics (submitted).
Lindahl, R., Evces, S., 1984. Aldehyde dehydrogenase II characterization of inducible isoenzymes. J. Biol. Chem. 259:11991-11996.
Rose, J. P., Hempel, J., Kuo, I, Lindahl, R., Wang, B.-C., 1990. Preliminary crystallographic analysis of Class 3 rat liver aldehyde dehydrogenase. Proteins: Structure, Function and Genetics (in press).

PROBING THE ACTIVE SITE OF ALDEHYDE DEHYDROGENASE BY SITE DIRECTED MUTAGENESIS

Henry Weiner, Jaume Farrés, Thomas T.Y. Wang, Suzanne J. Cunningham, Chao-Feng Zheng and Ghiorghis Ghenbot

Biochemistry Department
Purdue University
West Lafayette, IN 47907

INTRODUCTION

It has generally been accepted that aldehyde dehydrogenase functions through covalent catalysis where a nucleophilic amino acid residue attacks the carbonyl of the aldehyde substrate (Weiner et al., 1982). Since the first purification on ALDH by Feldman and Weiner in 1972, it has been argued that the active site nucleophile was a cysteine residue. The identity of this specific cysteine in the primary structure of the enzyme however, has not been unequivocally determined. Recently it has been suggested that the nucleophile is not cysteine but is serine (Loomis et al., 1990). A similar adduct could be formed with either amino acid; a thiohemiacetal in the former, a hemiacetal in the latter. The intermediate would then be oxidized in the presence of NAD to produce a thioacyl or an acyl intermediate respectively, as illustrated in Fig. 1.

$$
\begin{array}{ccc}
\overset{\text{OH}}{\underset{\text{H}}{\text{ES-}\overset{|}{\text{C}}\text{-CH}_3}} & \overset{\text{NAD}}{\longrightarrow} & \text{ES-}\overset{\overset{\text{O}}{||}}{\text{C}}\text{-CH}_3 \\
& & \mathbf{I}
\end{array}
$$

$$
\begin{array}{ccc}
\overset{\text{OH}}{\underset{\text{H}}{\text{EO-}\overset{|}{\text{C}}\text{-CH}_3}} & \overset{\text{NAD}}{\longrightarrow} & \text{EO-}\overset{\overset{\text{O}}{||}}{\text{C}}\text{-CH}_3 \\
& & \mathbf{II}
\end{array}
$$

Fig. 1. Possible acyl intermediates found after the oxidation of a covalently bound aldehyde. I shows a thioacyl intermediate if a cysteine were at the active site; II shows an acyl intermediate formed if serine were at the active site.

We proposed that the enzyme should also require general base catalysis in addition to nucleophilic catalysis (Weiner et al., 1985). This conclusion was partly based on the fact that acyl hydrolysis always is base catalyzed. The velocity-pH profile suggested that a group with a pK_a of ca 7 appeared to be involved (Takahashi et al., 1981). This is illustrated in Fig. 2.

Enzymology and Molecular Biology of Carbonyl Metabolism 3
Edited by H. Weiner et al., Plenum Press, New York, 1990

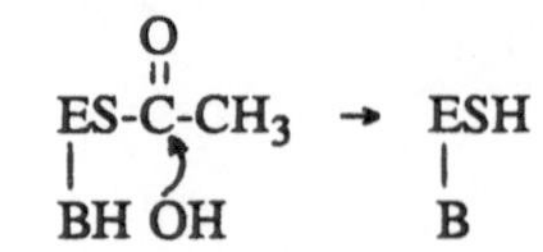

$$\underset{\underset{\text{BH OH}}{|\quad|}}{\text{ES-C-CH}_3} \rightarrow \underset{\underset{\text{B}}{|}}{\text{ESH}}$$

Fig. 2. A model showing how a general base (B) could be involved in the deacylation step catalyzed by aldehyde dehydrogenase.

Histidine often functions as the general base in hydrolytic reactions similar to the one illustrated in Fig. 2. At an earlier workshop we showed that histidine was not absolutely required for the enzyme to function, though chemical modifications of the residue caused partial inactivation of the enzyme (Weiner *et al.*, 1985). Now that the amino acid sequences of a number of ALDHs have been determined we find that there are a number of conserved histidine residues in the enzyme (Lindahl and Hempel 1991).

We recently inserted the cDNA coding for rat liver mitochondrial ALDH into an expression plasmid and successfully expressed active tetrameric enzyme in *E. coli*. We have subsequently initiated a project to probe for the active site residues by performing site directed mutagenesis (Farrés and Weiner, 1991). Residues altered were those implicated to be involved in catalysis either through chemical modification studies or which were conserved in many species. In this chapter we will discuss some preliminary results of our studies on the use of site directed mutagenesis for probing essential residues at the active site of rat liver mitochondrial ALDH.

RESULTS

The cDNA coding for rat liver aldehyde dehydrogenase was cloned into a pT7-7 plasmid behind the T-7 promotor. *E. coli* BL21(DE3) pLys S cells were used to express the enzyme. The promoter used in the *in vivo* expression vector was reported to allow for the efficient expression of foreign proteins in *E. coli*. In our system only a low level of expression of ALDH was found. From a liter culture approximately 4 mg of enzyme was expressed. We previously showed that the mRNA coding for mitochondrial aldehyde dehydrogenase could form secondary structures which appeared to cause a slow *in vitro* translation of the message (Guan and Weiner, 1989). Thus it is possible that the secondary structure of the mRNA prevented the maximum expression of ALDH to occur. In spite of the fact that liver mitochondrial ALDHs isolated from different species are acetylated on their N-terminal residue (Guan *et al.*, 1988; Weiner *et al.*, 1991), an active tetrameric enzyme was expressed in *E. coli*. The purified enzyme had a specific activity very similar to that of the native enzyme isolated from fresh rat liver.

Site directed mutagenesis using Bio-Rad's MUTA-GENE[TM] *IN VITRO* MUTAGENESIS kit was employed to alter three different cysteine residues reported to be possible candidates for the nucleophile at the active site of the enzyme. Cysteines 49, 162 and 302 were converted to alanines and the mutated proteins were expressed in *E. coli*. Though the three mutants were found to be expressed essentially as well as was the native enzyme, all were not found to be of equal stability. The 162 mutant was degraded much more rapidly during pulse chase experiments than were the other two mutants.

The mutant enzymes were purified to homogeneity as judged by SDS-PAGE. Each was a tetramer and had the same pI as did the native enzyme. The mutants' ability to function as a dehydrogenase and as an esterase was tested under Vmax conditions. Only the mutation at position 302 produced an enzyme void of catalytic activity. The results are summarized in Table 1.

Table 1

Catalytic Activity in Native and
Cysteine Mutants of Rat Liver
Mitochondrial Aldehyde Dehydrogenase

Mutant	Activity (Percent)	
	Dehydrogenase	Esterase
Native	100	100
49 Alanine	100	100
162 Alanine	100	100
302 Alanine	0	0

Based on these results we conclude that if a cysteine were involved at the active site of the enzyme it must be the residue at position 302 and not 49 or 162 as we previously reported based upon data obtained from chemical modification studies (Tu and Weiner, 1988a, 1988b). Evidence for cysteine 302 as the active site nucleophile has been obtained from chemical modification studies, as reviewed in the chapter by Pietruszko *et al.*, (1991). Recently it has been suggested that a serine, not a cysteine, is the actual nucleophile (Loomes *et al.*, 1990). Serine at position 74 in sheep liver cytosolic aldehyde dehydrogenase was reported to be covalently modified by a cinnamoyl substrate. To test for its involvement we changed the serine to an alanine. The enzyme has not yet purified to homogeneity but, at the time of writing, the serine-74 mutant was expressed and found to be essentially void of catalytic activity. Thus we still do not really know which is the actual nucleophile, serine or cysteine. Only cysteine at position 302 is conserved between all the known aldehyde dehydrogenases. Serine 74 is found only in the mammalian enzyme forms (Lindahl and Hempel, 1991).

Site directed mutagenesis also was used to determine if one of the conserved histidine residues could be essential for activity. The residues were converted to alanines and expressed in *E. coli*. Table 2 shows that none of the alanine mutants were inactive nor caused a drastic reduction of the activity of the enzyme. This means that none of the conserved histidine residues are essential for catalysis, though it can not be ruled out that one of them is not somehow involved in the active site. Results of pulse chase experiments indicate that the *in vivo* stability of the proteins were affected by the mutations. The largest decrease was found when the most conserved histidine (235) was changed. This finding leads us to suspect that this residue might be involved in the maintenance of the conformation of the enzyme.

Table 2

Effects of Mutation of Histidine
Residues in Rat Liver Aldehyde Dehydrogenase

Mutant	Stability[+]	Activity[#] (%)
Native	+++	100
29 Alanine	++	200
156 Alanine	++	100
235 Alanine	+	50
291 Alanine	+++	88

[+] Stability is based upon pulse chase analysis
[#] Activity is based upon a Vmax assay

Our previous chemical modification studies lead us to conclude that cysteine 49 and possibly 162 could be involved at the active site of the enzyme (Tu & Weiner, 1988a, 1988b). Modification with N-ethylmaleimide increased the size of the side chain of the cysteine residue, hence one can not tell if the inactivation was due to the loss of the functional group or to the increased mass. Since we found that replacement of cysteine 49 and 162 by alanine produced active enzymes we changed these residues to tryptophan to mimic what would be found after chemical modification with N-ethylmaleimide. The "essential" cysteine 302 was also mutated to a bulky tryptophan residue. All three enzymes were expressed, purified and found to be inactive. Thus replacing the -SH of cysteine 49 or 162 by a hydrogen did not affect activity but the addition of the bulky hydrophobic indole ring produced an inactive enzyme.

DISCUSSION

There are many approaches one can use to identify the residues which may be situated at the active site of an enzyme. One way is to perform chemical modifications with an affinity label. Alternatively, one can use a substrate or substrate analog to block the active site, modify the unprotected residues, remove the blocking group, and finally label the protected residue which is presumed to be at the active site. A totally different approach is to use sequence homology to identify residues conserved among a number of enzymes isolated from different species. Finally, after selecting a candidate for the active site residue one can use site directed mutagenesis to alter the suspected residue and determine the effect of the altered residue on the enzyme. We have used all three approaches and still are not certain as to whether or not we identified the active site components of mitochondrial aldehyde dehydrogenase.

Our earlier work with diethylpyrocarbonate showed that a modification of a histidine caused an 80% decrease but not an abolishment of activity of the horse liver enzyme (Weiner *et al.*, 1985). Since the residue was protected by a substrate analog we concluded that it most likely was in the environment of the active site, if not a component of the active site. In this study we show that the removal of any one of the three highly conserved histidines by site directed mutagenesis did not inactivate the enzyme. This is consistent with the conclusion that a histidine is not an obligatory component of the active site. Thus the role of a general base in the active site of aldehyde dehydrogenase is still uncertain.

Much more attention has been focused on the identification of the amino acid which could function as a nucleophile, as illustrated in Fig. 1. Prior to the recent study by Loomes *et al.*, (1990) it has been assumed that the residue was a cysteine. The site directed mutagenesis work described here leads us to suggest that if a cysteine is the active site nucleophile then it must be located at position 302, not 49 or 162 as we previously suggested. Our finding that the replacement of serine 74 by an alanine caused the enzyme to become inactive shows the potential importance of that residue. As this serine is not conserved in most aldehyde dehydrogenases makes it difficult to conclude the fact that this is really the essential residue.

It is difficult to determine why our previous attempt to identify the cysteine at the active site lead us to conclude that the essential one was at position 49 or 162. In that study the active site of the horse liver enzyme was blocked with either chloral hydrate, a substrate competitive inhibitor, or with o-nitrobenzaldehyde, a poor substrate for the enzyme (Tu and Weiner, 1988a, 1988b). The unprotected cysteines were labelled with N-ethylmaleimide; the blocking or protecting group was then removed and the enzyme was further treated with ^{14}[C]-N-ethylmaleimide. Only cysteine 49 and 162 were labelled. The dehydrogenase activity correlated well with the modification at position 49 while the esterase activity correlated with the loss of cysteine at position 162. These residues must somehow be located near the actual active site and were protected by the binding of substrate at the true active site residue.

Our finding that replacing the cysteines with a large tryptophan moiety produced an inactive enzyme shows that the size of the group at non-essential positions is critical for the integrity of the enzyme. Thus it is possible that the mis-identification was a result of adding bulk to a critical region of the enzyme, and not removing the reactive -SH group.

In spite of years of efforts many of the chemical aspects of aldehyde dehydrogenase are not known with certainty. It can only be assumed from a combination of chemical modification studies, site directed mutagenesis and sequence analysis that the active site components ultimately will be identified. Once a complete three dimensional structure of the enzyme is determined it may be possible to understand why so many different final conclusions were reached as to the nature of the active site.

ACKNOWLEDGEMENTS

This work was supported in parts from ADAMHA Grants AA08512 and P50-07611, Post-Doctoral Fellowship (T.T.Y.W.) AA05276 and Research Scientist Award (H.W.) AA00028.

REFERENCES

Feldman, R. I., and Weiner, H., 1972, Horse liver aldehyde dehydrogenase I, J. Biol. Chem., 247:260-266.

Farrés, J., and Weiner, H., 1991, To be submitted.

Guan, K-L, Pak, Y. K., Tu, G-C, Cao, Q-N, and Weiner, H., 1988, Purification and Characterization of Beef and Pig Liver Aldehyde dehydrogenase, Alcoholism Clin. Exp. Res., 12:713-719.

Guan K-L, and Weiner, H., 1989, Influence of the 5'-end region of aldehyde dehydrogenase mRNA on translational efficiency, J. Biol. Chem., 264:17764-17769.

Lindahl, R., and Hempel, J., 1991, Aldehyde dehydrogenase: What can be learned from a baker's dozen sequences?, In: These Proceedings.

Loomes, K. M., Midwinter, G. G., Blackwell, L. F., and Buckley, P. D., 1990, Evidence for reactivity of Serine-74 with trans-(N,N-dimethylamino)cinnamaldehyde during the oxidation by the cytoplasmic aldehyde dehydrogenase of sheep liver, Biochemistry, 29:2069-2075.

Pietruszko, R., Blatter, E., Abriola, D. P., and Prestwich, G., 1991, Localization of cysteine 302 at the active site of aldehyde dehydrogenase, In: These Proceedings.

Takahashi, K., Weiner, H., and Filmer D. L., 1981, Effect of pH on horse liver aldehyde dehydrogenase, Biochemistry 20:6225-6230.

Tu, G-C, and Weiner, H., 1988a, Identification of the cysteine residue at the active site of horse liver mitochondrial aldehyde dehydrogenase, J. Biol. Chem., 263:1212-1217.

Tu, G-C, and Weiner,H., 1988b, Evidence for two distinct active sites on aldehyde dehydrogenase, J. Biol. Chem., 263:1218-1222.

Weiner, H., 1982, Aldehyde dehydrogenase In "Enzymology of Carbonyl Metabolism," H. Weiner and B. Wermuth, eds., Alan R. Liss, New York.

Weiner, H., Cunningham, S. C., and Angelo, R. A., 1991, N-Terminal acetylated mitochondrial aldehyde dehydrogenase is found in fresh but not frozen liver tissue, In Press.

Weiner, H., Lin, F-P, and Sanny, C. G., 1985, Chemical probes for the active site of aldehyde dehydrogenase, In: Enzymology of Carbonyl Metabolism 2," T. G. Flynn and H. Weiner eds., Alan R. Liss, New York.

LOCALIZATION OF CYSTEINE 302 AT THE ACTIVE SITE OF ALDEHYDE DEHYDROGENASE

Regina Pietruszko[1], Erich Blatter[1], Darryl P. Abriola[1] and Glenn Prestwich[2]

[1]Center of Alcohol Studies, Rutgers University, Piscataway NJ 08855-0969 and [2]Chemistry Department, SUNY, Stony Brook NY 11794

INTRODUCTION

The superreactive cysteine was first identified in human cytoplasmic aldehyde dehydrogenase E1 isozyme, before its primary structure was known, as a part of 35 residue tryptic peptide (Hempel, 1981; Hempel and Pietruszko, 1981; Hempel et al., 1982) by employing iodoacetamide. When the primary structures of the E1 and E2 isozymes were established (Hempel et al., 1984, 1985; Hsu et al., 1985), this cysteine was found to occupy position 302 in a 500 amino acid residue polypeptide chain. Iodoacetamide fulfilled all criteria for an aldehyde-competitive, active-site-directed reagent with the exception of total inactivation of the mitochondrial E2 isozyme. Since that time, other investigators have also attempted to identify active site residues. Coenzyme-based affinity reagents (von Bahr-Lindstrom et al., 1985) identified cysteines 369 and 302, N-ethylmaleimide identified cysteine 49 and 162 (Tu and Weiner, 1988 a,b) and dimethylaminocinnamaldehyde identified serine 74 (Loomes et al., 1990). Our laboratory developed a substrate-based affinity reagent, bromo-acetophenone (MacKerell et al., 1986), which identified glutamate 268 (Abriola et al., 1987).

Two reagents have been employed during this investigation: (1) a vinyl ketone, (Z)-1,11-hexadecadiene-3-one and its tritium labelled form $[11,12-^3H_2]-(Z)$-1,11-hexadecadiene-3-one (Figure 1) and (2) bromo-acetophenone (2-bromo-1-phenylethanone) and its carbon C-14 labelled form, $carbonyl-[^{14}C]$-bromoacetophenone. Both are affinity reagents and function as analogues of aldehyde substrates.

The vinyl ketone, (Z)-1,11-hexadecadiene-3-one (Figure 1A,II), is a structural analogue of (Z)-11-hexadecenal (Figure 1A,I), a long-chain monounsaturated aldehyde which functions as a sex pheromone in the moth *Heliothis virescens* (Ding and Prestwich, 1988; Prestwich et al., 1989; Tasayco and Prestwich, 1990a,b). Selective irreversible inactivation of aldehyde dehydrogenase in the moth antenna can be achieved by the vinyl ketone analog of this aldehyde (Ding and Prestwich, 1988; Prestwich et al., 1989; Tasayco and Prestwich, 1990a,b). The vinyl ketone reacts with the enzyme sulfhydryl via its C=C double bond (Figure 1B) which is conjugated with C=O double bond of the ketone. The simple C=C double bond does not normally react with nucleophiles but serves as a source of

A

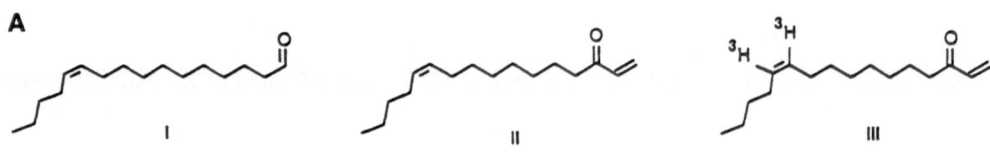

B

Michael Addition of Thiol:

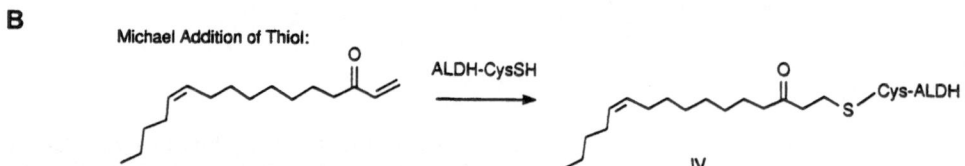

Figure 1. Insect Pheromone and its Vinyl Ketone Analogs. A. I - (Z)-11-hexadecenal, II - (Z)-1,11-hexadecadien-3-one, III - [11,12-^{3}H$_2$]-(Z)-11-hexadecadien-3-one. B. Proposed mechanism of interaction of compound II with human liver aldehyde dehydrogenase; IV - Michael addition product.

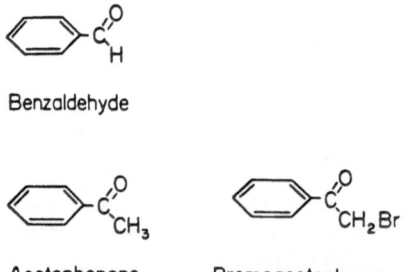

Benzaldehyde

Acetophenone   Bromoacetophenone

Figure 2. Chemical Formulae of Benzaldehyde, Acetophenone, and Bromoacetophenone.

electrons for electrophilic reagents. However, the presence of electron withdrawing groups such as -C=O, -COOH, -COOR and -CN in the vicinity of the double bond affects availability of electrons to electrophilic reagents and makes the double bond susceptible to attack by reagents that are electron rich. Thus, the conjugated C=C bond of the ketone undergoes nucleophilic addition which is not common to simple alkenes. At neutral pH values sulfhydryl groups of proteins are most nucleophilic and therefore most susceptible to the reaction with $\alpha\beta$-unsaturated carbonyl compounds.

The chemical structure of bromoacetophenone (MacKerell et al., 1986) resembles that of the substrate, benzaldehyde, and also that of the substrate-competitive inhibitor, acetophenone (Figure 2). The side chain contains a carbonyl group which probably aligns at the enzyme active site in a similar fashion to that of the substrate, benzaldehyde. The powerful electron withdrawing effect of the adjacent carbonyl makes the bromine extremely reactive. In fact, so positioned, bromine will react with any protein nucleophile in its vicinity, including -NH$_2$, -OH and -SH groups. Carboxyl groups would not normally react with this reagent; in fact, acetophenone is normally brominated in anhydrous acetic acid used as a solvent. It has been demonstrated that crown ether catalysts are necessary for carboxyl groups to react with haloketones (Pedersen and Frendsdorff, 1972; Durst et al., 1975). Thus, the carboxyl group has to be located in an extremely hydrophobic environment and be present as a "naked anion" (Carey and Sundberg (1984) in order to react with bromo-acetophenone. This fact, in addition to all other criteria (Shaw, 1970), enabled localization of glutamate 268 at the active site of aldehyde dehydrogenase (Abriola et al., 1987) and has also made the reagent useful for any other residue it may derivatize.

MATERIALS AND METHODS

Materials employed, enzyme preparation, determination of enzyme activity as well as labelling strategies were previously described (Pietruszko and Yonetani, 1981; MacKerell et al., 1986; Abriola et al., 1987, 1990; Blatter et al., 1990).

Synthesis of Pheromone Aldehydes and Vinyl Ketones Analogs. (Z)-11-Hexadecenal (Figure 1A,I) and (Z)-1,11-hexadecadien-3-one (Figure 1A,II) were synthesized as described by Ding and Prestwich (1988), the [11,12-^{3}H$_2$]-(Z)-1,11-hexadecadien-3-one (Figure 1A,III) was described by Prestwich (1989) and Tasayco and Prestwich (1990b).

Preparation of Bromoacetophenone and Synthesis of Carbonyl-[^{14}C]-bromoacetophenone. Bromoacetophenone (Aldrich) was recrystallized twice from methanol:water (8:2) prior to use. Carbonyl-[^{14}C]-bromoacetophenone was obtained by bromination of carbonyl-[^{14}C]-acetophenone (Amersham) (MacKerell et al., 1986).

Peptide Mapping and Sequencing. Following derivatization, the enzymes were reduced, carboxymethylated and digested by trypsin as described by Hempel et al. (1984). After purification via HPLC (Waters) the labelled material from the final column was applied to an Applied Biosystems Model 477A Protein Sequencer as described by Abriola et al., 1990, and Blatter et al., 1990.

RESULTS AND DISCUSSION

(Z)-11-Hexadecenal - An Insect Pheromone is a Substrate and (Z)-11-Hexadecadiene-3-one is an Irreversible Inhibitor of Human Aldehyde Dehydrogenase

In Table 1, the Michaelis constants, catalytic rate constants and k_{cat}/Km ratios obtained with (Z)-11-hexadecenal are compared with those of propanal (which is a good substrate for both isozymes) and o-nitrobenzaldehyde (which is a poor substrate). In general, the E1 isozyme exhibited smaller k_{cat} and larger Km values for a given substrate than the E2 isozyme. This is even more apparent when the ratio of k_{cat}/Km is compared, showing that the E1 isozyme is a less efficient catalyst than the E2 isozyme. For the E1 isozyme, the k_{cat}/Km value for (Z)-11-hexadecenal was larger than the corresponding one for propanal at both pH's, indicating that the pheromone was actually a better substrate. With the E2 isozyme, the pheromone k_{cat}/Km values were about the same as for propanal but much higher than for the poor substrate, o-nitrobenzaldehyde.

Table 1. Comparison of Michaelis and Catalytic Rate of Constants of the E1 and E2 Isozymes with Propanal, (Z)-11-Hexadecenal and o-Nitrobenzaldehyde as Substrates.

Enzyme	Substrate	pH 7.0			pH 9.0		
		Km (μM)	k_{cat}*	$\frac{k_{cat}}{Km}$	Km (μM)	k_{cat}*	$\frac{k_{cat}}{Km}$
E1	Propanal	4.8	0.5	0.1	11.0	0.7	0.06
	Hexadecenal	2.0	0.4	0.2	1.0	0.3	0.3
E2	Propanal	0.7	0.25	0.36	1.2	1.4	1.2
	Hexadecenal	0.6	0.24	0.4	1.1	0.7	0.6
	o-Nitrobenz-						
	aldehyde	-	-	-	1.6	0.16	0.1

- not determined. * = μmoles NADH/min/mg enzyme protein.

The vinyl ketone, (Z)-1,11-hexadecadien-3-one (a structural analogue of the pheromone, (Z)-11-hexadecenal) (Figure 1), when incubated at stoichiometric concentrations with human aldehyde dehydrogenase, produced rapid enzyme inactivation. Complete inactivation was achieved with both the E1 and E2 isozymes. This inactivation was not reversible by dialysis or gel filtration, indicating covalent bond formation. The loss of esterase activity paralleled that of dehydrogenase activity. Incubation of [³H]-vinyl ketone derivatized E1 or E2 isozymes with 2-mercaptoethanol did not restore activity and no loss of label occurred.

The Effect of Modification of the E2 Isozyme by the Vinyl Ketone on the Burst Amplitude

At high enzyme concentrations, at pH 5.5, a presteady state burst was observed when the E2 isozyme was mixed with o-nitrobenzaldehyde and NAD+. When 3.8 μM E2 isozyme (specific activity = 0.7 μmoles/min/mg) was mixed with 100 μM o-nitrobenzaldehyde and 525 μM NAD+, a burst of 3 μM NADH

(0.79 μmoles/μmole E2 isozyme) was observed. After incubation of 6.4 μM E2 isozyme with 65 μM vinyl ketone for 30 minutes, the enzyme activity decreased to 26% of control. After stopping the reaction with 2-mercaptoethanol (325 mM), and gel filtration at pH 5.5 to remove reagents, the modified enzyme was concentrated to 7.4 μM and then mixed with 100 μM o-nitrobenzaldehyde and 525 μM NAD+. The burst was diminished to 0.56 μM NADH (0.076 μmoles/μmole E2 isozyme); e.g., 10% of control. The control E2 isozyme treated with 2-mercaptoethanol and subjected to gel filtration retained the full burst of 0.8 μmoles/μmol E2 isozyme. Thus, modification with the vinyl ketone affected not only the steady state velocity but also the burst amplitude, suggesting that the amino acid residue modified by the vinyl ketone is important to catalysis.

Compliance With Criteria for Active Site-Directed Reagents

Both the vinyl ketone and bromoacetophenone complied with all criteria for active site-directed reagents. These criteria include:

(1) Complete irreversible inactivation.
(2) Protection by substrates against inactivation.
(3) Stoichiometry of label incorporation of 2 molecules of label per mole of enzyme.
(4) Specificity of incorporation.

The first three criteria were fulfilled in a similar manner by both compounds. Specificity of incorporation was examined after the modified enzyme was fragmented with trypsin, and the resulting mixture of tryptic peptides was separated on a C18 reverse-phase HPLC column with a gradient of 0.1% trifluoroacetic acid to 100% methanol. In these conditions a single radioactive peptide peak was obtained with the vinyl ketone (Figure 3); multiple radioactive peaks were visualized with carbonyl-[^{14}C]-bromo-acetophenone (Figure 4).

Peptide Mapping of Vinyl Ketone-Labelled E2 Isozyme

Following a differential labelling strategy with the vinyl ketone (Blatter et al., 1990), the E2 isozyme was fragmented with trypsin and chromatographed on a C18 column. The peptide containing the label eluted in a single peak at the end of the chromatogram at a position where peptides do not normally elute. This was apparently due to the extremely hydrophobic nature of the vinyl ketone. After further digestion with chymotrypsin and additional chromatographic steps, amino acid sequence analysis showed various fragments of the tryptic peptide comprising residues 273-307. The only extraneous material, which was the smallest component, consisted of amino acid residues 93-109. Thus, the peptides that bore the label all consisted of various cleavage products of the 273-307 tryptic peptide that contains cysteines 301, 302, and 303. The peptide present as an impurity contained no cysteine, indicating that one of the above-mentioned residues was labelled by the vinyl ketone. Due to difficulty in separation of the individual partial cleavage products, the resulting multiple overlapping sequences prevented the exact position of the label from being identified with confidence in the first experiment. Therefore, the experiment was repeated employing direct labelling with the vinyl ketone followed by a single enzymatic hydrolysis with trypsin. Results from sequencing showed two components; the major component (74%) comprised residues 296 to 307 in the primary structure of the E2 isozyme. When the material from the sequencing cycles was counted for radioactivity, cycle 7, corresponding to cysteine 302, was found to contain the label (Figure 5). The impurity (26%) was a part of the same tryptic peptide (residues 273-289) and contained a serine residue in cycle 7. This residue, however, cannot react with the vinyl ketone, and

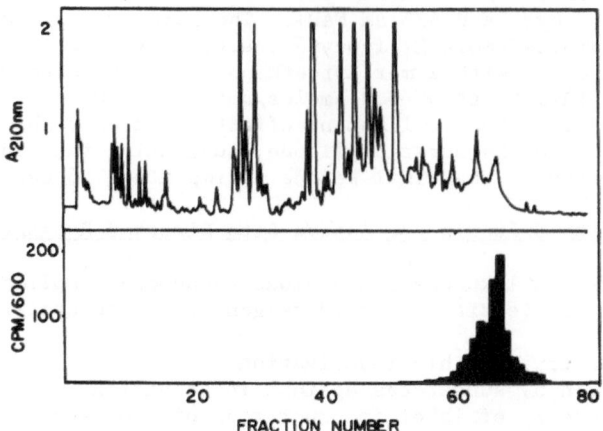

Figure 3. Identification of the [³H]-tryptic Peptide from Differentially Labelled E2 Isozyme. HPLC on Waters C-18 µBondapak with a linear gradient from 0-100% methanol (solvent A = 0.1% aqueous TFA; solvent B = 100% methanol. The upper part of the Figure shows absorbance at 210 nm; the lower part shows the distribution of radioactivity.

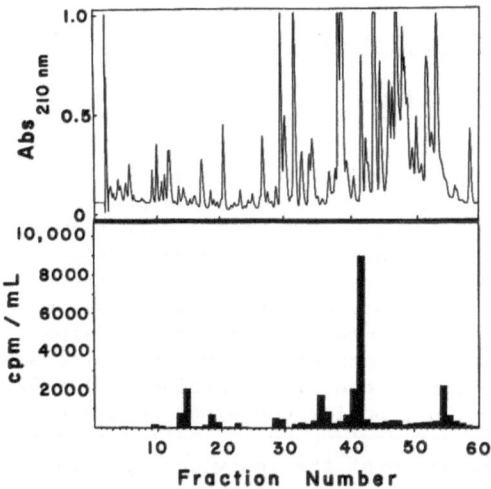

Figure 4. Peptide Mapping of *Carbonyl*[¹⁴C]bromoacetophenone-modified E1 Isozyme. HPLC on Waters C-18 µBondapak with a linear gradient from 0-100% methanol (solvent A = 0.1% aqueous TFA; solvent B = 100% methanol. Upper figure: column elution profile followed by absorbance at 210 nm; lower figure: column elution profile followed by radioactivity.

likelihood of its derivatization by the vinyl ketone was also eliminated by prior sequencing experiments (Blatter et al., 1990). Thus, the results of both direct and differential labelling showed that the label was located exclusively in cysteine 302. No label was detected in any other cysteine residue of the mitochondrial E2 isozyme.

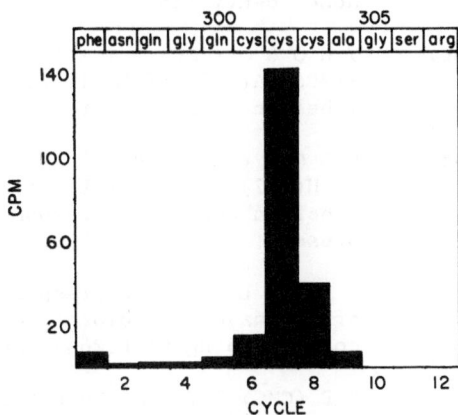

Figure 5. Sequence Analysis of the Directly-labelled E2 Peptide. [³H] radioactivity of the PTH-amino acid derivatives obtained from automated Edman degradation are shown on one axis and amino acid identity and cycle number of the other.

Peptide Mapping of Bromoacetophenone Labelled E1 Isozyme

From data shown in Figure 4 it can be seen that at low stoichiometries the incorporation of the majority of *carbonyl*[¹⁴C]-bromoacetophenone occurred into peptides eluting in fractions 40-42 where it reacted with Glu 268 (Abriola et al., 1987). The incorporation was, however, not completely specific and became less specific with increasing stoichiometry. At low stoichiometries incorporation of radioactive label consistently occurred into fractions 14-15, 35-37 and 54-56, in addition to the main peak. Since these could have represented partial or other cleavage products containing the same reactive residue as the main peak (eluted in fractions 40-42) or other residues localized in the vicinity of Glu 268, they were selected for identification via amino acid sequence analysis. Purification of the radioactive peaks occurring in fractions 19-20 and 29-30 was also considered but had to be abandoned because these materials were absent from the subsequently labelled E1 isozyme. Thus, it appears that they were cleavage products produced by that particular batch of trypsin.

Identification of Other Residues Labelled by Bromoacetophenone

The first attempt at purification of the peptide eluting at 55 min tentatively established label localization in the tryptic peptide comprising residues 273-306 with the reactive residue in the 30th cycle (Table 2). There was also a discrepancy in quantitation by specific radioactivity and 254 nm absorbance of phenylthiohydantoin derivatives, which necessitated a second experiment. The second attempt at purification confirmed the peptide's identity and comparison with the primary structure allowed determination of Cys-302 as the reactive amino acid in this tryptic peptide.

Table 2. Results of Sequence Analysis of Peptide Fractions Obtained By Tryptic Digestion of *Carbonyl*[^{14}C]bromoacetophenone-labelled E1 Isozyme

Fraction Number	Identity of Labelled Peptide
15	None identified
35A (major component)	Y H Q G Q (CMC) ? ? A A (corresponding to Y(296)-A(305) of E1 isozyme); radioactive label in cycle 7 corresponding to C(302).
35B (minor component)	H Q G Q (CMC) (CMC) I A A S (corresponding to H(297)-S(306) of E1 isozyme); radioactive label in cycle 6 corresponding to C(302); also present; P(334)-K(352).
42	V T L E L G G K (corresponding to V(265)-K(272) of E1 enzyme); radioactive label in cycle 4, corresponding to E(268) (Abriola et al., 1987).
55	S P (CMC) I V L A D A D L D N A V E F A H H G V F Y H Q G Q (CMC) (CMC) I A A S R (corresponding to S(273)-R(307) of E1 enzyme); radioactive label in cycle 30, corresponding to C(302).
55	Same as previously obtained; also present: E(210)-K(240) (at approx. 20% level); radioactive label in cycle 30, corresponding to C(302).

Material from fractions eluted at approximately 35 min was resolved into two distinct components, each of which was purified separately by different schemes. Sequencing of the major component of this fraction showed that the labelled peptide was a shorter version of the same peptide that occurred in fraction 55, beginning at Tyr-296 rather than at Ser-274 (Table 2). Independent purification of the minor component led to a double sequence, containing one peptide that was an even further shortened portion of the fraction 55 peptide, starting at His-297. Thus both fraction 35 peptides have been identified as further cleavage products of the fraction 55 peptide, with the label appearing with Cys-302 in each case.

Sequencing of material in fractions 14-15 from the first HPLC chromatography was carried out on approx. 1 nmole of labelled derivative, but no sequence was obtained corresponding to this quantity, and no radioactivity was detected either in material from any of the cycles or on the filter disc of the sequencer. A different sample of fraction 15 was then purified by isocratic separations on two columns and subjected to amino acid composition analysis. None of the amino acids commonly occurring in proteins was detected above the background level. Steps were taken to purify remaining material further on an additional HPLC column, and both desorption-chemical-ionization and fast-atom-bombardment mass spectrometry were done on three separate occasions, giving inconclusive results each time. Since the radioactive species in fractions 14-15 had no identifiable amino acid it appeared to be derived from the reagent, but the exact nature of the derivative could not be determined.

The vinyl ketone is a reagent that is relatively specific for sulfhydryls. It cannot react with other nucleophiles, especially with carboxyl groups. Out of 9 cysteines present per subunit of the E2 isozyme, it reacted exclusively with cysteine 302. Cysteines 49 and 162 (Tu and Weiner, 1988) and cysteine 369 (von Bahr-Lindstrom et al., 1985) were not derivatized by this reagent. Bromoacetophenone, on the other hand, can react with any nucleophile (-SH, -NH$_2$, -OH, or a suitably-positioned -COOH) in its vicinity and therefore lacks the absolute specificity of the vinyl ketone. It can be seen from the results presented that, in addition to glutamate 268, the only other residue derivatized by carbonyl-[^{14}C]-bromoacetophenone is cysteine 302. None of the other residues such as cysteines 49 and 162 (Tu and Weiner, 1988), cysteine 369 (von Bahr-Lindstrom et al., 1985), or serine 74 (Loomes et al., 1990) were derivatized by this reagent.

Information about the exact nature of the derivatized residue is contained in the characteristics of reagents employed for chemical modification. Both the vinyl ketone and bromoacetophenone are affinity reagents which differ considerably in their reactivity with protein nucleophiles; despite these differences, with the vinyl ketone, cysteine 302 is the only residue derivatized and with bromoacetophenone, cysteine 302 is derivatized in addition to glutamate 268. No other cysteine residues are labelled by either reagent. On the basis of the above evidence we conclude that cysteine 302 is located at the active site of aldehyde dehydrogenase: the results with carbonyl-[^{14}C]-bromoacetophenone (Abriola et al., 1990) also suggest its localization in close proximity to glutamate 268.

Table 3. Sequence Comparison of Residues 265-307 in Aldehyde Dehydro-genases.

Enzyme Source	Sequence
1. Human cytosolic	VTLELGGKSPCIVLADADLDNAVEFAHHGVFYHQGQCCIAASR
2. Horse cytosolic	VTLELGGKSPFIVFADADLETALEVTHQALFYHQGQCCVAASR
3. Human mitochondrial	VTLELGGKSPNIIMSDADMDWAVEQAHFALFFNQGQCCCAGSR
4. Horse mitochondrial	VTLELGGKSPNIIVSDADMDWAVEQAHFALFFNQGQCCGAGSR
5. Bovine mitochondrial	VTLEIGGKSPNIIMSDADMDWAVEQAHFALFFNQGQCCCAGSR
6. Rat mitochondrial	VTLELGGKSPNIIMSDADMDWAVEQAHFALFFNQGQCCCAGSR
7. Rat phenobarbitol induced	VTLELGGKSPCIVFADADLDSAVEFAHQGVFFHQGQICVAASR
8. Rat hepatoma	VTLELGGKSPCYVDKDCDLDVACRRIAWGKFMNSGQTCVAPDY
9. Pseudomonas oleovorans	VTLELGGKSPTIIGPTANLPKAARNIVWGKFSNNGQTCIAPDH
10. Aspergillus nidulans	VTLELGGKSPNIVFDDADIDNAISWANFGIFFNHGQCCCAGSR
11. Spinach chloroplasts	VTLELGGKSPIVVFEDVDIDKVVEWTIFGCFWTNGQICSATSR
12. Saccharomyces cerevisiae*	IIGETGGKNFHLVHPSANISHAVLSTIRGTFEFQGQKCSAASR

1. Hempel, J. et al., Eur. J. Biochem. 141: 21-35, 1984; 2. von Bahr-Lindstrom, H. et al., Eur. J. Biochem. 141: 37-42, 1984; 3. Hempel, J. et al., Eur. J. Biochem. 153: 13-28, 1985; 4. Johansson, J. et al., Eur. J. Biochem. 172: 527-533, 1988; 5. & 6. Farres, J. et al., Eur. J. Biochem. 180: 67-74, 1989; 7. Dunn, T. et al., J. Biol. Chem. 264: 13057-13065, 1989; 8. Jones, D. et al., PNAS 85: 1782-1785, 1988; 9. Kok, M. et al., J. Biol. Chem. 264: 5442-5451, 1989; 10. Pickett, M. et al., Gene 51: 217-226, 1987; 11. Weretilnyk, E.A. and Hanson, A.D., Proc. Natl. Acad. Sci. USA, 87:2745, 1990; 12. Krzywicki, K. and Brandriss, M.C., Mol. Cell. Biol. 4: 2837-2842, 1984. *glutamic-γ-semialdehyde dehydrogenase

Up to the present time, twelve aldehyde dehydrogenases have been sequenced (von Bahr-Lindstrom et al., 1984; Hempel et al., 1984, 1987; Krzywicki and Brandriss, 1984; Pickett et al., 1987; Johansson et al., 1988; Jones et al., 1988; Dunn et al., 1989, Farres et al., 1989; Kok et al., 1989; Weretilnyk and Hanson, 1990). Of all the cysteine residues present in human aldehyde dehydrogenase, only cysteine 302 is conserved. Cysteine 302 always aligns 34 residues downstream from glutamate 268 which is localized in a conserved region that spans from residue 265-274 (previously identified in our laboratory employing *carbonyl*-[^{14}C]-bromoacetophenone, Abriola et al., 1987). Glutamate 268 is also conserved in every aldehyde dehydrogenase so far sequenced.

Since glutamate 268 and cysteine 302 are both located at the active site of aldehyde dehydrogenase, both are likely candidates for the catalytic residue. The amino acid sequence comparison of the region of the molecule containing both residues (Table 3) demonstrates that the peptide containing glutamate 268 is considerably more conserved than the one containing cysteine 302. Thus, the argument for glutamate being the catalytic residue stems from greater conservation of amino acids around it, while the argument for cysteine is in its greater nucleophilicity. In the absence of concrete evidence, assuming that cysteine 302 is the catalytic residue that forms a covalent intermediate with substrates, glutamate 268 presumably functions in the ionization of this residue.

Acknowledgements

The authors wish to thank Mrs. Patricia LaSasso for typing and lay out of this manuscript. Financial support of USPHS Grant AA00186, Research Scientist Award K05 AA0046 and Charles and Johanna Busch Memorial Fund is gratefully acknowledged.

REFERENCES

Abriola, D.P., Fields, R., Stein, S., MacKerell, A.D., Jr. and Pietruszko, R., 1987, Active site of human aldehyde dehydrogenase, Biochemistry, 26:5679.

Abriola, D.P., MacKerell, A.D., Jr. and Pietruszko, R., 1990, Correlation of loss of activity of human aldehyde dehydrogenase with reaction of bromoacetophenone with glutamic acid-268 and cysteine-302 residues. Partial-sites reactivity of aldehyde dehydrogenase, Biochem. J., 266:179.

von Bahr-Lindstrom, Hempel, J. and Jornvall, H., 1984, The cytoplasmic isoenzyme of horse liver aldehyde dehydrogenase. Relationship to the corresponding human isoenzyme, Eur. J. Biochem., 141:37.

von Bahr-Lindstrom, Jeck, R., Woenckhaus, C., Sohn, S., Hempel, J. and Jornvall, H., 1985, Characterization of coenzyme binding site of liver aldehyde dehydrogenase: Differential reactivity of coenzyme analogues, Biochemistry, 24:5847.

Blatter, E.E., Tasayco, M.L., Prestwich, G. and Pietruszko, R., 1990, Chemical modification of aldehyde dehydrogenase by a vinyl ketone analog of an insect pheromone, Biochem. J., *in press*.

Carey, F.A. and Sundberg, R.J., 1984, Advanced Organic Chemistry. Second Edition. Part A: Structure and Mechanisms, Plenum Press, New York and London, p. 207.

Ding, Y.-S. and Prestwich, G.D., 1988, Chemical studies of proteins that degrade pheromones: cyclopropanated, fluorinated, and electrophilic analogs of unsaturated aldehyde pheromones, J. Chem. Ecol., 14:2033.

Dunn, T.J., Koleske, A.J., Lindahl, R. and Pitot, H.C., 1989, Phenobarbital-inducible aldehyde dehydrogenase in the rat. cDNA sequence and regulation of the mRNA by phenobarbital in responsive rats, J. Biol. Chem., 264:13057.

Durst, D., Milano, M., Kikta, E.J., Jr., Connelly, S.A. and Gruska, E., 1975, Phenacyl esters of fatty acids via crown ether catalysts for enhanced ultraviolet detection in liquid chromatography, Analyt. Chem., 47:1797.

Farres, J., Guan, K.-L. and Weiner, H., 1989, Primary structure of rat and bovine liver mitochondrial aldehyde dehydrogenases deduced from cDNA sequences, Eur. J. Biochem., 180:67.

Fukaya, M., Tayama, K., Tamaki, T., Tagami, H., Okumura, H., Kawamura, Y.. and Beppu, T., 1989, Cloning of the membrane-bound aldehyde dehydrogenase gene of *Acetobacter polyoxogenes* and improvement of acetic acid production by use of the cloned gene, Appl. Environ. Microbiol., 55:171.

Hempel, J.D., 1981, Chemical modification of human liver aldehyde dehydrogenase isoenzymes E1 and E2. Doctoral Dissertation, Rutgers University. Dissertation Abstracts International 42:3664B, University Microfilms No. DA80204216.

Hempel, J.D. and Pietruszko, R., 1981, Selective chemical modification of human liver aldehyde dehydrogenases E1 and E2 by iodoacetamide, J. Biol. Chem., 256:10889.

Hempel, J.D., Pietruszko, R., Fietzek, P. and Jornvall, H., 1982, Identification of a segment containing a reactive, cysteine residue in human liver cytoplasmic aldehyde dehydrogenase (isoenzyme E1)., Biochemistry, 21:6834.

Hempel, J., von Bahr-Lindstrom, H. and Jornvall, H., 1984, Aldehyde dehydrogenase from human liver. Primary structure of the cytoplasmic isoenzyme, Eur. J. Biochem., 141:21.

Hempel, J., Kaiser, R. and Jornvall, J., 1985, Mitochondrial aldehyde dehydrogenase from human liver. Primary structure, differences in relation to the cytosolic enzyme, and functional correlations, Eur. J. Biochem., 153:13.

Hsu, L.C., Tani, K., Fujiyoshi, T., Kurachi, K. and Yoshida, A., 1985, Cloning of cDNAs for human aldehyde dehydrogenases 1 and 2, Proc. Natl. Acad. Sci., USA, 82:3771.

Johansson, J., von Bahr-Lindstrom, Jeck, R., Woenckhaus, C. and Jornvall, H., 1988, Mitochondrial aldehyde dehydrogenase from horse liver. Correlations of the same species variants for both the cytosolic and the mitochondrial forms of an enzyme, Eur. J. Biochem., 172:527.

Jones, D.E., Jr., Brennan, M.D., Hempel, J. and Lindahl, R., 1988, Cloning and complete nucleotide sequence of a full-length cDNA encoding a catalytically functional tumor-associated aldehyde dehydrogenase, Proc. Natl. Acad. Sci. USA, 85:1782.

Kok, M., Oldenhuis, R., van der Linden, M.P.G., Meulenberg, C.H.C., Kingma, J. and Witholt, B., 1989, The *Pseudonomas olevorans* alkBAC operon encodes two structurally related rubredoxins and an aldehyde dehydrogenase, J. Biol. Chem., 264:5442.

Krzywicki, K.A. and Brandriss, M.C., 1984, Primary structure of the nuclear PUT2 gene involved in the mitochondrial pathway for proline utilization in *Saccharomyces cerevisiae*, Mol. Cell. Biol., 1984, 4;2837.

Loomes, K.M., Midwinter, G.G., Blackwell, L.F. and Buckley, P.D., 1990, Evidence for reactivity of serine-74 with trans-4-(N,N-dimethyl-amino)cinnamaldehyde during oxidation by the cytoplasmic aldehyde dehydrogenase from sheep liver, Biochemistry, 29:2070.

MacKerell, A.D., Jr., MacWright, R.S. and Pietruszko, R., 1986, Bromoacetophenone as an affinity reagent for human liver aldehyde dehydrogenase, Biochemistry 25:5182.

Pedersen, C.J. and Frensdorff, K.H., 1972, Macrocyclic polyethers and their complexes, Angew. Chem., Int. Ed. Engl. 11:16.

Pickett, M., Gwynne, D.I., Buxton, F.P., Elliott, R., Davies, R.W., Lockington, R.A., Scazzocchio, C. and Sealy-Lewis, H.M., 1987, Cloning and characterization of the aldA gene of *Aspergillus nidulans*, Gene, 51:217.

Pietruszko, R. and Yonetani, T., 1980, Aldehyde dehydrogenase from liver, In: Methods in Enzymology, Collowick and Kaplan (Eds.), 71:772.

Prestwich, G.D., Graham, S. McG., Handley, M., Latli, B., Streinz, L. and Tasayco J., M.L., 1988, Enzymatic processing of pheromones and pheromone analogs, Experientia, 45:267.

Shaw, E., 1970, Chemical modification by active-site-directed reagents. In: The Enzymes (Student Edition), Vol. 1, Structure and Control, Academic Press, p. 91.

Tasayco J., M.L. and Prestwich, G.D., 1990a, A specific affinity reagent to distinguish aldehyde dehydrogenases and oxidases, J. Biochem. Chem., 265:3094.

Tasayco J., M.L. and Prestwich, G.D., 1990b, Aldehyde-oxidizing enzymes in a adult moth: in vitro study of aldehyde metabolism in *Heliothis virescens*, Arch. Biochem. Biophys., 276:444.

Tu, G.-C. and Weiner, H., 1988a, Identification of the cysteine residue in the active site of horse liver mitochondrial aldehyde dehydrogenase, J. Biol. Chem., 263:1212.

Tu, G.-C. and Weiner, H., 1988b, Evidence for two distinct active sites on aldehyde dehydrogenase, J. Biol. Chem., 263:1218.

Weretilnyk, E.A. and Hanson, A.D., 1990, Molecular cloning of a plant betaine-aldehyde dehydrogenase, an enzyme implicated in adaptation to salinity and drought, Proc. Natl. Acad. Sci. USA, 87:2745.

PH EFFECTS ON CYTOPLASMIC ALDEHYDE DEHYDROGENASE FROM

SHEEP LIVER

Paul D. Buckley, +Rosemary L. Motion, Leonard F. Blackwell
and Jeremy P. Hill

Department of Chemistry and Biochemistry, Massey University
Palmerston North, New Zealand
+New Zealand Dairy Research Institute, Palmerston North
New Zealand

INTRODUCTION

Although studies on the effect of pH on aldehyde dehydrogenase have
been carried out in a number of laboratories , the results of these studies do
not give a complete picture of the mechanism of the aldehyde dehydrogenase
catalyzed oxidation of propionaldehyde over an extended pH range. The
enzyme catalyzed oxidation reaction at pH 7.0 and 7.6 is generally agreed to
occur by the following ordered mechanism:

$$E \rightleftharpoons E.NAD^+ \rightleftharpoons E.NAD^+ . ALD \rightleftharpoons E.NADH.acyl \rightarrow {}^*E.NADH \rightleftharpoons E.NADH \rightleftharpoons E$$

Scheme I

At least at low propionaldehyde concentrations, the rate of the reaction is
controlled by the slow conformational change which occurs during the release
of NADH from the enzyme (Blackwell et al., 1987; Dickinson, 1985).

EXPERIMENTAL METHODS

Materials

NADH (grade III) and NAD$^+$ (grade III) were obtained from Sigma
Chemical Co. (St. Louis, MO, U.S.A.), propionaldehyde solutions were
prepared as described by MacGibbon et al. (1977a).

Methods

Sheep liver cytosolic aldehyde dehydrogenase was prepared essentially
by the method of MacGibbon et al. (1979) with the addition of a pH gradient
chromatography step (Dickinson et al., 1981) to remove mitochondrial aldehyde dehydrogenase contamination. The method of Motion (1986) was used
to determine enzyme active-site concentrations.

Gel filtration chromatography was performed on a Pharmacia FPLC
system using a calibrated Superose 6HR 10/30 column. Enzyme assays were
carried out as described by Bennett et al. (1982).

RESULTS AND DISCUSSION

Detailed steady-state kinetic studies of propionaldehyde oxidation with aldehyde dehydrogenase have been performed at pH 7.0 (Hart and Dickinson, 1982) and pH 7.6 (MacGibbon et al., 1977b). Both groups report similar results with Lineweaver-Burke plots being linear at low aldehyde concentrations but curved at high aldehyde concentrations indicating substrate activation. Dickinson (1985) has shown that the substrate activation effect results from aldehyde binding to the E.NADH binary complex which increases the rate-limiting release of NADH from the enzyme.

In order to test the hypothesis that NADH release from the enzyme was rate-limiting throughout the pH range 5 to 9, displacement experiments were carried out in the stopped flow apparatus (MacGibbon et al., 1977b). At all pH values, a biphasic displacement reaction was observed, as was seen previously at pH 7.6 (MacGibbon et al., 1977b) and pH 7.0 (Dickinson, 1985). The apparent rate constants derived for the two processes, λ_{fast} and λ_{slow}, were measured at each pH and the results are shown in Figure 1.

Both λ_{slow} and λ_{fast} have their highest values at pH 5, decrease markedly between pH 5 and 6 and then decrease steadily from pH 6 to 9. In Figure 2, λ_{slow} is compared with the k_{cat} value obtained by extrapolating the linear portion of the Lineweaver-Burke plot obtained at low propionaldehyde, (the non-activated k_{cat} value.) Between pH 7 and 9, the pH dependencies of λ_{slow} and the non-activated k_{cat} closely parallel one another and in this pH range the conformation change controlling NADH release is rate limiting. The wide divergence between λ_{slow} and k_{cat} (non-activated) at below pH 7 suggests that some other step is rate limiting in this pH range.

However, even at pH 5.0, a low amplitude presteady-state burst in fluorescence was observed (approximately 4% of the burst at pH 7.6) upon rapidly mixing enzyme and NAD$^+$ with propionaldehyde in the stopped-flow apparatus (Figure 3). This burst is thought to be associated with the acyl-

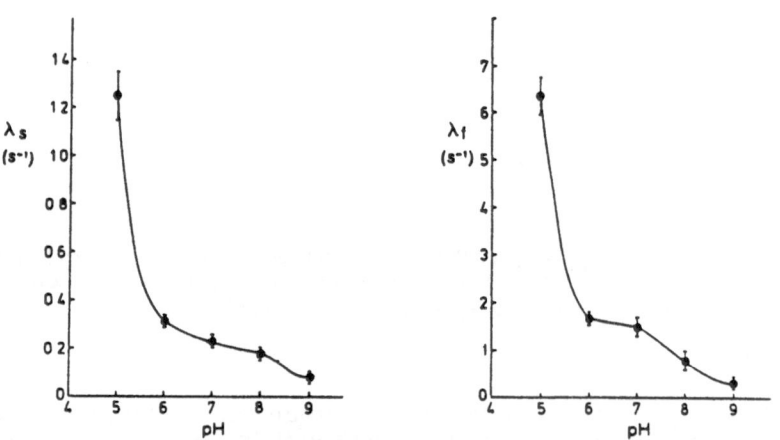

Fig. 1. The Effect of pH on the Displacement of NADH. NADH displacement experiments were carried out over the pH range 5-9 using buffers of constant ionic strength (I = 0.1 M). Solutions containing enzyme (9.0 µM) and NADH (40 µM) in 3.3 mM pH 7.3 phosphate buffer were rapidly mixed with solutions containing NAD$^+$ (2.5 mM) in the buffer of the appropriate pH.

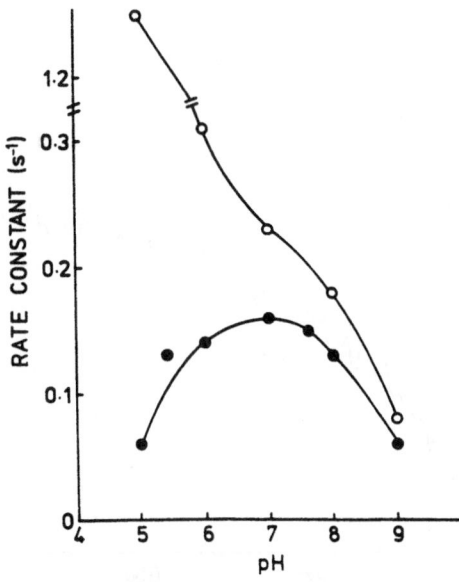

Fig. 2. Relationship Between λ_{slow} and k_{cat} (non activated) for Propionaldehyde. The profile for k_{cat} (●) was determined for assays containing aldehyde dehydrogenase (2.1 µM) and NAD⁺ (1 mM) with propionaldehyde in constant ionic strength buffers (0.1 M). For the displacement experiments (o), one syringe contained aldehyde dehydrogenase (9.0 µM) and NADH (40 µM) and the other contained NAD⁺ (2.5 mM).

enzyme hydrolysis step in the mechanism (Hart and Dickinson, 1982; Dickinson and Haywood, 1986). Thus, as the enzyme complex produced by the step associated with the fluorescence burst is *E.NADH of Scheme I , then the release of NADH from this complex should still be rate-limiting at low pH as was found to be the case at pH values above 7.0 (Dickinson, 1985; Blackwell et al., 1987).

Figure 4 shows the variation of the burst rate constant with pH. The fluorescence burst pH profile we report here and that shown by Dickinson (1986) are very similar. The fluorescence burst profile mirrors the pH profile for the steady-state phase of the oxidation of 4-N,N-dimethylcinnamaldehyde (Dunn and Buckley, 1985), as is to be expected since both these processes are believed to reflect the acyl-enzyme hydrolysis step in the mechanism.

The burst data seems to require that NADH release be rate limiting even at low pH even though the NADH displacement rates at low pH are much greater than k_{cat} (non-activated). How can this apparent contradiction be resolved?

The results of Blackwell et al. (1987) showed that predilution of the enzyme in the absence of propionaldehyde gave lower oxidation rates in assays at pH 7.6 when compared with enzyme which had not been prediluted. Dilution was therefore somehow inactivating the enzyme which, however, could be protected from the effect by predilution in the presence of propionaldehyde.

Figure 5 shows the effect of enzyme concentration on the oxidation of propionaldehyde at pH 5.0. At high enzyme concentrations, the plot is linear (data not shown), however, at low concentrations of enzyme (<0.3 µM) the

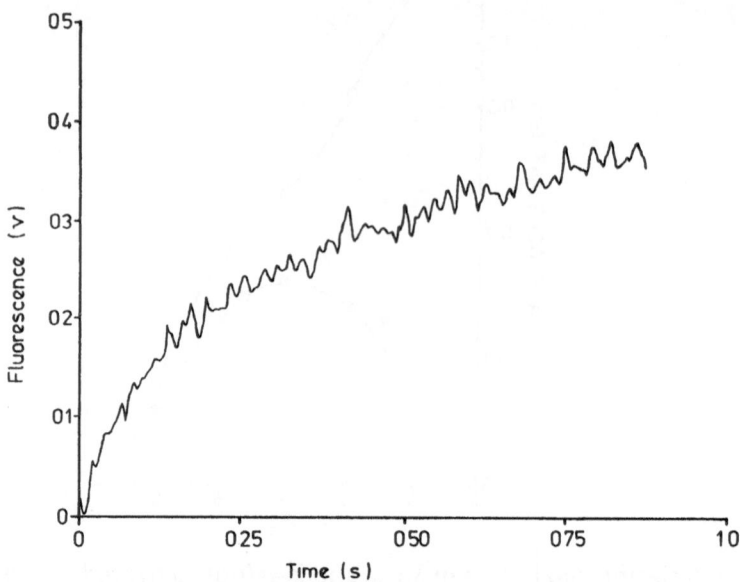

Fig. 3. Burst in NADH Fluorescence with 20 mM Propionaldehyde at pH 5.0.
Enzyme (2.3 μM) and NAD⁺ (2 mM) in distilled water was rapidly mixed
with propionaldehyde (20 mM) in constant ionic strength sodium
acetate buffer, pH 5.0 (200 mM).

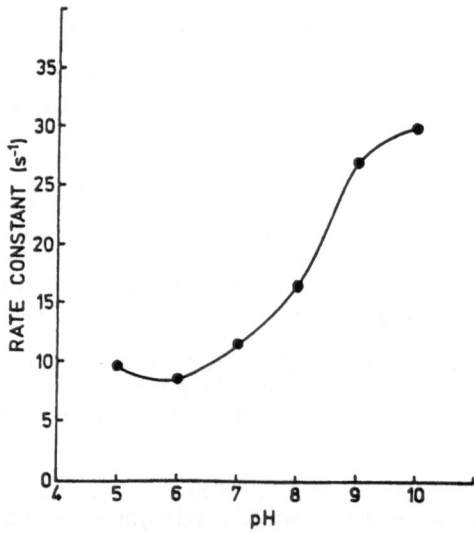

Fig. 4. The Effect of pH on the NADH Burst at High Concentrations of
Propionaldehyde. The burst pH profile was determined by rapidly
mixing solutions of aldehyde dehydrogenase (2.3 μM) and NAD⁺
(2 mM) in water with a second solution containing propionaldehyde
(40 mM) dissolved in the buffer of the appropriate pH (0.2 M).

plot is not linear, showing that inactivation of enzyme by predilution is also occurring at low pH.

The simplest explanation for such a result is that on predilution the enzyme is dissociating into an inactive species. In order to explore this possibility, we carried out gel filtration chromatography on the enzyme and the results for pH 7.4 and 5.0 are shown in Figure 6. In both cases, two distinct peaks in absorbance at 280 nm were eluted from the column. At pH 7.4 and 44 μM enzyme, one major enzymatically active peak (peak B, molecular weight 200,000) and one minor enzymatically inactive peak (peak A, molecular weight 50,000 to 100,000) were observed with the amount of peak A increasing when the enzyme was ten-fold more dilute. At pH 5.0 the amount of the enzymatically inactive peak A had increased markedly compared to pH 7.4 at the expense of the enzymatically active peak B both with 44 μM enzyme and 4.4 μM enzyme. The presence of magnesium favored the enzymatically active form.

The important kinetic consequences of this dissociation phenomena can be demonstrated in a number of ways.

The observed rate of production of NADH for an assay at pH 5.0 with 20 mM propionaldehyde and 2 mM NAD$^+$ depends on the order of mixing and on the times between addition of enzyme, substrate and cofactor. If, when enzyme is added last to an assay mixture, the rate is 0.010 μM s^{-1}, the same rate can be obtained by adding the enzyme to the buffer ten seconds before adding both NAD$^+$ and propionaldehyde. However, if enzyme is prediluted in the buffer for 15 minutes before addition of substrate and cofactor, the rate of NADH production is only 20 percent of the previous value. Premixing enzyme, buffer and propionaldehyde (20 mM) for 15 minutes before addition of NAD$^+$

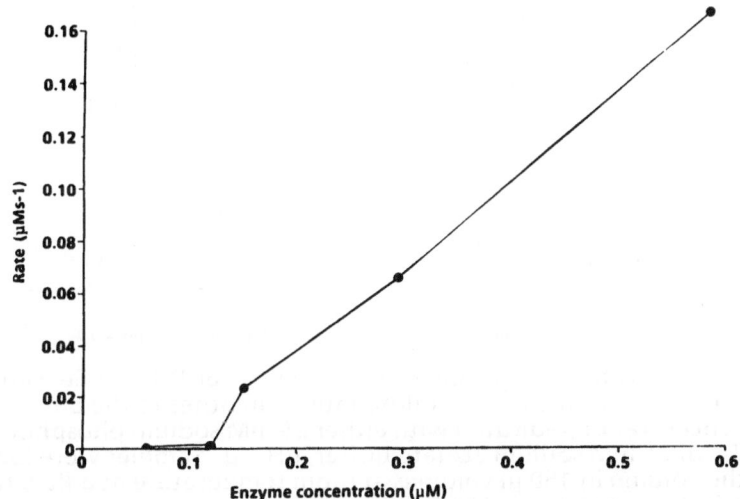

Fig. 5. Plot of Propionaldehyde Oxidation Rate Versus Enzyme Concentration at pH 5.0. Enzyme, buffer (25 mM sodium acetate, 5.0) and 1 mM NAD$^+$ were incubated for 5 min at 25°C prior to the addition of 20 mM propionaldehyde. The steady-state production of NADH was monitored by absorbance on a Hewlett Packard 8452 Diode Array Spectrophotometer.

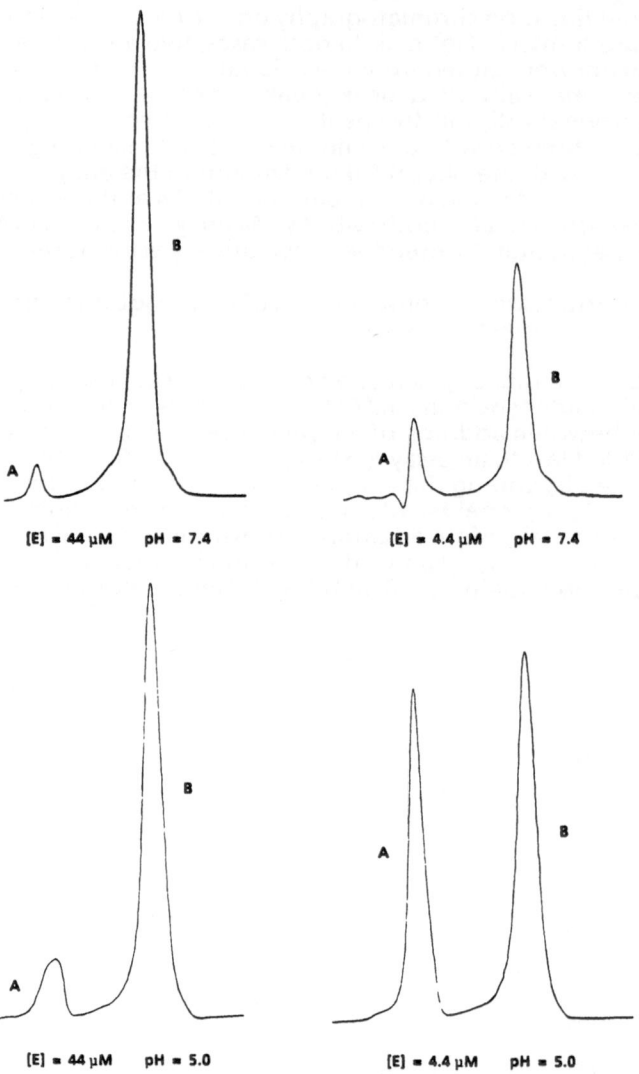

Fig. 6. Elution Profiles of Enzyme from a Superose 6HR 10/30 Gel Filtration Column. Enzyme was prediluted for 30 minutes to the desired preload concentration indicated with either 25 mM sodium phosphate buffer, pH 7.4 or 25 mM sodium acetate buffer, pH 5.0. Samples were loaded onto the column in 150 µl volumes at room temperature at a flow rate of 0.5 ml min^{-1} in a Pharmacia FPLC system.

gave the same rate as the control assay when enzyme was added last (0.010 µM s^{-1}). Although 20 mM propionaldehyde provides 100 percent protection against the partial dissociation of the tetramer, premixing with NAD$^+$ (2 mM) for the same length of time results in an NADH production rate about 50% of the control rate.

Another example of the importance of these effects is shown in Figure 7. A low concentration of enzyme (0.075 µM) was preincubated in pH 5.22 and 7.6 buffers at buffer concentrations between 2.5 mM and 200 mM. Predilution at pH 5.22 is, again, more inactivating that at pH 7.6; however, lower ionic strength causes a greater inactivation of the enzyme at pH 7.6 than at pH 5.22. Clearly, there are a number of factors which determine the functional concentration of enzyme in assays: predilution, predilution period, pH, ionic strength, presence of substrate (propionaldehyde) and enzyme concentration.

Figure 8 shows a Lineweaver-Burke plot for the oxidation of propionaldehyde by the enzyme at pH 5.22 in which the effect of dissociation to an inactive form is minimized by adding enzyme last to the assay mixture.

In contrast to the plots at pH 7.0 (Hart and Dickinson, 1982) and 7.6 (Blackwell et al., 1977b), the plot is curved at low propionaldehyde concentrations but linear at high propionaldehyde concentrations. Using the linear section of the plot V_{max}, K_m, and k_{cat} values can be determined. The K_m for propionaldehyde was approximately 58 µM and has thus increased significantly from a value of approximately 2 µM at pH 7.0 (Hart & Dickinson, 1982). The k_{cat} value determined from the plot was found to be 0.493 s^{-1}. According to Blackwell et al. (1987), in the absence of substrate activation effects k_{cat} can be determined from the values of the apparent first order NADH displacement rate constants by the expression $\lambda_s \lambda_f/(\lambda_s + \lambda_f)$. At pH 5.22, this gives a calculated value for k_{cat} of 0.52 s^{-1}. The similarity of this value with the measured k_{cat} value strongly suggests that, as proposed on the basis of fluorescence burst data, NADH release from the enzyme is rate limiting at low pH values, and that substrate activation of NADH release from the enzyme no longer occurs at this pH. This is in contrast to the situation at pH 7.0 and 7.6 where substrate activation at approximately 20 mM propionaldehyde produces a three-fold enhancement in the rate of NADH release (Hart and Dickinson, 1982; MacGibbon et al., 1977b). This concentration of propionaldehyde gives oxidation rates approaching V_{max} at pH 5.22. If such rates were three-fold activated, they would no longer relate to the calculated k_{cat} value from the displacement rate constants.

The Lineweaver-Burke plot is presumably nonlinear at low propionaldehyde concentrations because such concentrations of propionaldehyde are insufficient to suppress the dissociation to the inactive form.

We also wish to report that pyrophosphate buffers used in the pH range 7 to 9 have a specific effect on the rate at which the enzyme can oxidize propionaldehyde. Figure 9 shows how pH effects the steady-state turnover of the enzyme using a mixture of nonpyrophosphate (phosphate, carbonate, glycine buffers) and pyrophosphate buffers. The discrepancy between the two profiles of Figure 9 can be explained in terms of the enhancement of the rate-limiting release of NADH from the enzyme in pyrophosphate buffers (Figure 10). Such a result indicates that atypical pH/activity profiles can result from the use of pyrophosphate buffers to study this enzyme.

In conclusion, there are two main factors which appear to effect the rate of aldehyde oxidation by sheep liver cytosolic aldehyde dehydrogenase; firstly, there is a mechanistic control through NADH release, and secondly, there is a

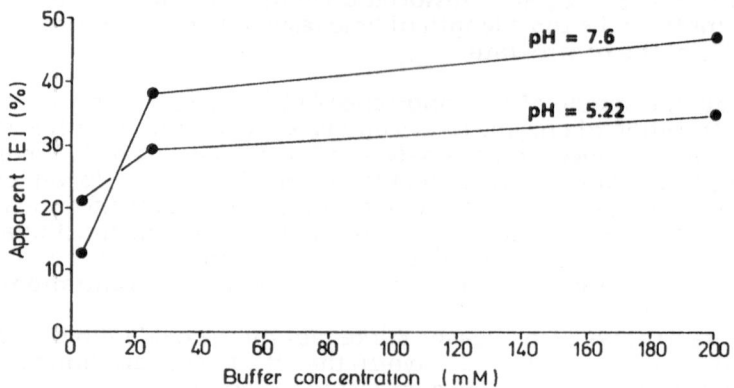

Fig. 7. Effect of Buffer Concentration on the Inactivation Effect which Occurs on the Predilution of Aldehyde Dehydrogenase at pH 5.22 and 7.6. The aldehyde dehydrogenase active site concentration (3 µM) was determined at pH 7.6 in 35 mM phosphate buffer by assaying with NAD$^+$ (3 mM) and propionaldehyde (20 mM). The enzyme was then diluted 40-fold into the buffer system containing 1 mM NAD$^+$, and after 5 minutes propionalde-hyde (6 mM) was added. Activity was measured by monitor-ing NADH production in absorbance. The apparent enzyme concentra-tion (%) was then determined by dividing the observed activity by the activity expected from 40-fold dilution of the original enzyme, and multiplying by one hundred.

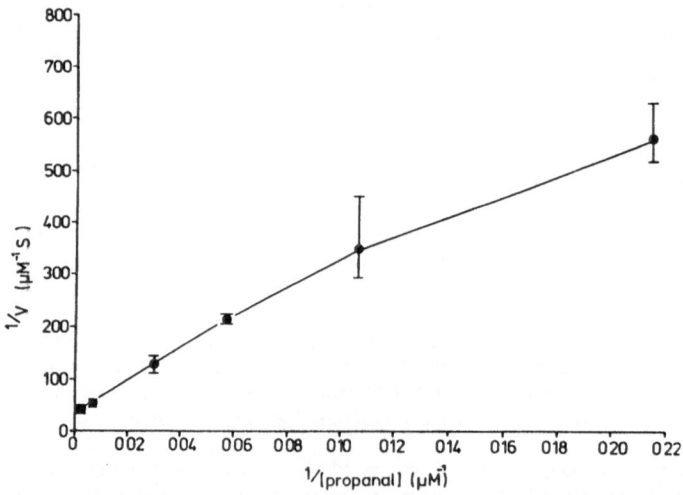

Fig. 8. Double Reciprocal Plot for the Oxidation of Propionaldehyde by the Enzyme at pH 5.22. Enzyme (40 µM) was added last to assay containing 1 mM NAD$^+$ and propionaldehyde at 25°C. The rate of NADH production was followed spectrophotometrically in a Hewlett Packard HP8452 spectrophotometer.

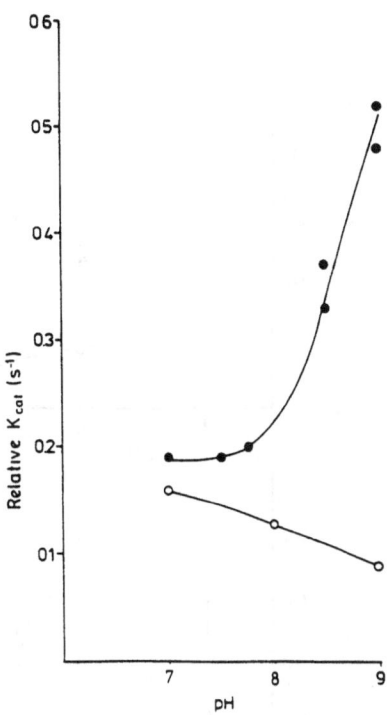

Fig. 9. Variation of k_{cat} with pH for Pyrophosphate and Nonpyrophosphate
Buffer Systems. Enzyme (3 μM) was assayed in both pyrophosphate
buffers I = 100 mM (●) and nonpyrophosphate buffers (O) using
phosphate, glycine and carbonate buffers, I = 100 mM at 25°C, spectrophotometrically in a Hewlett Packard HP 9452 spectrophotometer.

dissociation effect which leads to reduction in the effective enzyme concentration. As has been shown, a number of conditions can effect each or both of the
above factors. Under nonsubstrate activating conditions, k_{cat} is controlled by
the release of NADH from the enzyme throughout the pH range 5 to 9. Activation of the release of NADH at high pH values by substrate leads to enhanced
turnover rates. At low pH, apparent substrate activation may arise because of
an increase in the amount of active or associated enzyme on the addition of high
concentrations of propionaldehyde. On this latter point it was found that the
addition of 3.5 mM Mg^{2+} ions to assays performed in the linear propionaldehyde concentration range of Figure 8 resulted in up to 85% inhibition of rates
while addition of Mg^{2+} to assays in the nonlinear propionaldehyde concentration range caused up to 50% activation of rates. Such a concentration of
magnesium was also found to result in a total association of diluted enzyme as
shown by FPLC gel filtration on Superose 6 columns. Thus, when the enzyme is
in an associated form, induced by propionaldehyde at pH 5.22, magnesium
addition inhibits the enzyme through a slowing in the rate-determining release
of NADH as has been reported at pH 7.6 (Bennett et al., 1983). When the
propionaldehyde concentration is too low to cause significant association of the
enzyme, magnesium addition activates the enzyme by causing it to associate.

Finally, the sheep liver cytosolic aldehyde dehydrogenase has now been
crystallized and X-ray diffraction studies have been initiated with preliminary
results looking promising.

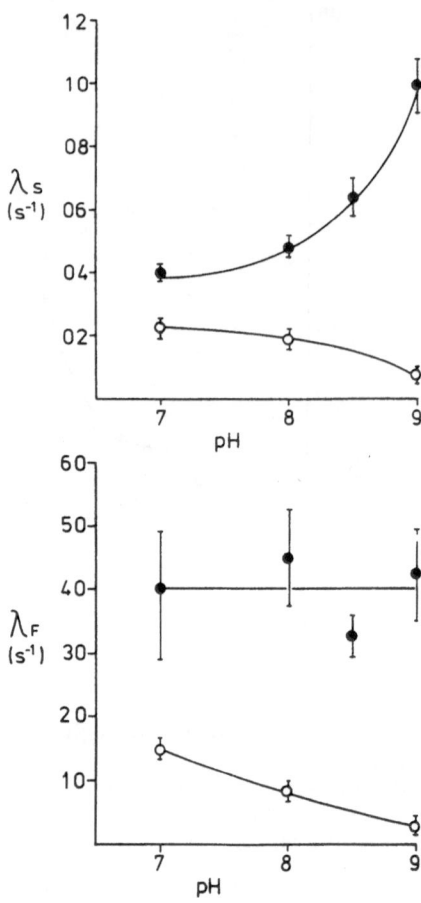

Fig. 10. Effect of pH on the Displacement of NADH from the Enzyme in Pyrophosphate (●) and Nonpyrophosphate (O) Buffers Over the pH Range 7-9. Experiments were performed as described in Figure 1.

REFERENCES

Bennett, A. F., Buckley, P. D. and Blackwell, L. F., 1982, Proton release during the pre-steady-state oxidation of aldehydes by aldehyde dehydrogenase. Evidence for a rate-limiting conformational change, Biochemistry, 21:4407-4413.
Bennett, A. F., Buckley, P. D. and Blackwell, L. F., 1983, Inhibition of the dehydrogenase activity of sheep liver cytoplasmic aldehyde dehydrogenase by magnesium ions, Biochemistry, 22:776-784.
Blackwell, L. F., Motion, R. L., MacGibbon, A. K. H., Hardman, M. J. and Buckley, P. D., 1987, Evidence that the slow conformation change controlling NADH release from the enzyme during the oxidation of propionaldehyde by aldehyde dehydrogenase, Biochem. J., 242:803-808.
Dickinson, F. M., 1985, Studies on the mechanism of sheep liver cytosolic aldehyde dehydrogenase, Biochem. J., 242:803-808.
Dickinson, F. M., 1986, Studies on the mechanism of sheep liver cytosolic aldehyde dehydrogenase. The effect of pH on the aldehyde binding steps and a reexamination of the problem of the site of proton release in the mechanism, Biochem. J., 238:75-82.

Dickinson, F. M. and Haywood, G. W., 1986, The effects of Mg^{2+} on certain steps in the mechanisms of the dehydrogenase and esterase reactions catalyzed by sheep liver aldehyde dehydrogenase, *Biochem. J.,* **222**:877-883.

Dickinson, F. M., Hart, G.J. and Kitson, T. M., 1981, The use of pH gradient ion-exchange chromatography to separate sheep liver cytoplasmic aldehyde dehydrogenase from mitochondrial contamination, and observations on the interaction between the pure cytoplasmic enzyme and disulfiram, Biochem. J., 199:573-579.

Dunn, M. F. and Buckley, P. D., 1985, Kinetic and spectroscopic characterization of the sheep liver aldehyde dehydrogenase acyl-enzyme, Enzymology of Carbonyl Metabolism 2: Aldehyde Dehydrogenase, Aldo-Keto Reductase and Alcohol Dehydrogenase, pp 15-27, A. R. Liss Inc., New York.

Hart, G. J. and Dickinson, F. M., 1982, Kinetic properties of highly purified preparations of sheep liver cytoplasmic aldehyde dehydrogenase, Biochem. J., 203:617-627.

MacGibbon, A. K. H., Blackwell, L. F. and Buckley, P. D., 1977a, Pre-steady-state kinetic studies on cytoplasmic sheep liver aldehyde dehydrogenase, Biochem. J., 167:469-477.

MacGibbon, A. K. H., Blackwell, L. F. and Buckley, P. D., 1977b, Kinetics of sheep-liver cytoplasmic aldehyde dehydrogenase, Eur. J. Biochem., 77:93-100.

MacGibbon, A. K. H., Motion, R. L., Crow, K. E., Buckley, P. D. and Blackwell, L. F., 1979, Purification and properties of sheep-liver aldehyde dehydrogenases, Eur. J. Biochem., 96:585-595.

Motion, R. L., 1986, Structural and mechanistic studies of sheep liver aldehyde dehydrogenase, Ph.D. Thesis, Massey University, New Zealand.

BIOCHEMICAL, IMMUNOLOGICAL, AND MOLECULAR CHARACTERIZATION OF A "HIGH KM"

ALDEHYDE DEHYDROGENASE*

Rolf Eckey, Rüdiger Timmann, John Hempel**,
Dharam P. Agarwal and H. Werner Goedde

Institute of Human Genetics
University of Hamburg, Hamburg, F.R.G.
**Department of Microbiology, Biochemistry
and Molecular Biology, University of Pittsburgh
Medical School, Pittsburgh, PA 15261

INTRODUCTION

Aldehyde dehydrogenase (ALDH), the enzyme mainly responsible for the oxidation of acetaldehyde and other aliphatic and aromatic aldehydes, plays an important role in the toxic consequences of a deranged acetaldehyde metabolism in alcohol-related disorders (Agarwal and Goedde, 1989, 1990; Goedde and Agarwal, 1989). Two broadly defined groups of ALDHs have been recognized based upon their Michaelis constants ("low Km" and "high Km" isozymes). The mammalian liver ALDHs differ in their electrophoretic mobility, isoelectric point, molecular size, kinetic properties, inhibition with disulfiram, subunit structure, as well as in their cellular and tissue distribution, and chromosomal assignment.

At least two "low Km" ALDHs (ALDH1 and ALDH2) have been purified to homogeneity from human liver and their catalytic and structural properties reported (Harada et al., 1980; Pietruszko et al., 1987). The liver ALDH1 (cytosolic) and ALDH2 (mitochondrial) isozymes differ in their electrophoretic properties and catalytic efficiencies. Sequence studies at the protein and gene level have also revealed differences at several positions in their primary and genomic structures (Hempel et al., 1984, 1985; Hsu et al., 1988, 1989; Agarwal et al., 1989a). The so-called "high Km" ALDH enzymes (designated here as ALDH3 and ALDH4) show much higher Km values with aliphatic aldehyde substrates (Pietruszko et al., 1987; Ryzlak and Pietruszko, 1989). A stomach-specific high Km ALDH isozyme (ALDH3) has attracted considerable attention due to its unique kinetic properties (Teng, 1983; Meier-Tackmann et al., 1984; Duley et al., 1985; Santisteban et al., 1985). This unique isozyme also exhibits complex multiple forms due to a genetic polymorphism (Yin et al., 1989).

In a previous study (Agarwal et al., 1989b), we have shown that an ALDH3-like activity was present in a human hepatocarcinoma in high amounts, but not detectable in normal liver. This tumor-associated ALDH shares many properties with the constitutive form of human stomach

---

*This work will be a part of the MD thesis of Rüdiger Timmann to be submitted to the Faculty of Medicine, University of Hamburg.

Enzymology and Molecular Biology of Carbonyl Metabolism 3
Edited by H. Weiner *et al.*, Plenum Press, New York, 1990

ALDH3, as well as with the tumor-associated class 3 ALDH expressed during rat hepatocarcinogenesis (Lindahl, 1982). The human and rat liver ALDH activity associated with hepatocellular carcinoma appears to be distinct from the low Km cytosolic and mitochondrial ALDHs (Agarwal et al., 1989b; Hempel et al., 1989). This unusual ALDH enzyme form is also detectable in some normal tissues from various species (Lindahl, 1986; Chieco et al., 1988).

The purpose of the present study was to purify and characterize the human stomach ALDH3 form, and to establish the structural and functional relationship to the other ALDH isozymes from man and rat. Specifically, we have investigated the electrophoretic, kinetic, and immunological properties of a human stomach-specific ALDH isozyme (ALDH3) and compared them with other human and rat ALDHs. We have also determined the amino acid sequences of selected peptides of ALDH3. These sequences are compared with known sequences of other human ALDH isozymes and rat class 3 ALDH form.

MATERIAL AND METHODS

Human autopsy stomach samples were obtained from the Department of Legal Medicine, University of Hamburg. The tissue samples were washed in water and kept frozen at -20°C until used. Extracts were prepared by homogenizing the samples in 30 mM sodium phosphate buffer, pH 5.6 (1 ml buffer/g tissue), containing 1 mM EDTA and 1 mM dithioerythritol.

Aldehyde dehydrogenase activity was measured in tissue extracts and purified preparations at 25°C in semimicro cuvettes using a double beam spectrophotometer equipped with a recorder (Kontron, Model Uvicon 810). The NADH production was recorded at 340 nm. The reaction mixture contained 0.1 M tetrasodium pyrophosphate buffer, pH 9.0, 2.0 mM NAD+, 10 mM pyrazole and various aldehyde substrates (5 mM).

Purification Procedure

Autopsy stomach tissue extract, prepared from 300-400 g tissue as described above, was centrifuged for 15 min at 27,000g. The supernatant was dialysed against 5 volumes of the extraction buffer and centrifuged to remove denatured proteins. ALDH3 was purified by successive chromatography with CM-Sephadex C50, Red Sepharose CL-6B, and 5'-AMP Sepharose 4B. Extraction buffer was used to equilibrate the columns and to remove unbound proteins. ALDH3 was eluted from the CM-Sephadex and Red Sepharose columns by a linear salt gradient (0-1M NaCl). For enzyme elution from the 5'-AMP Sepharose column a 100 mM sodium phosphate buffer, pH 8.0 containing 1 mM EDTA, 1 mM dithioerythritol, and 0.5 mg NADH/ml was used. Further purification was achieved by preparative isoelectric focusing (IEF) in agarose gels using a pH range of 3.5-9.5 (Eckey et al., 1988).

Protein Determination

Soluble protein content in different preparations was determined with Coomassie Brilliant Blue G 250 by the method of Macart and Gerbaut (1982) using bovine serum albumin as a standard.

Preparation of Anti-ALDH Antibodies

Specific antisera against human ALDH isozymes were raised in rabbits by immunizing the animals with the purified enzyme preparations as described before (Agarwal et al., 1984). One part of antigen (200-400 µg protein/ ml 0.9% NaCl) and one part of Freund's complete adjuvant were injected subcutaneously in the animals. Subsequent booster injections were given in incomplete adjuvant. In all, 4 injections of 1 ml each were given in weekly intervals. After 4 weeks, 2 ml blood was collected from the ear vein of the animal and antiserum tested for titer and specificity. The IgG fractions of the antisera were salted out by adding 2.8 g ammonium sulphate to 10 ml of antiserum, and the sediment was washed several times with 2.12 M ammonium sulphate. The hemoglobin free sediment was redissolved in 3 ml of distilled water, dialysed against distilled water for 24 h followed by dialysis against 20 mM sodium barbital buffer, pH 8.6, containing 0.1% sodium azide. After dialysis, any insoluble material was removed by centritrifugation at 27,000g for 20 min, and the clear supernatant was stored at −20°C in 0.2 ml portions until used.

Peptide and Amino Acid Sequence Analysis

The purified ALDH3 protein was S-(14C) carboxymethylated and cleaved with trypsin as described for the human liver ALDH1 and ALDH2 (Hempel et al., 1984, 1985). The tryptic digest was prefractionated in 30% acetic acid over 1.6 x 190 cm Sephadex G-75 column. The eluate was applied to a C-18 reversed phase HPLC column (Ultrapack, LKB, 4.6 x 250 mm) in 0.1% trifluoroacetic acid and eluted with a linear gradient of acetonitrile, with UV-monitoring at 214 and 280 nm. Aliquots from UV-absorbing peaks were taken for hydrolysis (6N HCl, 0.5% phenol, 110°C in vacuo, 24 h) and compositional analysis (Beckman 6300 amino acid analyzer) to judge peptide purity. Material from selected peaks was submitted to automated liquid phase Edman degradation (Beckman 890 M sequenator) with identification of the derivatives by HPLC.

RESULTS

Purification of Human Stomach ALDH3

Table 1 summarizes the results of a purification of ALDH3, giving details of yields and specific activities at different stages of purification. The purified preparations were found to be homogeneous by SDS-electrophoresis, as well as by a lack of cross-reaction with antisera raised against other proteins.

Electrophoretic and Kinetic Properties

Some of the main physico-chemical properties of the human stomach ALDH3 isozyme as compared to other human ALDH isozymes are summarized in Table 2. In contrast to other ALDH isozymes, ALDH3 shows a higher affinity

Table 1. Purification of Human Stomach Aldehyde Dehydrogenase (ALDH3)

Purification Step	Protein mg	Activity mU	Spec. Act. mU/mg protein	Yield %	Purification Factor
Crude extract	3100.0	24614	7.9	100.0	1.0
CM-Sephadex	594.0	23258	39.2	94.5	5.0
Red Sepharose	108.8	22824	209.8	92.7	26.6
5'-AMP Sepharose	18.0	15831	879.5	64.3	111.3
IEF	9.7	9331	962.0	37.9	121.8

Enzyme activity was determined with 5 mM 3-nitrobenzaldehyde as substrate and 2 mM NAD+ as coenzyme.

Table 2. Physico-Chemical Properties of Major Human ALDH Isozymes

Isozyme	pI	Number of Subunits	Subunit MW	Preferred Aldehyde Substrate	Coenzyme
ALDH1	5.1	4	54 kD	aliphatic	NAD+
ALDH2	4.9	4	54 kD	aliphatic	NAD+
ALDH3	6.0-6.5	2	60 kD	aromatic	NAD+/NADP+
ALDH4	6.8-7.0	2	63 kD	aromatic	NAD+/NADP+

for aromatic aldehydes such as benzaldehyde and nitrobenzaldehyde than for aliphatic aldehydes such as propionaldehyde and acetaldehyde, and accepts both NAD+ and NADP+ as coenzymes. Differences in substrate and coenzyme affinity of ALDH isozymes could also be demonstrated by using different substrates and coenzymes in the staining gels after separation by isoelectric focusing (Fig. 1). While the best staining results for ALDH1 and ALDH2 were obtained with propionaldehyde and NAD+, ALDH3 showed highest staining intensities with NADP+ and 3-nitrobenzaldehyde. Inhibition studies with disulfiram at a concentration of 20 μmol/l, pH 7.0, revealed a lower inhibition for ALDH3 (44%) as compared to the liver cytosolic ALDH (88%). Optimum pH for the enzymatic activity was determined to be in the range of 9.0-9.5. The purified ALDH3 preparation was unstable when kept beyond 45°C for more than three minutes. The enzyme was most stable in a ph range of 5.5-6.0. Table 3 summarizes some of the kinetic properties of human stomach ALDH3 with a range of aldehydes and coenzymes. It was apparent that acetaldehyde was a relatively poor substrate, whereas medium-chain aliphatic and aromatic aldehydes exhibited good activity. The apparent Km values for NAD+ and NADP+ as coenzymes (substrate: 3-nitrobenzaldehyde) were found to be 0.005 and 0.255 mM, respectively.

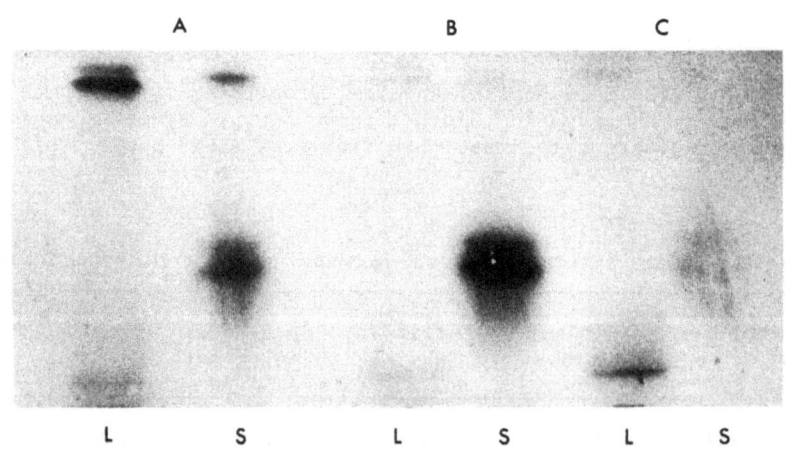

Fig. 1. Staining intensity of ALDH isozymes with various aldehyde substrates and coenzymes after separation by isoelectric focusing on agarose gel. L: liver extract; S: stomach extract. Substrates and coenzymes: A, propionaldehyde and NAD+; B, 3-nitrobenzaldehyde and NADP+; C, glutamic γ-semialdehyde and NADP+.

Table 3. Substrate Specificity of the Human Stomach-Specific ALDH3

Substrate	Km (mM)		Vmax (U/mg protein)	
	NAD+	NADP+	NAD+	NADP+
Acetaldehyde	11.0		0.5	
Propionaldehyde	5.3		0.75	
Benzaldehyde	0.41		0.9	
3-Nitrobenzaldehyde	0.1	0.032	9.8	42.2
4-Nitrobenzaldehyde	0.11		2.8	
Furfuraldehyde	5.5		0.35	
Glutaraldehyde	4.2		0.1	
Glyceraldehyde	17.6		0.15	
Cinnamic aldehyde	0.05	0.03	5.45	42.0
Glyceraldehyde-3-phosphate	0.42	2.0	0.2	2.1

Molecular Weight and Subunit Composition

On SDS-polyacrylamide slab gel electrophoresis, all purified ALDH isozymes migrated as single bands indicating that each isozyme is composed of
equal subunits (Fig. 2). The molecular weights of the subunits were determined to be 54 kD for ALDH1 and ALDH2, 60 kD for ALDH3, and 63 kD for ALDH4.
Migration distances of the native isozymes in polyacrylamide gradient gel
electrophoresis indicated an apparent native molecular weight for ALDH3
between 120 kD and 150 kD.

Immunological Relationship

Rabbit antibodies raised against human ALDH1, 2, 3, and 4 were highly
specific for their antigens (Fig. 3). In double diffusion tests using liver
and stomach extracts of different mammalian species, the liver extracts of

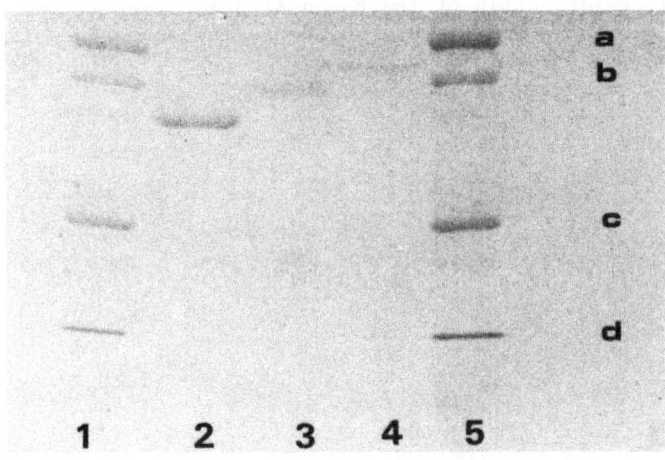

Fig. 2. SDS-electrophoresis of purified human low and high Km ALDH
isozymes for the determination of subunit molecular weights.
Lane 1 and 5: calibration kit; lane 2: ALDH2; lane 3: ALDH3;
lane 4: ALDH4. a: albumin, MW 67 kD; b: catalase, MW 60 kD;
c: lactate dehydrogenase, MW 36 kD; d: ferritin, MW 18.5 kD.

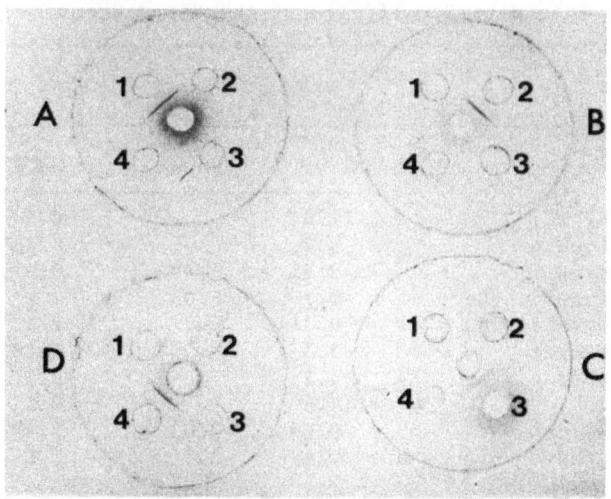

Fig. 3. Ouchterlony double diffusion tests showing specificity
of antibodies to purified human ALDH isozymes. A, B,
C, and D: antibodies to ALDH1, 2, 3, and 4 isozymes,
respectively (center wells). 1, 2, 3, and 4: purified
human ALDH1, 2, 3, and 4 isozymes, respectively.

these animals cross-reacted only with antibodies to human ALDH2 and ALDH4
(Fig. 4). No cross-reaction was detected with anti-ALDH3 antibodies.
Except for rat ALDH2, the precipitin lines obtained with isozymes from
other mammalian species did not completely fuse with the precipitin lines
of the corresponding human enzyme proteins, but showed the so-called
"spurs". Only the immunoprecipitates from rat and human ALDH2 formed a
completely fused line supporting a very close structural relationship of
the two isozymes from rat and man (Farres et al., 1989). By contrast, when
antibodies to human ALDH1 or ALDH3 were used, no cross-reacting material
was found with the liver or stomach homogenates from the animals. The
immunological relationship of the human ALDH isozymes to the corresponding
isozymes of rat, pig, cow, sheep, and horse are summarized in Table 4.

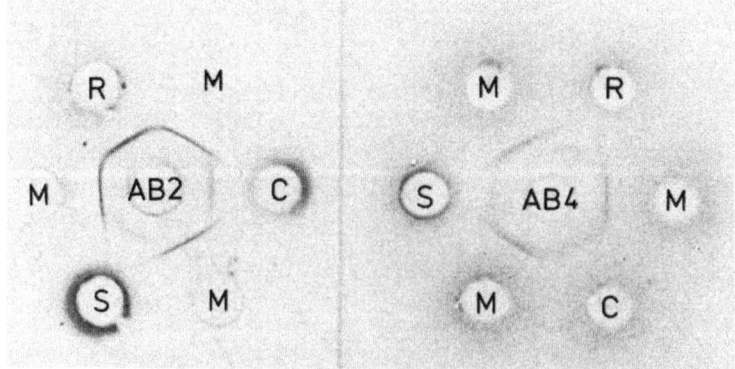

Fig. 4. Immunological cross-reaction of antibodies to
human ALDH2 (AB2) and ALDH4 (AB4) with liver
homogenates of man (M), rat (R), cow (C), and
sheep (S).

Table 4. Immunological relationship of human ALDH
isozymes to the corresponding isozymes of
other mammals

Species	ALDH1	ALDH2	ALDH3	ALDH4
Rat	n	c	n	p
Pig	n	p	n	p
Cow	n	p	n	p
Sheep	n	p	n	p
Horse	n	p	n	p

n: no immunological identity, p: partial identity,
c: complete identity.

Structural Relationships

The sequence of five peptides from human ALDH3 are summarized in
Fig. 5. Alignment of the sequence of these peptides with the corresponding
sequence from the human low Km ALDH isozymes and the rat class 3 ALDH
revealed a strong homology with the rat liver ALDH. Peptides from the human
stomach ALDH3 match the ALDH structure from rat class 3 ALDH at all positions examined except for a few species variations. Two segments corresponding exactly to residues 26-34 and 202-213 in the rat sequence are very
different from the corresponding parts in ALDH1 and ALDH2. Another segment

```
26   ILE GLN GLN LEU GLU ALA LEU GLN ARG                        Rat class 3
     ILE GLN GLN LEU GLU ALA LEU GLN ARG                        Human ALDH3
85   GLY ARG LEU LEU TYR LYS LEU ALA ASP                        Human ALDH1

202  HIS LEU THR PRO VAL THR LEU GLU LEU GLY GLY LYS
     HIS LEU THR PRO VAL THR LEU GLU LEU GLY GLY LYS
261  ASN LEU LYS ARG VAL THR LEU GLU LEU GLY GLY LYS

293  VAL LYS GLY LEU ILE  -   -   -   -   -  ASP ASN GLN LYS
     VAL MET GLY LEU ILE  -   -   -   -   -  ASN GLY GLN LYS
353  ILE LEU ASP LEU ILE GLU SER GLY LYS LYS GLU GLY ALA LYS

314  TYR ILE ALA PRO THR ILE LEU VAL ASP VAL ASP PRO GLN
     TYR ILE ALA PRO THR ILE LEU GLU ASP VAL ASP PRO GLN
380  PHE VAL GLN PRO THR VAL PHE SER ASN VAL THR ASP GLU

376  MET ILE ALA GLU THR SER SER GLY GLY VAL THR ALA ASN ASP
     MET ILE ALA GLU THR SER SER GLY GLY VAL ALA ALA ASN ASP
439  ALA ILE THR ILE SER SER ALA LEU GLN ALA GLY THR VAL TRP

390  VAL ILE VAL HIS ILE THR VAL PRO THR LEU  -  PRO PHE GLY GLY
     VAL ILE VAL(HIS)ILE THR LEU(HIS)THR)LEU  -  PRO PHE GLY GLY
453  VAL ASN CYS TYR GLY VAL VAL SER ALA GLN CYS PRO PHE GLY GLY
```

Fig. 5. Alignment of rat class 3 aldehyde dehydrogenase sequence
(Jones et al., 1988) with the human stomach ALDH3 (present
study) and human liver class 1 aldehyde dehydrogenase
(Hempel et al., 1989). Residues conserved between the three
structures are boxed. Upper numbers refer to rat class 3
ALDH, numbers below to human ALDH1. Positions with gaps are
supported from the total alignment of ALDH3 vs. ALDH1.

that corresponds to residues 293-301 in the rat, represents exchanges from Lys, Asp, and Asn to Met, Asn, and Gly, respectively. The segment 314-326, also unique for class 3 structure, shows a single species variation: Val exchanged with Glu. Residues 376-403 give three species variations: Thr, Val, and Pro exchanged with Ala, Leu, and His, respectively.

DISCUSSION

We have previously shown that a human hepatocarcinoma-associated ALDH isozyme was identical with the human stomach-specific ALDH3 isozyme form concerning its isoelectric point, Km values, substrate and coenzyme specificity, as well as in its immunological cross-reaction (Agarwal et al., 1989b). In the present study, the human stomach aldehyde dehydrogenase isozyme (ALDH3) has been purified to homogeneity, and characterized regarding electrophoretic, kinetic, immunological and structural properties. No evidence for multiple forms of ALDH3 was detected on isoelectric focusing of the crude stomach extracts and purified ALDH3 preparations.

Consistent with earlier observations, the results in the present study also suggest that ALDH3 isozyme is a homodimer in contrast to the low Km isozymes (ALDH1 and ALDH2) which are known to be homotetramers. These results suggested a parallel to the tumor-associated class 3 ALDH in rat liver. The rat class 3 ALDH structure shows a clear but distant relationship to class 1 and 2 ALDHs (Jones et al., 1988; Hempel et al., 1989). The peptide and amino acid analysis reported here clearly shows that the human stomach ALDH3 isozyme is a class 3 enzyme and strongly supports its identity to the tumor-associated rat liver enzyme. Further evidence for a common identity of the human ALDH3 and rat class 3 enzyme may be obtained by hybridization studies using rat liver tumor-associated ALDH cDNA. Such a study is currently in progress in our laboratory in collaboration with Ronald Lindahl.

Taken together, the results obtained in the present study - combined with those from our previous studies - suggest that low and high Km ALDH isozymes apparently represent different enzymes with distinct physiological functions and distant evolutionary relationship.

ACKNOWLEDGEMENTS

This study was supported by grants from the Deutsche Forschungsgemeinschaft, and the Bundesministerium für Forschung und Technologie, Bonn.

REFERENCES

Agarwal, D. P., Eckey, R., Harada, S. and Goedde, H. W., 1984, Basis of aldehyde dehydrogenase deficiency in Orientals: immunochemical studies. Alcohol, 1:111.

Agarwal, D. P., Cohn, P., Goedde, H. W. and Hempel, J., 1989a, Aldehyde dehydrogenase from human erythrocytes: structural relationship to the liver cytosolic enzyme. Enzyme, 42:47.

Agarwal, D. P., Eckey, R., Rudnay, A. C., Volkens, T. and Goedde, H. W., 1989b, "High Km" aldehyde dehydrogenase isozymes in human tissues: constitutive and tumor-associated forms, in: "Enzymology and Molecular Biology of Carbonyl Metabolism 2," H. Weiner and T. G. Flynn, eds., Alan R. Liss, New York.

Agarwal, D. P. and Goedde, H. W., 1989, Human aldehyde dehydrogenases: their role in alcoholism. Alcohol 6:517.

Agarwal, D. P. and Goedde, H. W., 1990, "Alcohol Metabolism, Alcohol Into-
     lerance and Alcoholism. Biochemical and Pharmacogenetic Approaches",
     Springer-Verlag, Berlin, Heidelberg.
Chieco, P., Normanni, P. and Moslen, M. T., 1988, Localization of high
     benzaldehyde dehydrogenase activity in rat upper gastrointestinal
     tract mucosa: a quantitative histochemical study. J. Histochem.
     Cytochem., 36:245.
Duley, J. A., Harris, O. and Holmes, R. S., 1985, Analysis of human
     alcohol- and aldehyde-metabolizing isozymes by electrophoresis and
     isoelectric focusing. Alcoholism: Clin. Exp. Res., 9:263.
Eckey, R., Agarwal, D. P., Volkens, T. and Goedde, H. W., 1988, A simple
     and rapid method for the purification of aldehyde dehydrogenase
     isozymes from human liver, in: "Alcohol Toxicity and Free Radi-
     cals Mechanisms", R. Nordmann, C. Ribiere, and H. Rouach, eds.,
     Pergamon Press, Oxford, New York.
Farrés, J., Guan K.-L. and Weiner, H., 1989, Primary structure of rat
     and bovine liver mitochondrial aldehyde dehydrogenases deduced
     from cDNA sequences. Eur. J. Biochem., 180:67.
Goedde, H. W. and Agarwal, D. P., 1989, Acetaldehyde metabolism: genetic
     variation and physiological implications, in: "Alcoholism:
     Biomedical and Genetic Aspects," H. W. Goedde and D. P. Agarwal,
     eds., Pergamon Press, New York.
Harada, S., Agarwal, D. P. and Goedde, H. W., 1980, Electrophoretic and
     biochemical studies of human aldehyde dehydrogenase isozymes in
     various tissues. Life Sci., 26:1771.
Hempel, J., von Bahr-Lindström, H. and Jörnvall, H., 1984, Aldehyde
     dehydrogenase from human liver. Primary structure of the cyto-
     plasmic isoenzyme. Eur. J. Biochem., 141:21.
Hempel, J., Kaiser R. and Jörnvall, H., 1985, Mitochondrial aldehyde
     dehydrogenase from human liver: primary structure, differences in
     relation to the cytosolic enzyme and functional correlations.
     Eur. J. Biochem., 153:13.
Hempel, J., Harper, K. and Lindahl, R., 1989, Inducible, class 3 aldehyde
     dehydrogenase from rat hepatocellular carcinoma and 2,3,7,8-tetra-
     chlorodibenzo-p-dioxin treated liver: distant relationship to the
     class 1 and 2 enzymes from mammalian liver cytosol/mitochondria.
     Biochemistry, 28:1160.
Hsu, L. C., Bendel, R. E. and Yoshida, A., 1988, Genomic structure of the
     human mitochondrial aldehyde dehydrogenase gene. Genomics, 2:57.
Hsu, L. C., Chang W. C. and Yoshida, A., 1989, Genomic structure of the
     human cytosolic aldehyde dehydrogenase gene. Genomics, 5:857.
Jörnvall, H., Hempel, J. and Vallee, B., 1987, Structures of human
     alcohol and aldehyde dehydrogenases. Enzyme, 37:5.
Jones, D. E., Brennan, M. D., Hempel, J. and Lindahl, R., 1988, Cloning
     and complete nucleotide sequence of a full-length cDNA encoding a
     catalytically functional tumor-associated aldehyde dehydrogenase.
     Proc. Natl. Acad. Sci. USA, 85:1782.
Lindahl, R., 1982, Induction of aldehyde dehydrogenase during hepatocar-
     cinogenesis, in: "Enzymology of Carbonyl Metabolism: Aldehyde
     Dehydrogenase and Aldo/Keto Reductase," H. Weiner and B. Wermuth,
     eds., Alan R. Liss, New York.
Lindahl, R., 1986, Identification of hepatocarcinogenesis-associated
     aldehyde dehydrogenase in normal rat urinary bladder. Cancer
     Res., 46:2502.
Macart, M. and Gerbaut, L., 1982, An improvement of the Coomassie Blue
     dye binding method allowing an equal sensitivity to various
     proteins. Application to cerebrospinal fluid. Clin. Chim. Acta,
     122:93.
Meier-Tackmann, D., Agarwal, D. P., Saha, N. and Goedde, H. W., 1984,
     Aldehyde dehydrogenase isozymes in stomach autopsy specimens from
     Germans and Chinese. Enzyme, 32:170.

Pietruszko, R., Ryzlak, M. T. and Forte-McRobbie, C. M., 1987, Multiplicity and identity of human aldehyde dehydrogenases. Alcohol Alcoholism, Suppl.1:175.

Ryzlak, M. T. and Pietruszko, R., 1989, Human brain glyceraldehyde-3-phosphate dehydrogenase, succinic semialdehyde dehydrogenase and aldehyde dehydrogenase isozymes: substrate specificity and sensitivity to disulfiram. Alcoholism: Clin. Exp. Res., 13:755.

Santisteban, I., Povey, S., West, L. F., Parrington, J. M. and Hopkinson, D. A., 1985, Chromosome assignment, biochemical and immunological studies on a human aldehyde dehydrogenase, ALDH3. Ann. Hum. Genet., 49:87.

Teng, Y.S., 1983, Stomach aldehyde dehydrogenase: Report of a new locus. Hum. Hered., 31:74.

Yin, S. J., Liao, C. S., Wang, S. L., Chen, Y. J. and Wu, C. W., 1989, Kinetic evidence for human liver and stomach aldehyde dehydrogenase-3 representing an unique class of isozymes. Biochem. Genet., 27:321.

PURIFICATION AND PROPERTIES OF BABOON CORNEAL ALDEHYDE DEHYDROGENASE:

PROPOSED UVR PROTECTIVE ROLE

Elizabeth M. Algar[1], Mahin Abedinia[1], John L. VandeBerg[2]
and Roger S. Holmes[1-3]

[1]Division of Science and Technology, Griffith University
Nathan, Brisbane, Qld 4111, Australia [2]Department of
Genetics, Southwest Foundation for Biomedical Research
PO Box 28147, San Antonio, TX, USA [3] Address for
correspondence

INTRODUCTION

The mammalian cornea plays an important role in protecting the
eye from ultra-violet radiation (UVR) induced damage, particularly
photosensitive retinal cells and the lens, by the absorption of UVR in
the 290-320 nm wavelength range (UV-B) (Boettner and Wolters, 1962;
Zigman, 1983). The mechanisms of corneal photoreception of UV-B light
have not been determined, as yet, although studies examining the action
spectra for the cornea in the induction of photokeratitis, have
suggested the major involvement of a soluble protein in this process
(Cogan and Kinsey, 1946; see Ringvold, 1980). Photobiological processes
in the cornea, arising from UV-B absorption, have also not been
described. Recent studies by Stephens et al (1989) have indicated,
however, that lipid peroxidation may account for phototoxicity in the
retina, via the generation of peroxidic aldehydes, which are the
cytotoxic products of lipid peroxidation (see Esterbauer et al,
1988).

A recent study from this laboratory reported the purification and
properties of bovine corneal aldehyde dehydrogenase (ALDH; E.C.
1.2.1.3), and proposed a dual role for this enzyme in protecting the
eye against UV-B light (Abedinia et al, 1990). This was based upon
the very high soluble protein levels for bovine corneal ALDH (around
0.5 percent wet weight of tissue), constituting 20-40 percent of total
soluble corneal protein; the earlier studies by Cogan and Kinsey (1946)
and others, which had proposed soluble corneal protein as the likely
UV-B photoreceptor; and the defined specificity of this enzyme towards
peroxidic aldehydes. These proposals were that corneal ALDH may serve
both as UV-B photoreceptor in the cornea, and in the disposal of
cytotoxic by-products generated by light absorption. The latter
catalytic role was supported by earlier biochemical studies of Holmes
and VandeBerg (1986) and Evces and Lindahl (1989), who had reported
very high levels of a Class 3 ALDH in mammalian cornea, with a
preference for peroxidic aldehydes.

In this study, we describe the isolation and characterisation of corneal ALDH from a primate species, the baboon (Papio hamadryas anubis), and report further evidence that this enzyme exists in very high levels in this tissue and apparently functions in peroxidic aldehyde metabolism.

MATERIALS AND METHODS

Chemicals. CM-sepharose CL-6B, agarose (IFF grade), Pharmalyte and isoelectric point calibration kits (3-10 and 5-8 ranges) were obtained from Pharmacia Fine Chemicals (Uppsala, Sweden). Biochemicals, including aldehyde substrates and buffers were from Sigma Chemical Co., St. Louis, MO., USA. 4-Hydroxynonenal was a gift from Dr H. Esterbauer, University of Graz, Austria.

Homogenate Preparation. Corneas were obtained from healthy baboons (Papio hamadryas anubis) at the Southwest Foundation for Biomedical Research, San Antonio, TEXAS, USA, and stored at -70^0C prior to homogenate preparation. Corneal extracts were prepared by homogenization of the cornea in an Ultra Turrax homogenizer (Janke and Kunkel, FRG) in 50 mM Tricine-sodium hydroxide buffer (pH8.0), containing 0.25 percent sodium deoxycholate and 0.1mM EDTA, in a ratio of 5 volumes of buffer per gram of tissue. The average weight of individual corneas was $75^\pm15mg$. Corneal homogenates were centrifuged at 45,000g for 30 mins at 4^0C prior to electrophoretic analysis and enzyme purification.

Isoelectric Focusing and Staining. Crude corneal extracts and purified baboon corneal ALDH were subjected to isoelectric focusing (IEF) according to the method of Pharmacia Fine Chemicals (Uppsala, Sweden) using Pharmalyte in the range 3-10. Samples were focused for 1500 Vhr at 10^0C and the gels histochemically stained for ALDH activity (Holmes, 1978) or protein, using Coomassie Brilliant Blue G-250. The activity staining mixture contained 33mg ml^{-1} agar, 65mM Tris-HCl (pH8.0), 0.25mM NAD, 2.5mM sodium pyruvate, 0.5mM phenazine methosulfate, 3mM methyl thiazolyl blue and 10mM benzaldehyde as substrate.

Spectrophotometric Assays. Corneal ALDH was routinely assayed in 100mM Tris-HCl (pH8.5) buffer, containing 0.5mM NAD, 20mM pyrazole and 0.01 mg ml^{-1} bovine serum albumin. Benzaldehyde (made up in methanol) was added last to a final concentration of 10mM to initiate the reaction. All enzyme assays were performed at 30^0C on an LKB UV/visible spectrophotometer and the production of NADH was followed at 340nm. Enzyme activity was expressed in International Units (μmole min^{-1}) per ml.

Purification of Baboon Corneal ALDH. Approximately 2g of corneal tissue comprising 26 corneas was routinely used for the purification of baboon corneal ALDH. Corneal extraction was performed as earlier described, following which the pH of the extract was lowered to 6.0 with dilute acetic acid. After standing on ice for 20 mins, denatured protein was removed by centrifugation at 45,000g for 15 mins. The clear supernatant was applied to a column of CM-Sepharose CL-6B ($2cm^2$x20cm), which had been pre-equilibrated in 20mM Mes(pH6.0) buffer, containing 0.1mM EDTA, 20 percent glycerol and 1mM dithiothreotel (DTT). The column was washed with the same buffer until the A280 of the eluate dropped to baseline level, following which the column was washed with 100mM Mops buffer (pH7.5), containing 0.1mM EDTA, 20 percent glycerol and 1mM DTT, and the eluting ALDH activity was pooled. This was then applied to a 5'AMP-Sepharose column ($1cm^2$ x 10cm), pre-equilibrated with 20mM Mops (pH7.0) buffer, containing 0.1mM EDTA, 20 percent glycerol and 1mM DTT. The column was washed with equilibrating buffer until a low A280 reading was reached, and the ALDH activity was eluted with 0.5mM NAD in the Mops buffer. Fractions containing corneal ALDH activity were then pooled, concentrated and stored at -70°C until required for analysis.

Protein Estimation. Protein concentration was determined by the method of Sedmak and Grossberg (1977) using Coomassie Brilliant Blue G250 to detect microgram amounts of protein.

Sodium Dodecyl Sulfate (SDS) Gel Electrophoresis. Polyacrylamide gels (10 percent resolving and 4 percent stacking) containing 1 percent SDS were prepared in a Protean TM Slab Cell system (Ames, 1974). Electrophoresis was performed vertically at 10°C, following which the gels were fixed in a solution of 30 percent methanol/10 percent acetic acid, and stained for protein with 0.25 percent Coomassie Brilliant Blue G-250 in 50 percent methanol/10 percent acetic acid for 2 hours and destained in the fixing solution.

Kinetic Studies. Aldehydes which were relatively insoluble in aqueous solution, such as benzaldehyde and trans-2-hexenal, were dissolved in methanol. Malondialdehyde was prepared from acid hydrolysis of malonaldehyde tetramethyl acetal (Hjelle et al, 1982). Kinetic studies were performed at pH8.5 in 100mM Tris-HCl buffer, containing 20mM pyrazole, 0.5mM NAD and 0.01mg ml^{-1} bovine serum albumin. The pH optimum of the corneal ALDH was tested using the following buffers and pH ranges: 100mM Mes (6.0-6.5); 100mM Mops (7.0-7.5); 100mM Tris (8.0-8.5); and 100mM glycine (9.0-10.0). For the latter study, ALDH was assayed using 10mM benzaldehyde as substrate.

RESULTS

Purification of baboon corneal ALDH. Baboon corneal ALDH was purified by extraction into buffer, acid precipitation treatment at pH6.0, CM-sepharose chromatography, and affinity chromatography using 5'-AMP Sepharose to achieve final purification. Table 1 summarises the results of this procedure, and provides details of yields and specific activities of ALDH at the various steps during purification. The procedure resulted in a homogeneous enzyme preparation in reasonable yield (25 percent), providing >200 µg of pure enzyme from 2g original cornea, which was free of alcohol dehydrogenase activity.

Table 1

Purification of Baboon Cornea Aldehyde Dehydrogenase (ALDHIII)

Purification Step	Protein	Activity	Specific Activity	Recovery
	mg	IU	IUmg^{-1}	%
Crude extract*	27.9	47.0	1.68	100
Cm-Sepharose	16.0	24.3	1.52	52
5'AMP Sepharose	0.21	12.0	57.1	25

* obtained from 2g cornea

Purity/Subunit Molecular Weight. SDS gel electrophoretic analysis (Fig.1) illustrated that corneal ALDH was in a highly purified state, and migrated with an apparent subunit molecular weight of 63,000. In addition, a comparison with the electrophoretic pattern for crude extract protein from baboon cornea indicated that ALDH represents the second most prevalent soluble protein in this tissue. This was confirmed

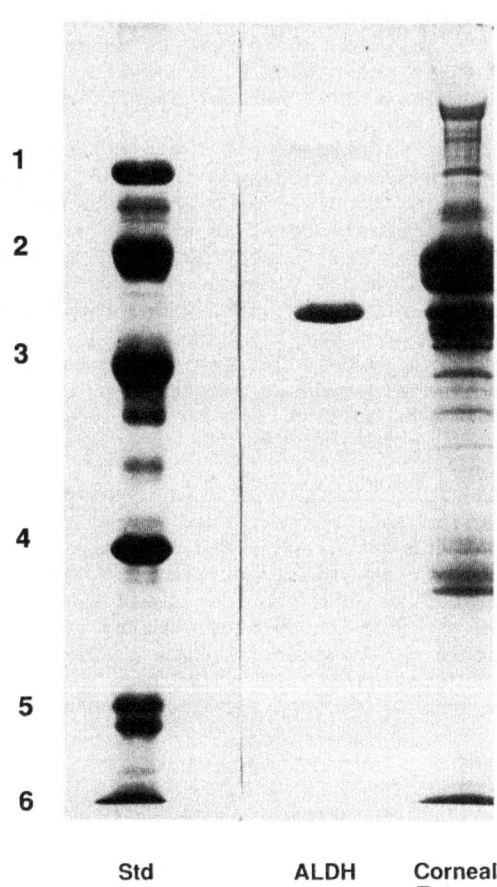

1

2

3

4

5

6

Std   ALDH Corneal
          Extract

Figure 1. Sodium dodecyl sulphate polyacrylamide gel of purified
baboon corneal ALDH (middle zone) and a crude extract of
baboon cornea (right zone). Protein subunit standards (Std)
include 1, phosphorylase b (94,000); 2, serum albumin
(67,000); 3, ovalbumin (43,000); 4, carbonic anhydrase
(30,000); 5, trypsin inhibitor (20,100); and 6, α-
lactalbumin (14,400). A subunit MW for the enzyme was
deduced from plots of log 10 MW versus mobility as 63,000.

using agarose-IEF gels resolving soluble corneal protein from crude extracts and the purified corneal ALDH preparation, for which 3-4 bands of protein (and activity) were readily distinguished (Fig.2).

Kinetic Analyses. The aldehyde substrates were chosen to represent a range of naturally occurring peroxidic aldehydes, which may be encountered in corneal tissue, as well as acetaldehyde, the primary product of ethanol metabolism, and an aromatic aldehyde, which has been previously reported as a good substrate for corneal ALDHs (Evces and Lindahl, 1989; Abedinia et al 1990). The enzyme was active with both NAD and NADP as coenzyme, but exhibited a ten fold lower Km with NAD (Table 2). The activity of baboon corneal ALDH was highest with benzaldehyde and 4-hydroxynonenal among the substrates studied, and the enzyme exhibited low activity and a high Km value with acetaldehyde. No detectable activity was observed using malondialdehyde or glyceraldehyde 3-phosphate as substrates. Substrate turnover (kcat) among the peroxidic aldehydes (malondialdehyde, trans-2-hexenal and 4-hydroxynonenal) was highest for the latter substrate, suggesting that longer chain, unsaturated aldehydes may be the preferred substrates for this enzyme.

The purified enzyme was optimally active at pH8.5, with that activity being approximately 4 times the activity at pH7.0

Table 2

Kinetic Properties of Baboon Corneal ALDH

Substrate	Michaelis Constant Km	Maximum Velocity Vmax	Turnover Number kcat*	$\frac{kcat}{Km}$
Aldehyde	mM	$IUmg^{-1}$	min^{-1}	$mM^{-1}min^{-1}$
acetaldehyde	40.0	12.8	800	20
benzaldehyde	0.56	75.2	4700	8393
trans-2-hexenal	0.29	45.0	2800	9655
malondialdehyde	-	0	0	-
4-hydroxynonenal	0.22	70.0	4400	20000
Coenzyme				
NAD	0.017	-	-	-
NADP	0.17	-	-	-

* moles substrate used $min^{-1}mole^{-1}$ active site (63,000 MW)

DISCUSSION

Baboon corneal aldehyde dehydrogenase, designated ALDH III by Holmes and VandeBerg (1986), has been purified to homogeneity, and characterized in terms of subunit MW, substrate and coenzyme specificity, and pH optimum. In many of these properties, the enzyme was similar to the mouse stomach ALDH (Algar and Holmes, 1989a), rat corneal ALDH (Evces and Lindahl, 1989) and the bovine corneal ALDH (Abedinia et al, 1990), all of which have been classified as Class 3 ALDHs. This class of enzymes has been readily distinguished from Class 1 and Class 2

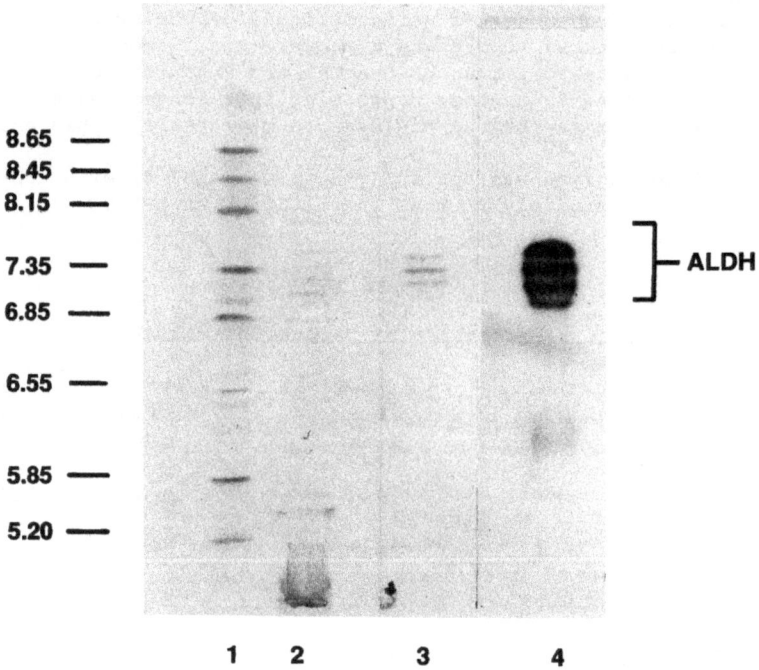

Figure 2. Agarose isoelectric focusing patterns of baboon corneal soluble protein (channel 2); and purified baboon corneal ALDH protein (channel 3) and activity (channel 4). Note the correspondence of the 3-4 zones of protein and activity in crude and purified preparations. Channel 1 illustrates isoelectric point protein standards.

ALDHs, which are predominantly localized in liver cytosol and mitochondrial sources, and exhibit characteristically different kinetic and biochemical properties (see Algar and Holmes, 1989b). Class 3 ALDHs exhibit high Km values with acetaldehyde as substrate (mM), and show a dual coenzyme specificity with NAD and NADP, and are highly active with benzaldehyde as substrate (see Koivula and Koivusalo, 1975; Evces and Lindahl, 1989).

The properties of baboon corneal ALDH can be most directly compared with those recently reported for rat corneal ALDH (Evces and Lindahl, 1989) and for bovine corneal ALDH (Abedinia et al, 1990). The baboon and bovine enzymes have similar subunit molecular weights, 63,000 and 65,000 respectively, whereas the rat corneal enzyme, in common with the tumour-associated Class 3 ALDH from this species, has a smaller subunit size of 51,000 daltons (Evces and Lindahl, 1989). The larger subunit size for the baboon and bovine ALDHs is consistent with that observed for the Class 3 mouse stomach ALDH, recently described (Algar and Holmes, 1989 b), which has a subunit MW of 65,000. In terms of substrate preference, however, the corneal ALDH enzymes from baboon, bovine and rat show close similarity. They exhibit high Km values with acetaldehyde, and share a preference for aromatic and medium-chain length aliphatic aldehydes. Indeed, comparative substrate kcat/Km values suggest that baboon corneal ALDH may function "in vivo" in the detoxification of UVR-induced peroxidic aldehydes, such as trans-2-hexenal and 4-hydroxynonenal (Table 2).

Studies on bovine corneal ALDH (Abedinia et al 1990) have led to the proposal that bovine corneal ALDH may play a dual role in protecting the eye against UVR-induced tissue damage, by serving both as a photoreceptor for UV-B light in corneal tissue, as well as in the detoxification process, via the NAD-dependent oxidation of peroxidic aldehydes. This hypothesis was supported by the observation that bovine corneal ALDH is the major soluble protein (accounting for 20-40% of all soluble protein), and by earlier studies, from a number of groups examining the action spectrum for UVR-induced photokeratitis in mammalian cornea, which showed a peak sensitivity in the 270-290 nm wavelength range, consistent with absorption by a major corneal protein (Cogan and Kinsey, 1946; reviewed in Ringvold, 1980). In contrast, the baboon corneal ALDH constitutes <5 percent of the total soluble protein in the cornea, although preliminary analysis of corneal extracts on SDS-gel electrophoresis (Fig.1) suggested that this enzyme is the second most major soluble protein in this tissue, and may therefore contribute to UV-B photoreception. In addition, the purified baboon corneal ALDH exhibited a 50-fold higher specific activity compared with the bovine enzyme, thereby providing higher specific activity activity levels (expressed as international units g^{-1} tissue) for baboon corneal ALDH, as compared with the bovine enzyme, even though the latter is present at a higher protein content

In summary, we have purified and characterized a high activity ALDH isozyme in baboon cornea, and compared the properties of this enzyme with those previously reported for rat (Evces and Lindahl, 1989) and bovine corneal ALDHs (Abedinia, et al 1990). These enzymes exhibited Class 3 ALDH properties, showing 'high' Km values with acetaldehyde as substrate, dual coenzyme specificities, and a preference for 'peroxidic' medium chain aliphatic and aromatic aldehydes. A substantive role in protecting the cornea from cytotoxic UVR-induced peroxidic aldehydes is suggested by this study. In addition, the enzyme may also contribute to UV-B photoreception, although levels of corneal ALDH protein are significantly lower in the baboon cornea, as compared to the very high levels reported in bovine cornea (Abedinia et al, 1990).

ACKNOWLEDGEMENTS

This research was supported in part by a grant from the National Health and Medical Research Council of Australia.

REFERENCES

Abedinia, M., Pain, T., Algar, E.M. and Holmes, R.S., 1990, Bovine corneal aldehyde dehydrogenase: the major soluble corneal protein with a possible dual protective role for the eye. Exp. Eye Research (in press).

Algar, E.M and Holmes, R.S., 1989a, Purification and properties of mouse stomach aldehyde dehydrogenase. Evidence for a role in the oxidation of peroxidic and aromatic aldehydes, Biochim. Biophys. Acta, 995:168.

Algar, E.M. and Holmes, R.S., 1989b, Kinetic properties of murine liver aldehyde dehydrogenases, in "Enzymology and Molecular Biology at Carbonyl Metabolism 2, H.Weiner and T.G. Flynn, eds., Alan R. Liss, N.Y., 93.

Ames, G. F-L., 1974, Resolution of bacterial proteins by poly-acrylamide gel electrophoresis on slabs. Membrane, soluble and periplasmic fractions. J.Biol. Chem., 249:634.

Boettner, E.A. and Wolters, J.R., 1962, Transmittance of the ocular media, Invest. Ophthalmol., 1:776.

Cogan, D.G. and Kinsey, V.E., 1946, Action spectrum of keratitis produced by ultraviolet radiation, Arch. Ophthalmol.55:670

Esterbauer, H. Zollner, H. and Schaur, R.J., 1988, Hydroxyalkenals:cytotoxic products of lipid peroxidation, ISI Atlas of Science: Biochemistry, 1:311.

Evces, S. and Lindahl, R., 1989, Characterization of rat cornea aldehyde dehydrogenase, Arch. Biochem. Biophys., 274:518,

Holmes, R.S., 1978, Electrophoretic analyses of alcohol dehydrogenase, aldehyde dehydrogenase, aldehyde oxidase, sorbitol dehydrogenase and xanthine oxidase from mouse tissues. Comp. Biochem. Physiol., 61B:339.

Holmes, R.S. and Vandeberg, J.L., 1986, Ocular NAD-dependent alcohol dehydrogenase and aldehyde dehydrogenase in the baboon, Exp. Eye Res., 43:383.

Hjelle, J.J., Grubbs, J.H. and Petersen, D.R., 1982, Inhibition of mitochondrial aldehyde dehydrogenase by malondialdehyde, Toxicol. lett., 14:35.

Koivula, T. and Koivusalo, M., 1975, Different forms of rat liver aldehyde dehydrogenase and their subcellular distribution, Biochim. Biophys. Acta, 397:9.

Ringvold, A., 1980, Cornea and ultraviolet radiation, Acta Ophthalmol. 58:63.

Sedmak, J.J. and Grossberg, S.E., 1977, A rapid, sensitive and versatile assay for protein using Coomassie brilliant blue G250, Anal. Biochem., 79:544.

Stephens, R.J., Negi, D.S., Short, S.M. van Kuijk, J.G.M., Dratz, E.A. and Thomas, D.W., 1989, lipid peroxidation and retinal phototoxic degeneration, in "Oxygen Radicals in Biology and Medicine", M.G. Simic, K.A. Taylor, J.F. Ward and C. von Sonntag, eds., Plenum, N.Y., :283.

Zigman, S., 1983, The role of sunlight in human cataract formation, Surv. Ophthalmol. 27:317.

A POLYMORPHISM IN THE RAT LIVER MITOCHONDRIAL ALDH2

GENE IS ASSOCIATED WITH ALCOHOL DRINKING BEHAVIOR

L. Carr, B. Mellencamp,
D. Crabb, L. Lumeng, T.-K. Li

Indiana University School of Medicine
and VA Medical Center
Indianapolis, IN 46202   USA

INTRODUCTION

Liver mitochondrial aldehyde dehydrogenase (ALDH2) has a low Km for acetaldehyde and plays a major role in acetaldehyde oxidation in vivo following alcohol consumption. In humans, a single base pair mutation produces a catalytically inactive enzyme (Yoshida et al., 1984). Approximately 50% of Asians have this defective ALDH2 allele; upon ingestion of ethanol, these individuals experience classic symptoms of acetaldehyde accumulation (Mizoi et al., 1979) which deters alcohol-drinking and thus reduces the risk of becoming an alcoholic. Japanese alcoholics have a low incidence of the ALDH2-deficient phenotype (2.3%) compared with the general Japanese population (41%) suggesting that polymorphism of ALDH2 could play a role in an individual's susceptibility to alcoholism (Harada et al., 1982).

Since there is a mutation in the human ALDH2 gene that influences alcohol drinking, we were interested in studying ALDH2 in the alcohol-preferring (P) and -nonpreferring (NP) rat lines. These rats have been selectively bred for high and low alcohol-seeking behavior, respectively, and the P line satisfies all the perceived criteria for an animal model of alcoholism (Li et al., 1987). We found that the ALDH2 enzymes from the P and NP rat lines exhibit different patterns on isoelectric focusing gels. To establish that this difference is due to a mutation in the coding region of the gene, we sequenced the ALDH2 cDNA from the P and NP rat lines. We report here a nucleotide substitution which can account for the difference seen on the isoelectric focusing gels.

Enzymology and Molecular Biology of Carbonyl Metabolism 3
Edited by H. Weiner *et al.*, Plenum Press, New York, 1990

MATERIALS AND METHODS

Isoelectric focusing

Isolated mitochondria were prepared from P and NP rat livers (Johnson and Lardy, 1967). Isoelectric focusing gels (1% Isogel agarose) were prepared according to Rex et al., (1985). Five milliunits of ALDH activity were loaded for each sample, and the gels were run for 1.5 hours at 1500 volts. ALDH activity was detected by staining as described (Rex et al., 1985).

RNA extraction, amplification, and sequencing

Total cellular RNA was prepared from P and NP rat liver using the single-step RNA isolation method (Chomczynski and Sacchi, 1987). First strand cDNAs were generated from 30 μg of total RNA using MMLV reverse transcriptase (Bethesda Research Laboratories) in 50 mM Tris-HCl (pH 8.3), 75 mM KCl, 10 mM DTT, 3 mM $MgCl_2$, and 0.5 mM dNTPs with specific antisense oligonucleotide primers. The cDNAs were then subjected to 30 cycles of enzymatic amplification (Saiki et al., 1988). The amplified cDNAs were cloned into pUC18 and sequenced using Sequenase (United States Biochemical Corp., Cleveland, OH).

RESULTS AND DISCUSSION

We determined that P and NP rats differ in mitochondrial ALDH isozyme patterns on isoelectric focusing gels (Fig. 1). The P rats exhibited a predominant and a weak anodic band (lanes 3 and 4), whereas the NP rats showed two cathodic bands of approximately equal intensity (lanes 1 and 2). For comparison, the cytosolic ALDH isozyme pattern for NP rats is also shown (lane 5). P and NP rats did not differ in cytosolic ALDH banding patterns.

To determine if the different electrophoretic patterns for ALDH2 detected in the P and NP rats were due to a mutation in the coding region of the gene, we compared the cDNAs of the P and NP rats. We cloned and sequenced the ALDH2 cDNAs of the P and NP rats by reverse transcription of RNA followed by PCR to specifically amplify the ALDH2 cDNA. The complete cDNA was generated in five overlapping cDNA fragments using five sets of sense/antisense oligonucleotide primers based on the rat cDNA sequence (Farres et al., 1989). Therefore, the complete coding region of P and NP rat ALDH2 genes was amplified in overlapping fragments, subcloned into pUC18, and sequenced.

Preliminary experiments showed a single nucleotide difference between the ALDH2 cDNAs from P and NP rats. The reported cDNA sequence for rat ALDH2 (obtained using a library constructed from liver RNA from a Sprague-Dawley rat) shows a glutamine (CAG) at position 67 (Farres et al., 1989). This codon was also found in the ALDH2 cDNA from P rats. In the NP rat cDNA, the codon for residue 67 was CGG, which encodes arginine (Fig. 2). The Arg for Gln substitution would make the enzyme more basic and could account for the difference in isoelectric points. Knowing the nucleotide change will now

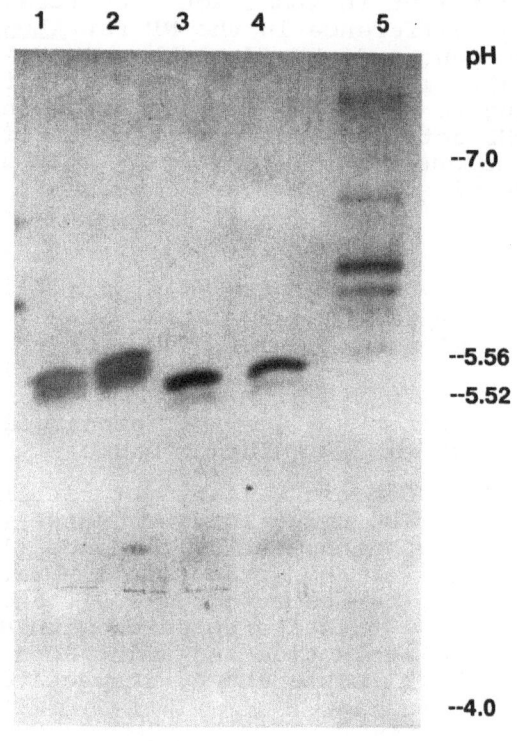

Fig. 1.   Agarose isoelectric focusing of
mitochondrial ALDH from P and NP
rats.   Mitochondrial proteins were
separated in 1% Isogel agarose and
stained for enzyme activity (see
Materials and Methods).   Lanes
1 and 2 - NP rat mitochondrial
extract.   Lanes 3 and 4 - P rat
mitochondrial extract.   Lane 5 - NP
cytosolic extract.

allow us to genotype the P and NP rats to determine if they are homozygous or heterozygous at the <u>ALDH2</u> locus. This information could help explain the multiple bands of ALDH2 activity seen on isoelectric focusing gels.

In the human <u>ALDH2</u> mutation, the nucleotide change is toward the 3' of the coding region (Exon 12; lys for glu_{487}) and produces an inactive, or partially inactive, enzyme. Individuals both homozygous and heterozygous for the mutant allele are deficient in ALDH2 activity (Crabb et al., 1989). The nucleotide difference in the NP rat <u>ALDH2</u> is in the 5' region of the gene and, based on the human <u>ALDH2</u> sequence (Yoshida et al., 1984), would be in the third exon. It will be interesting to determine if this mutation causes an alteration in ALDH2 activity and substrate specificity. The $glutamine_{67}$ residue has not previously been implicated as part of the active site of ALDH2.

```
                       67
               Ala    Gln    Ala
     P rat    --GCT   CAG   GCA--
    NP rat    --GCT   CGG   GCA--
               Ala    Arg    Ala

     P rat - glutamine (uncharged)
    NP rat - arginine (basic)
```

Fig. 2.   Genetic variants of rat ALDH2 –
          The amino acid sequence (line 1)
          and· nucleotide sequence (line 2)
          for the Sprague-Dawley rats
          (Farres et al., 1989) and the
          P rats are compared with NP rats
          nucleotide and amino acid sequences
          (lines 3 and 4, respectively).

ACKNOWLEDGEMENT

We thank Melinda Steel for excellent technical assistance. This research was supported by PHS AA 07611.

REFERENCES

Yoshida, A., Huang, I.-Y., and Ikawa, M., 1984, Molecular abnormality of an inactive aldehyde dehydrogenase gene results in loss of enzyme activity, Proc. Natl. Acad. Sci., 81:258-161, USA.
Mizoi, Y., Ijiri, I., Tatsuno, Y., Kijima, T., Fujiwara, S., Adachi, J., 1979, Relationship between facial flushing and blood acetaldehyde levels after alcohol intake, Pharmacol. Biochem. Behav., 10:303-311.
Harada, S., Agarwal, D. P., Goedde, H. W., and Ishikawa, B., 1983, Aldehyde dehydrogenase isoenzyme variation and alcoholism in Japan, Pharmacol. Biochem. Behav., 18 (Suppl. 1): 151-153.
Li, T.-K, Lumeng, L., McBride, W. J., Murphy, J. M., 1987, Rodent lines selected for factors affecting alcohol

sumption, Alcohol and Alcoholism, Suppl. 1:91-97.

Johnson, D., and Lardy, H. A., 1967, IN Methods in Enzymology (Estabrook, R. W., and Pullman, M. E., eds.), 10:94-96.

Rex, D. K., Bosron, W. F., Smialek, J. E., and Li, T.-K., 1985, Alcohol and aldehyde dehydrogenase isoenzymes in North American Indians, Alcohol. Clin. Exp. Res., 9:147-152.

Chomczynski, P., and Sacchi, N., 1987, Single-step method of RNA isolation of acid guanidinium thiocyanate-phenol-chloroform extraction, Anal. Biochem., 162:157-159.

Saiki, R. K., Scharf, S., Faloona, F., Mullis, K. B., Horn, G. T., Erlich, H. A., Arnheim, N., 1985, Enzymatic amplification of B-globin genomic sequences and restriction site analysis for diagnosis of sickle cell anemia, Science, 230:1350-1354.

Farres, J., Guan, K.-L., Weiner, H., 1989, Primary structures of rat and bovine liver mitochondrial aldehyde dehydrogenases reduced from cDNA sequences, Eur. J. Biochem., 180:67-74.

Crabb, D. W., Edenberg, H. J., Bosron, W. F., Li, T.-K., 1989. Genotypes for aldehyde dehydrogenase deficiency and alcohol sensitivity, J. Clin. Invest., 83:314-316.

ALDEHYDIC PRODUCTS OF LIPID PEROXIDATION: SUBSTRATES OR

INHIBITORS OF HEPATIC ALDEHYDE DEHYDROGENASE?

D.R. Petersen, J.J. Hjelle and D.Y. Mitchell

School of Pharmacy, Alcohol and Hepatobiliary Research Centers
University of Colorado Health Sciences Center
Denver, CO

INTRODUCTION

Peroxidation of membrane lipids is thought be a dynamic process which is ongoing in virtually all cells. Under normal conditions, cellular lipid peroxidation is well regulated and, as noted in a recent review (Esterbauer. et al., 1990), is always associated with the formation of numerous and chemically diverse aldehydic products. Malondialdehyde (MDA) and *trans*-4-hydroxy-2-nonenal (HNE) are classified as major products of lipid peroxidation since they are present in the greatest quantities during peroxidation of cellular membrane lipids while *trans*-2-hexenal, acrolein and crotonaldehdye are representative of minor products of lipid peroxidation thought to be formed in significantly smaller quantities.

Carbon tetrachloride is a well documented example of a chemical which, as a result of hepatic biotransformation, acts as a pro-oxidant and is a potent stimulator of hepatocellular lipid peroxidation (Recknagle 1967 and Slater, 1972). Agents such as CCl_4 or ADP-iron which are potent stimulators of lipid peroxidation have been documented to produce millimolar concentrations of carbonyl products of lipid peroxidation within isolated hepatocytes (Poli et al., 1985). These investigators noted that HNE concentrations within CCl_4-intoxicated hepatocytes approached 10 μM indicating that chemical-induced lipid peroxidation, or other events leading to cellular oxidative stress, result in production of relatively high intracellular concentrations of aldehydic products which have been proposed to be "second toxic messenger" (Esterbauer et al., 1990). Although somewhat controversial, it has also been proposed that ethanol toxicity is associated with a stimulation of increased production of reactive oxygen intermediates which initiate uncontrolled hepatocellular lipid peroxidation (see Cederbaum, 1989 for a comprehensive review).

The occurrence of relatively high concentrations of aldehydic products of lipid peroxidation within cells raise obvious questions about the role of aldehyde dehydrogenases in detoxification of these potentially toxic biogenic aldehydes. The proposition that ethanol is capable of stimulating hepatic lipid peroxidation, and is oxidized to acetaldehyde, has prompted a series of investigations in our laboratory concerning co-metabolism of aldehydic products of lipid peroxidation and acetaldehyde by the high-affinity form of aldehyde dehydrogenase (ALDH) present in rat liver mitochondria. This communication summarizes the results of these studies.

METHODS AND PROCEDURES

The studies described here were designed to characterize the oxidative metabolism of specific aldehydic products of lipid peroxidation by the high-affinity form of ALDH present in rat liver mitochondria. The rational for studying this specific Class II ALDH is its well documented importance in the *in vivo* oxidation of ethanol-derived acetaldehyde (Weiner, 1979). Livers were obtained from male Sprague-Dawley rats and processed for isolation of mitochondria as described previously (Mitchell and Petersen, 1987). The high-affinity form of mitochondrial ALDH was isolated and partially purified according to the method of Siew et. al., 1975 with additional modifications as outlined by Hjelle and Petersen 1983 and Mitchell and Petersen, 1987.

Crotonaldehyde, acrolein and *trans*-2-hexenal were purchased from commercial sources. Malondialdehyde was obtained commercially in the diacetal form and liberated prior to use as described elsewhere (Hjelle and Petersen, 1983). Four-hydroxynonenal is not commercially available and was synthesized as described previously (Esterbauer and Weger, 1967; Esterbauer et al., 1985) and stored in its diacetal form at -20°C. Malondialdehyde and 4-hydroxynonenal were liberated from their respective diacetal form immediately prior to use. Only 100% pure 4-hydroxynonenal, as determined by GC/MS analysis was used in the enzymatic assays.

Aldehyde dehydrogenase activity was assessed at pH 7.4 using a spectrophotometric procedure outlined previously (Mitchell and Petersen, 1987). The kinetic data were analyzed by Michaelis-Menton plots to estimate the kinetic parameters, V/K and V_{max} in order to eliminate error resulting from transformation of the data to fit double reciprocal Lineweaver-Burk plots. The kinetic parameter V/K describes the affinity of the enzyme for substrate at low substrate concentrations. However, the K_m values are also reported for the purpose of reference to previous literature. In the instances where certain α,β-unsaturated aldehydes were not substrates for this Class II rat liver ALDH, the respective adehydes were evaluated as inhibitors of ALDH-mediated acetaldehyde oxidation. The type of ALDH inhibition by specific α,β-unsaturated aldehydes (competitive or noncompetitive) was determined by double reciprocal plots of enzyme activity in the presence of variable substrate/inhibitor concentrations. Inhibition constants (K_i values) were determined by replots of line slopes determined from double reciprocal plots. The reversibility of inhibition was tested by plotting V_{max} values, as a function of the amount of enzyme added, in the presence and absence of various concentrations of inhibitor (Akermann and Potter, 1949). The inhibition constant for irreversible inhibition (K_{mi}) was estimated from a replot of line slopes derived from rates of the log% ALDH activity remaining at various time points after addition of different concentrations of inhibitor (Mares-Guia and Shaw, 1967).

RESULTS

The kinetic parameters of rat liver mitochondrial high-affinity ALDH for acetaldehyde and the five aldehydic products of lipid peroxidation evaluated are presented in Table 1. Based on the V/K parameter, these data demonstrate that the high-affinity mitochondrial ALDH displays optimal affinity for acetaldehyde which 8-fold greater than that observed for *trans*-2-hexenal and 28-fold greater than the value estimated for 4-hydroxynonenal. Interestingly, no ALDH-mediated activity could be detected when malondialdehyde, acrolein or crotonaldehyde were evaluated as substrates for the mitochondrial high-affinity form of the enzyme.

Table 1. Kinetic Parameters for oxidation of acetaldehyde and aldehydic products of lipid peroxidation by purified rat liver high-affinity aldehyde dehydrogenase[a].

Kinetic Parameter

Substrate	V_{max}[b]	V/K[c]	K_m (μM)
Acetaldehyde	10.6$\pm$0.8	6.780$\pm$0.50	1.6$\pm$0.1
trans-2-hexenal	1.3$\pm$0.2	0.839$\pm$0.14	1.7$\pm$0.3
4-hydroxynonenal	3.5$\pm$1.0	0.243$\pm$0.50	14.3$\pm$4.0
Malondialdehyde	-[d]	-	-
Acrolein	-	-	-
Crotonaldehyde	-	-	-

[a]Data are expressed as mean $\pm$ SEM for 4-6 experiments.
[b]V_{max} values are expressed as nmoles NADH formed/min/mg protein.
[c]V/K values are expressed as nmoles NADH formed/min/mg protein/μmole respective substrate/ml.
[d]No enzymztic activity detected

The inability of mitochondrial ALDH to oxidize malondialdehyde, acrolein and crotonaldehyde prompted us to examine the possibility that these aldehydes may function as inhibitors. Likewise, the potential of trans-2-hexenal and 4-hydroxynonenal to competitively inhibit high-affinity mitochondrial ALDH-mediated acetaldehyde oxidation was also assessed. The results of these experiments are summarized in Table 2.

Table 2. Characteristics and apparent inhibition constants of aldehyde dehydrogenase-mediated acetaldehyde oxidation by aldehydic products of lipid peroxidation.[a]

Aldehydic Product	Mechanism of Inhibition	Apparent K_i (μM)
trans-2-hexenal	Competitive	0.43
4-hydroxynonenal	Competitive	0.57
Malondialdehyde	Irreversible	2.60[b]
Acrolein	Irreversible	<1.00[b]
Crotonaldehyde	Noncompetitive	7.00

[a]Data are expressed as mean of 3-5 individual experiments.
[b]Represents K_{mi} value calculated from a replot of line slopes derived from rates of log% ALDH activity remaining at various time points after addition of different concentrations of inhibitor.

Malondialdehyde was found to be a relatively potent irreversible inhibitor of the mitochondrial high-affinity ALDH as evidenced by an apparent K_{mi} of 2.6 μM. Acrolein displayed the same pattern of inhibition which was characterized by a very rapid rate of inactivation making it difficult to calculate an absolute K_{mi} value. However, the data obtained suggest an estimated K_{mi} value of somewhat less than 1.0 μM. Crotonaldehyde inhibition was characterized as noncompetitive reversible with an estimated K_i value of 7.0 μM. Given the observation that *trans*-2-hexenal and 4-hydroxynonenal are relatively good substrates for this Clas II ALDH, it is not surprising that they functions as competitive inhibitors. However, the potency of inhibition is very impressive with both unsaturated aldehydes displaying K_i values of 0.4 to 0.5 μM.

DISCUSSION

The data presented in Tables 1 and 2 demonstrate that, depending on chemical structure, the α,β-unsaturated aldehydic products of lipid peroxidation may be either substrates or inhibitors of the rat liver mitochondrial high-affinity ALDH. Four-hydroxynonenal and *trans*-2-hexenal are readily oxidized by this Class II ALDH while malondialdehyde, acrolein and crotonaldehyde function as relatively potent inhibitors. An interesting feature of these data is that all of the aldehydic products of lipid peroxidation evaluated have the ability to inhibit acetaldehyde oxidation by this high-affinity form of ALDH. In the case of malondialdehyde and acrolein, the inhibition was irreversible. Inhibition of ALDH-mediated acetaldehyde oxidation by crotonaldehyde was noncompetitive and competitive in the case of *trans*-2-hexenal and 4-hydroxynonenal. Given these different mechanisms of inhibition, a striking feature of these data are the low inhibition constants ranging from approximately 0.5 μM for *trans*-2-hexenal and 4-hydroxynonenal to 7.0 μM for crotonaldehyde. The kinetic significance of these data can best be envisioned in terms of the inhibitory effect of these aldehydes on the V/K parameter for acetaldehyde oxidation by this Class II ALDH. For example, in the presence of either 0.5 μM 4-hydroxynonenal or 2.6 μM malondialdehyde, the V/K value for acetaldehyde oxidation by this high-affinity form of ALDH would be decreased by 50%.

At present, there are no published reports presenting quantitative data concerning hepatic concentrations of these various aldehydic products of lipid peroxidation following acute or chronic alcohol administration. However, experiments with CCl_4 or ADP-iron induced lipid peroxidation in isolated rat hepatocytes have produced some insight into the possible *in vivo* concentrations of certain aldehydic products of lipid peroxidation (Poli *et al*., 1985). From their data it can be estimated that control hepatocytes contain approximately 10 μM malondialdehyde and 0.4 μM 4-hydroxynonenal. Stimulation of lipid peroxidation by addition CCl_4 or ADP-iron increases the concentrations of these aldehydes approximately 2- to 10-fold. These calculations suggest that chemical-induced lipid peroxidation, or certain cellular events leading to oxidative stress, raise intracellular concentrations of these aldehydes which could result in inhibition of this high-affinity form of mitochondrial aldehyde dehydrogenase. The inhibition would be irreversible, as in the case of malondialdehyde, or competitive as was observed for 4-hydroxynonenal. Involvement of these proposed mechanisms in diminution of ethanol-derived acetaldehyde metabolism relys on demonstration that ethanol is capable of stimulating hepatic lipid peroxidation which results in accumulation of aldehydic products of lipid peroxidation with concomitant decreases in ALDH activity followed by increases of intracellular acetaldehyde.

The ability of chronic alcohol ingestion to decrease acetaldehyde metabolism (Korsten *et al.*, 1975; Palmer and Jenkins, 1982) and decrease hepatic ALDH activity (Jenkins and Peters, 1980; Jenkins *et al.*, 1984) has been documented in humans. While the specific mechanism responsible for these effects remain to be elucidated, it has been proposed that decreased hepatic acetaldehyde oxidation in human alcoholics is partially attributable to impaired mitochondrial function which in turn impacts the rate at which acetaldehyde is removed (Lieber, 1977). The accumulation of acetaldehyde would lead to further mitochondrial damage setting up a "viscous cycle" which initiates, or possibly promotes, alcohol-induced liver injury. Given our data presented in this communication, we propose the involvement of certain aldehydic products of lipid peroxidation in potentiation of acetaldehyde-mediated liver injury. This proposition is presented in Figure 1 as a component of the original scheme presented by Lieber, 1977.

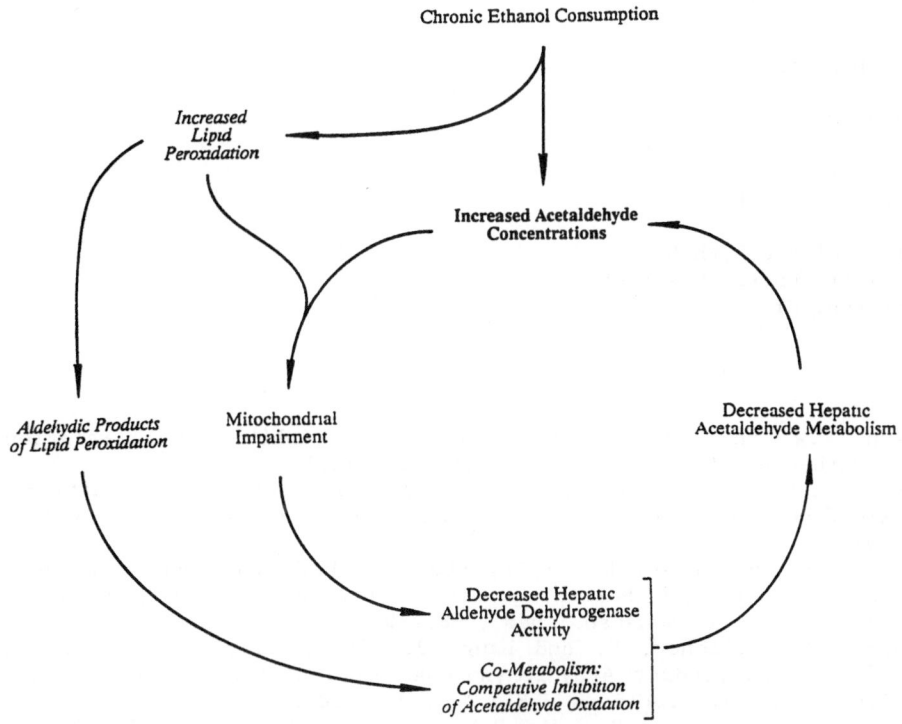

Figure 1. Possible role of aldehydic products of lipid peroxidation in disruption of acetaldehyde metabolism.

Our addition to this scheme is based on the assumption that chronic ethanol consumption promotes increased lipid peroxidation which may directly affect or modify mitochondrial membrane structure and in turn impair mitochondrial function. In addition, the aldehydic products of lipid peroxidation generated from this process could compromise ethanol-derived acetaldehyde metabolism by irreversibly inhibiting ALDH or through competitive inhibition of ALDH-mediated acetaldehyde oxidation. The combined result of these processes would be decreased acetaldehyde metabolism reflected in increased hepatic acetaldehyde concentrations which could further potentiate mitochondrial impairment or perhaps stimulate lipid peroxidation.

Presently, involvement of aldehydic products of lipid peroxidation in the events depicted in Figure 1 is speculative. However, the role of these chemically reactive and potentially toxic aldehydes in alcohol-induced liver injury is intriguing. Clearly, additional systematic studies are necessary to establish the association of prolonged ethanol ingestion with stimulation uncontrolled hepatic lipid peroxidation. Likewise, it is also important to identify and quantitate specific aldehydic products of lipid peroxidation produced in the liver during prolonged ethanol consumption. Mechanistic studies concerning the ability of various aldehydic products of lipid peroxidation to compromise acetaldehyde oxidation in isolated hepatocytes, perfused liver preparations and intact animals will be necessary to test the hypothetical concepts presented in this communication.

ACKNOWLEDGMENTS

This work was supported by Grants AA 03527 and AA00106 to DPP

CURRENT ADDRESSES

J.J. Hjelle, Ph.D.
Preclinical Research
NutraSweet Company
Deerfield, IL

D.Y. Mitchell, Ph.D.
Pfizer Central Research
Groton, CT

REFERENCES

Ackermann, W.W. and Potter, V.R., 1949, Enzyme inhibition in relation to chemotherapy, *Proc. Soc. Exp. Biol. Med.* 72:1.

Cederbaum, A.I., 1989, Introduction: Role of lipid peroxidation and oxidative stress in alcohol toxicity, *Free Radical Biol. & Medic.*, 7:537.

Esterbauer, H. and Weger, W., 1967, Uber die Wirkungen von aldehyden auf gesunde und maligne Sellen, 3. Mitt: Synthese von homologen 4-Hydroxy-2-alkenalen, II. *Monatsh. Chem.*, 98:1994.

Esterbauer, H., Zollener, H. and Lang, J., 1985, Metabolism of the lipid peroxidation product 4-hydroxynonenal by isolated hepatocytes and by liver cytosolic fractions. *Biochem. J.*, 228:363.

Esterbauer, H., Zollener, H., and Schuar, R.J., 1990, Aldehydes formed by lipid peroxidation: mechanisms of formation, occurrence and determination. In: Vigo-Pelfrey, C., ed. Membrane lipid oxidation. Vol I, Boca Raton Fl: CRC Press, Pg 239-268.

Hjelle, J.J. and Petersen, D.R., 1983, Metabolism of malondialdehyde by rat liver aldehyde dehydrogenase, *Toxicol. Appl. Pharmacol.*, 70:57.

Korsten, M.A., Matsuzaki, S., Fieinman, L., and Lieber, C.S., 1975, High blood acetaldehyde levels after ethanol administration. Differences between alcoholic and non-alcoholic subjects. *N. Engl. J. Med.*, 292:386.

Jenkins. W.J. and Peters, T.J., 1980, Selectively reduced hepatic aldehyde dehydrogenase in alcoholics. *Lancet*, 2:628.

Jenkins, W.J., Cakebread, K. and Palmer, K.R., 1984, Effect of alcohol consumption on hepatic aldehyde dehydrogenase activity in alcoholic patients, *Lancet*, 1:1048.

Lieber, C.S., 1977, *Metabolic Aspects of Alcoholism*, Bltimore, MD, Univ. Park Press, pg. 20.

Mares-Guia, M. and Shaw, E., 1967, The specific inactivation of trypsin by ethyl-p-guanidinobenzoate, *J. Biol. Chem.*, **242**:5782.

Mitchell, D.Y. and Petersen, D.R., 1987, The oxidation of α,β- unsaturated products of lipid peroxidation by rat liver aldehyde dehydrogenase, *Toxicol. Appl. Pharmacol.* **87**:403.

Palmer, K.R. and Jenkins, W.J., 1982, Impaired acetaldehyde oxidation in alcoholics, *Gut*, **23**:729.

Poli, G., Dianzani, M.U., Cheeseman, K.H., Slater, T.F., Lang, J. and Esterbauer, H., 1985, Separation and characterization of the aldehydic products of lipid peroxidation stimulated by carbon tetrachloride or ADP-iron in isolated rat hepatocytes and rat liver microsomal suspensions. *Biochem. J.*, **227**:629.

Recknagel, R.O., 1967, Carbon tetrachloride hepatotoxicity, *Pharmcol. Rev.* **19**:145.

Slater, T.F., 1972, *Free Radical Mechanisms in Tissue Injury*. Pion. London; Academic Press, New York.

Siew, C., Deitrich, R.A. and Erwin, V.G., 1976, Localization and characteristics of rat liver mitochondrial aldehyde dehydrogenases, *Arch. Biochem. Biophys.* **176**:638.

Weiner, H., 1979. Aldehyde dehydrogenase: Mechanism of action and possible physiological role. In Biochemistry and Pharmacology of Ethanol (E. Majchrowicz and E.P. Noble, Eds.) Vol 1, pp.107, Plenum Press New York.

METABOLISM OF 4-HYDROXYNONENAL IN HEPATOMA CELL LINES

R.A. Canuto, G. Muzio, A.M. Bassi[*], M.E. Biocca,
G. Poli, H. Esterbauer[**] and M. Ferro[*]

University of Turin and [*]Genoa (I),[**]University of Graz(A)
Dpt Experimental Medicine and Oncology, C.so Raffaello 30
10125 Torino, Italy

INTRODUCTION

The carbonyl compound 4-hydroxynonenal is one of the major and most toxic products of the oxidative lipid breakdown of biological membranes. Its formation has been quantified in rat liver homogenates, microsomes and whole parenchymal cells under various conditions of enhanced lipid peroxidation (Benedetti, et al., 1980; Esterbauer, et al., 1982; Poli, et al., 1985,1986).

4-hydroxynonenal is highly reactive towards thiol compounds (Esterbauer, 1982) and at concentrations between 10^{-4} M and 10^{-6} M it inhibits various biochemical processes in liver cells. A series of inhibitory effects on both basal and growth-related functions were observed in Ehrlich ascites tumour cells (Schauenstein, et al., 1977; Hauptlorenz, et al., 1985). These studies have generated renewed interest about the involvment of lipid peroxidation in cell growth during liver regeneration (Cheeseman, et al., 1986a), and during the formation of preneoplastic (Poli, et al., 1986) and neoplastic lesions (Poli, et al., 1986; Cheeseman, et al., 1986b). The actual implication of 4-hydroxynonenal-induced effects in both quiescent and rapidly dividing liver parenchimal cells is represented by the steady-state ratio between formation and removal through various catabolic pathways (Weiner and Wermuth, 1982). The metabolism of saturated aliphatic aldehydes and aromatic aldehydes have been quite well characterized in normal and neoplastic tissues (Tottmar, et al., 1973; Lindahl and Evces, 1984, 1987). From these studies it is evident that several differences are present in the compartmentalization and activity of major enzymes involved in the aldehyde metabolism. For example, the oxidation of aromatic aldehydes is strongly increased in the cytosolic fraction of some primary hepatomas and of hepatoma cell lines, while it is decreased or unchanged in highly deviated cells such as AH-130 Yoshida hepatoma cells (Canuto, et al., 1983).

More recently, we studied the aldehyde dehydrogenase activity during diethyl-nitrosamine-induced carcinogenesis in rat liver by using 4-hydroxynonenal as substrate. The NAD- and NADP-dependent aldehyde dehydrogenase activities are increased in nodules and in hepatoma. The distribution of these dehydrogenase activities in various subcellular fractions has shown that the dehydrogenase activity with 4-hydroxy-nonenal as substrate is concentrated in the cytosolic fraction in nodules and in hepatoma, whereas in the normal liver it is also significantly present in subcellular particles.

These results suggest that the increased aldehyde dehydrogenase with 4-hydroxynonenal as substrate can be considered a marker of neoplastic process, in the same way as can the increased level of aldehyde dehydrogenase with aromatic aldehydes.

In these experiments we measured the activity of oxidative and reductive enzymes involved in aldehyde metabolism, i.e. aldehyde dehydrogenase, aldehyde reductase and alcohol dehydrogenase, in normal hepatocytes and in two hepatoma cell lines, HTC and MH_1C_1 cells, respectively less and more differentiated. To identify the detoxifying role of the glutathione conjugation with 4-hydroxynonenal, the cytosolic activity of glutathione transferases were investigated in all cell types.

MATERIALS AND METHODS

Cell Preparations

Male Wistar rats (150-200 g body weight) were used for isolation of hepatocytes by the collagenase perfusion "in situ" technique described by Poli et al. (1979). Cell preparations with a viability index lower than 85% were discarded.

The hepatoma cell lines were grown routinely as monolayers in 25 cm^2 plastic flasks at 37 C in 95% air and 5% CO_2 atmosphere. MH_1C_1 cells were grown in presence of HAM F10 medium supplemented with 10% donor horse serum and 7.5% new-born calf serum; HTC cell line was grown with minimum essential medium supplemented with non essential aminoacids and 10% new-born calf serum. Media were supplemented with 100 U/ml penicillin and 0.1 mg/ml streptomycin. For the experiments described here, cells at an early stationary phase were harvested by treatment with a trypsin/EDTA (0.05-0.02%) solution followed by centrifugation at 600 x g for 10 min and three washings with cold HBSS.

Subcellular Fraction Preparation

Hepatocytes, MH_1C_1 and HTC cells were prepared as homogenates in a medium containing 70 mM sucrose, 220 mM mannitol, 20 mM Tris-HCl buffer (pH 7.4), 2 mM EGTA and 0.1 % (w:v) bovine albumin (fraction V, fatty acid free). The homogenate of hepatocytes was prepared in a Potter--Elvehjem homogenizer with 3 strokes of the tightly fitting teflon pestle in a volume of ice-cold isolation medium corresponding to 2.5 times their weight. The homogenates of MH_1C_1 and HTC cells were obtained by disruption with a hand-driven Tenbroeck glass homogenizer in a volume of 35 mM sucrose, 110 mM mannitol, 10 mM Tris-HCl buffer (pH 7.4), 1 mM EGTA corresponding to 2.5 times their weight. Then the homogenates were diluted to 20% with sucrose and mannitol so as to have an isotonic solution. A mild sonication was carried out.

Diluted homogenates were centrifuged at 1500 x g for 6.5 min (Beckman centrifuge J-6B). From the collected supernatants (cytoplasmic extract) the mitochondrial fraction was isolated at 28,000 x g for 2 min (rot. 30 Beckman ultracentrifuge L8-65) and washed 3 times. The supernatants obtained were centrifuged at 69,000 x g for 3 min (rot. 42.1 Beckman ultracentrifuge L8-65) to sediment the lysosomes. Microsomes were isolated from the post-lysosomal supernatant by centrifugation at 105,000 x g for 57 min (rot. 42.1 Beckman ultracentrifuge L8-65). Mitochondrial, lysosomal and microsomal fractions were finally resuspended in a medium containing 250 mM sucrose, 20 mM Tris-HCl so as to have 1 g of tissue per ml.

Post-microsomal supernatants were used as cytosolic fractions. The purity of subcellular fractions was checked by specific enzyme marker analysis.

Enzyme Assays

Aldehyde dehydrogenase activities (E.C. 1.2.1.5. and E.C. 1.2.1.3.) were measured both in subcellular particles (mitochondria and microsomes) and in cytosol. Other enzymatic activities: alcohol dehydrogenase (E.C. 1.1.1.1.), aldehyde reductase (E.C. 1.1.1.2.) and glutathione transferase (E.C. 2.5.1.18.) were measured only in the cytosol. 4-hydroxynonenal (0.1 mM) was used as substrate in the aldehyde dehydrogenase, alcohol dehydrogenase and aldehyde reductase assays.

The activity of aldehyde dehydrogenases was measured by monitoring the change in A_{340} due to NAD(P)H production during the oxidation of the substrate, as previously reported (Canuto, et al., 1983). The alcohol dehydrogenase and aldehyde reductase activities were measured by monitoring the change in A_{340}; the assay mixture (1 ml) for the measurement of these activities contained 0.1 mM NADH or NADPH, 200 mM K-phosphate buffer (pH 7.0), 0.1 mM aldehyde and an appropriate amount of the enzyme.

The activity of glutathione-transferase was measured by the method of Alin et al. (Alin, et al., 1985) with 0.05 mM 4-hydroxynonenal. All enzymatic assays were carried out at 30 C by a DU-7 Beckman spectrophotometer.

Proteins were determined by a biuret procedure (Gornall, et al., 1949).

Statistical Analysis

Standard deviations of the means were calculated and Student's t-test was used for determining significant differences from the mean.

RESULTS

Table 1 reports the aldehyde dehydrogenase activities detected in the cytoplasmic extracts of the three types of cells, using 0.1 mM

TABLE 1   4-HYDROXYNONENAL ALDEHYDE DEHYDROGENASE ACTIVITIES AND PROTEIN CONTENT IN THE CYTOPLASMIC EXTRACT FROM NORMAL HEPATOCYTES AND HEPATOMA CELLS

CELLS	ENZYME ACTIVITIES		PROTEIN
	NAD	NADP	
Hepatocytes	942.9 ± 45.9	300.9 ± 19.7	132.2 ± 12.9
MH_1C_1	432.7 ± 10.2**	174.1 ± 5.6**	70.3 ± 6.9**
HTC	2437.7 ± 131.7**	3794.5 ± 140.7**	72.7 ± 9.5**

Enzyme activities are expressed as nmoles of NAD(P)H produced/ min/g of wet wt of hepatocytes or hepatoma cells. Protein content is expressed as mg/g of wet wt of hepatocytes or hepatoma cells. Data are mean $\pm$ S.D. of 5 experiments. The cytoplasmic extract is the supernatant obtained from homogenate after nuclei and unbroken cell sedimentation, and contains mitochondria, lysosomes, microsomes and cytosol.
** = $p < 0.001$ (hepatoma cells vs hepatocytes).

TABLE 2. 4-HYDROXYNONENAL OXIDATIVE METABOLISM BY ALDEHYDE DEHYDROGENASES IN DIFFERENT SUBCELLULAR FRACTIONS OF NORMAL RAT HEPATOCYTES AND HEPATOMA CELLS

		Hepatocytes	MH_1C_1	HTC
Mitochondria	NAD	5.31 ± 0.81	2.24 ± 0.17*	5.69 ± 0.99
	NADP	1.91 ± 0.11	N.D.	3.68 ± 0.57*
Lysosomes	NAD	2.79 ± 0.31	0.49 ± 0.11**	6.29 ± 0.46**
	NADP	1.15 ± 0.24	0.04 ± 0.02**	5.87 ± 0.89**
Microsomes	NAD	10.51 ± 1.07	1.61 ± 0.22**	12.78 ± 1.89
	NADP	4.17 ± 0.56	0.25 ± 0.08**	23.02 ± 3.28**
Cytosol	NAD	7.67 ± 1.71	10.12 ± 4.16	43.60 ± 4.24**
	NADP	1.64 ± 0.29	3.87 ± 1.43*	68.09 ± 9.91**

Data are expressed as nmoles of NADH or NADPH produced/min/mg of protein. The values are the mean ± S.D. of 4 experiments.
N.D., not detectable.
* = $p < 0.05$; ** = $p < 0.001$ (hepatoma cells vs hepatocytes).

4-hydroxynonenal as substrate with either NAD or NADP as cofactors. MH_1C_1 cells show significant depression of the oxidative pathway. On the contrary, HTC hepatoma cells display enzymatic values considerably above those of control cells.

As illustrated in Table 2, the NAD- and NADP-dependent specific activities expressed in the cytosol fraction of MH_1C_1 cells are similar to or slightly higher than those of hepatocytes; the activities in the particulate fractions are lower than those of hepatocytes. The activities expressed by HTC were much higher than in hepatocytes, except in mitochondria with NAD as cofactor. The highest activity was found in the cytosolic fraction using NADP as cofactor.

The different subcellular distribution between hepatocytes and neoplastic cells can clearly be seen in Figure 1; the relative specific activities of 4-hydroxynonenal-dehydrogenase in cytosolic and particulate fractions clearly demonstrate the different metabolic condition of tumour cells with respect to normal cells. In tumor cells the values of relative specific activity and the percentage values of the histogram area compared with the total activities of the cytoplasmic extract show that both dehydrogenase activities are mainly localized in cytosolic fraction, whereas in the hepatocytes the activities of aldehyde dehydrogenase are equally distributed in cytosol and in subcellular particles. In fact, in hepatocytes the NAD-dependent activity was linked to mitochondria and microsomes for about 50% of the total activity, and to cytosol for the remaining amount; the NADP-dependent activity was linked to mitochondria and microsomes for about 60% of the total and to cytosol for about 40%. In hepatoma cells, both NAD- and NADP-dependent

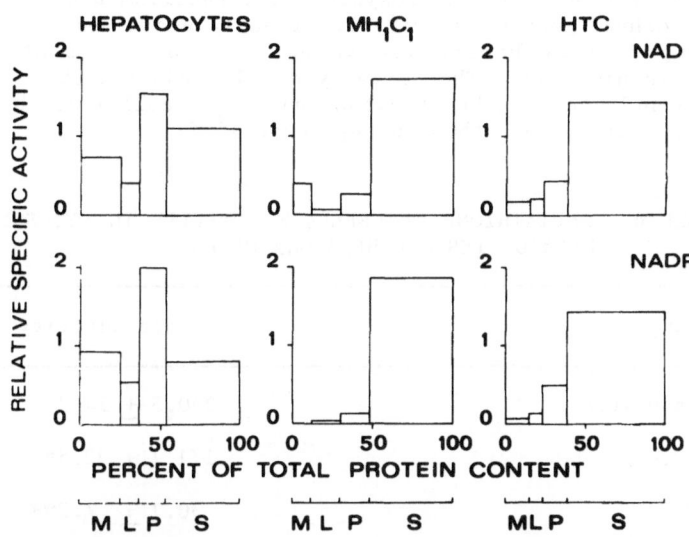

Figure 1. DISTRIBUTION OF 4-HYDROXYNONENAL DEHYDROGENASE ACTIVITIES IN THE SUBCELLULAR FRACTIONS FROM NORMAL HEPATOCYTES AND IN HEPATOMA CELLS

The ordinates indicate the specific activities on a protein basis, taking the specific activities of the total cytoplasmic extract (before fractionation) as one. The abscissas indicate the percentages of total cytoplasmic extract proteins in mitochondria (M), lysosomes (L), microsomes (P) and cytosol (S) subfractions.

TABLE 3. REDUCTIVE METABOLISM OF 4-HYDROXYNONENAL IN THE CYTOSOL OF NORMAL HEPATOCYTES AND HEPATOMA CELLS

CELLS	ALCOHOL DEHYDROGENASE	ALDEHYDE REDUCTASE
Hepatocytes	58.33 $\pm$ 15.30	7.62 $\pm$ 1.80
MH_1C_1	1.68 $\pm$ 0.51**	5.79 $\pm$ 1.66
HTC	28.48 $\pm$ 3.28**	26.24 $\pm$ 1.39**

Data are expressed as nmoles of NADH consumed per min per mg of protein for alcohol dehydrogenase and as nmoles of NADPH consumed per min per mg of protein for aldehyde reductase. The values are the mean $\pm$ S.D. of 5-6 experiments.
** = $p < 0.001$ (hepatoma cells vs hepatocytes).

activities were mostly present in the cytosolic fraction, less than 15% being linked to the subcellular organelles.

The specific activities of the NADH- and NADPH-dependent reductive pathways involved in 4-hydroxynonenal metabolism are shown in Table 3. Alcohol dehydrogenase activity is strongly decreased in MH_1C_1, where it drops to 2.9% of hepatocyte value, whereas it is about 50% of the control in HTC cells. The activity of aldehyde reductase did not differ significantly between hepatocytes and MH_1C_1 cell line, but in HTC cells it is 3.5-fold higher than in hepatocytes.

TABLE 4. GLUTATHIONE TRANSFERASE ACTIVITY IN CYTOSOL OF NORMAL HEPATOCYTES AND HEPATOMA CELLS

CELLS	Specific activity
Hepatocytes	280.5 ± 54.9
MH_1C_1	$171.1 \pm 32.5*$
HTC	$30.1 \pm 7.2**$

Data are expressed as nmoles of 4-hydroxynonenal consumed per min per mg of protein. The values are the mean $\pm$ S.D. of 4 experiments.
$* = p < 0.05$; $** = p < 0.001$ (hepatoma cells vs hepatocytes).

TABLE 5. PERCENTAGE DISTRIBUTION OF ENZYME ACTIVITIES METABOLIZING 4-HYDROXYNONENAL IN CYTOSOL FROM HEPATOCYTES AND HEPATOMA CELLS

Enzymes	Hepatocytes	MH_1C_1	HTC
NAD-ALDH	2.15	5.25	21.79
NADP-ALDH	0.46	2.00	34.64
ADH	16.39	0.87	14.49
ALRED	2.14	3.00	13.75
GST	78.84	88.85	15.32

The proportions of 4-hydroxynonenal metabolism due to each cytosolic enzyme activity are expressed as percentages of total 4-hydroxynonenal enzyme activities detected in cytosolic fractions. Enzyme activities were measured as nmoles of 4-hydroxynonenal consumed/min/g of tissue.
NAD-ALDH: NAD-dependent aldehyde dehydrogenase; NADP-ALDH: NADP-dependent aldehyde dehydrogenase; ADH: NAD-dependent alcohol dehydrogenase; ALRED, NADPH-dependent aldehyde reductase; GST: glutathione-S-transferase.

Table 4 displays the level of glutathione transferase activity detected in the three cell types. A progressive decrease of this enzymatic activity occurred in the hepatoma cells compared to hepatocytes, probably in direct relation to the degree of deviation. In fact, in well differentiated MH_1C_1 cells the glutathione transferase specific activity is about 60% of normal values, whereas in poorly differentiated HTC cells the activity is 10% of control values.

The percentage distribution of enzyme activities able to metabolize 4-hydroxynonenal in cytosol from the cells examined is reported in Table 5. In hepatocytes, glutathione transferase seems to be the more active enzyme in 4-hydroxynonenal metabolism, whereas the oxidative enzymes probably play a minor role. In MH_1C_1 cells glutathione transferase still maintains a remarkable importance in the aldehyde metabolic pathway; the aldehyde dehydrogenase activities,however, present higher percentage values than those of hepatocytes. In HTC cells, aldehyde dehydrogenase activities account for 56% of cytosolic metabolism of 4-hydroxynonenal; a strong decrease is present in percentage contribution of glutathione transferase in 4-hydroxynonenal disposition.

DISCUSSION

Under physiological conditions, hepatocytes display a defensive mechanism which has been shown to be extremely efficient against the accumulation of toxic aldehydes (Poli, et al., 1985; Weiner and Wermuth, 1982; Esterbauer, et al., 1985; Ferro, et al., 1988). In particular with regard to one of the most harmful classes of carbonyls, i.e. the hydroxyalkenals, rat liver cells can rapidly consume externally-added 4-hydroxynonenal up to concentrations 10-20 fold those actually recovered within the same cells undergoing oxidative stress (Poli, et al., 1985). The rapid removal of 4-hydroxynonenal by the hepatocytes is afforded through oxidative, reductive and transferase activity. The involvement of the cytosolic enzyme alcohol dehydrogenase in 4-hydroxynonenal metabolism has been demonstrated by Esterbauer et al. (1985). Mitchell and Petersen (1987) proved that 4-hydroxynonenal can also be actively oxidized by mitochondrial and cytosolic dehydrogenase. Further, 4-hydroxyalkenals have been shown to be good substrates for glutathione transferase isoenzymes (Alin, et al., 1985), and in particular the isoenzyme 8-8 is exceptionally active in the conjugation of 4-hydroxynonenal in the rat (Alin, et al., 1985; Danielson, et al., 1987). Our results show that glutathione transferase, but also alcohol dehydrogenase are the most active enzymes in the 4-hydroxynonenal metabolism by normal hepatocytes.
In relation to the oxidative metabolism of 4-hydroxynonenal, in hepatocytes the activity of aldehyde dehydrogenase is almost equally distributed between subcellular particles and cytosol.

The rate of oxidative metabolism of 4-hydroxynonenal in MH_1C_1 hepatoma line is 2-3 times lower than that of the control hepatocytes, with a significant decrease of both NAD- and NADP-dependent enzymatic pathways. These low activities of aldehyde dehydrogenase are in connection with a marked decrease of activities in the subcellular particles; in fact the enzyme specific activity of the MH_1C_1 cytosolic fraction is similar to that of hepatocytes. With regards to subcellular particles it is noteworthy that the intracellular production of 4-hydroxynonenal takes place in membranes.

In HTC cells, 4-hydroxynonenal oxidative consumption is greatly enhanced in comparison with hepatocytes and MH_1C_1 cells. As in the case of MH_1C_1 cells, in HTC cells there is also a marked shift of the normal percentage distribution of aldehyde dehydrogenase activities from particulate to cytosolic fraction.

The NADH-dependent reductive metabolism of 4-hydroxynonenal (through alcohol dehydrogenase) is almost totally depressed in MH_1C_1 cells, while it is only halved compared to control hepatocytes in HTC cells. The aldehyde reductase (NADPH-dependent) activity in MH_1C_1 shows a specific activity which does not differ from controls; in HTC cells it increases about 4 fold compared to the control. Glutathione transferase activities in control and neoplastic cells decrease in direct and linear relation to the degree of differentiation.

In conclusion, in the hepatoma cells, the metabolic pathway for aldehyde removal is markedly different from hepatocytes. The whole bulk of data indicates that the liver biochemical pathways demonstrated to be physiologically involved in aldehyde metabolism undergo a complex variety of changes during neoplastic transformation, so that hepatoma cells with different degrees of deviation show different metabolic patterns with respect to 4-hydroxynonenal. The analysis of aldehyde enzymatic activities suggests that MH_1C_1 cells have decreased levels of all the enzymes involved in the catabolism of 4-hydroxynonenal except for NAD and NADP aldehyde dehydrogenase in the cytosol, compared to hepatocytes and the other cell line. 4-Hydroxynonenal can however be metabolized through a sufficiently high activity of glutathione transferase. NAD and NADP dependent aldehyde dehydrogenase are of great importance in HTC cells for 4-hydroxynonenal metabolism. In these hepatoma cells the metabolic efficiency is due mainly to oxidative enzymes, whereas in normal hepatocytes the metabolic efficiency is significantly related to reductive metabolism and to glutathione-mediated conjugation. The differences between the two types of cultured hepatoma cells studied might be due to genetic changes occurring during the adaptation to growth in vitro, as suggested also by other authors (Lin, et al., 1986).

The oxidative route is probably the most active metabolic mechanism by which defined hepatoma cells can avoid the accumulation in the intracellular environment of antiproliferating aldehydes.
Some neoplastic cells have a recognized increased susceptibility to aldehydes (Dianzani, 1982; Esterbauer, 1985; Canuto, et al., 1985); a more detailed investigation of both aldehyde production and removal in experimental hepatomas in general might lead to a better understanding of the antiproliferative behaviour and efficacy of aldehydes.

Acknowledgements

This paper was supported by grants from Association for International Cancer Research, St. Andrews, United Kingdom and Ministry of the University and Scientific Research, Rome, Italy.

REFERENCES

Alin, P., Danielson, H. and Mannervik, B., 1985, 4-hydroxyalk-2-enals are substrates for glutathione transferase, FEBS Lett., 179:267.

Benedetti, A., Comporti, M. and Esterbauer, H., 1980, Identification of 4-hydroxynonenal as a cytotoxic product originating from the peroxidation of liver microsomal lipids, Biochim. Biophys. Acta, 620:281.

Canuto, R.A., Garcea, R., Biocca, M.E., Pascale, R., Pirisi, L. and Feo, F., 1983, The subcellular distribution and properties of aldehyde dehydrogenase of hepatoma AH-130, Eur. J. Cancer Clin. Oncol., 19:389.

Canuto, R.A., Biocca, M.E., Muzio, G., Garcea, R. and Dianzani, M.U., 1985, The effect of various aldehydes on the respiration of rat liver and hepatoma AH-130 cells, Cell Biochem. Funct., 3:3.

Cheeseman, K.H., Collins, M., Maddix, S., Milia, A., Proudfoot, K., Slater, T.F., Burton, G.W., Webb, A.C. and Ingold, K.U., 1986a, Lipid peroxidation in regenerating rat liver, FEBS Lett., 209:191.

Cheeseman, K.H., Collins, M., Proudfoot, K., Slater, T.F., Burton, G.W., Webb, A.C. and Ingold, K.U., 1986b, Studies on lipid peroxidation in normal and tumour tissues. The Novikoff rat liver tumour, Biochem. J., 235:507.

Danielson, U.H., Esterbauer, H. and Mannervik, B., 1987, Biochem. J., Structure-activity relationships of 4-hydroxyalkenals in the conjugation catalysed by mammalian glutathione transferases, 247:707.

Dianzani, M.U., 1982, Biochemical effects of saturated and unsaturated aldehydes, in: "Free radicals, Lipid peroxidation and Cancer", D.C.H. Mc Brian and T.F.Slater, eds, Academic Press, New York/London.

Esterbauer, H., Cheeseman, K.H., Dianzani, M.U., Poli, G. and Slater, T.F., 1982a, Separation and characterization of the aldehydic products of lipid peroxidation stimulated by ADP-Fe^{2+} in rat liver microsomes, Biochem. J., 208:129.

Esterbauer, H., 1982b, Aldehydic products of lipid peroxidation, in: "Free radicals, Lipid peroxidation and Cancer", D.C.H. Mc Brian and T.F.Slater, eds, Academic Press, New York/London.

Esterbauer, H., Zollner, H. and Lang, J., 1985a, Metabolism of the lipid peroxidation product 4-hydroxynonenal by isolated hepatocytes and by liver cytosolic fractions, Biochem. J., 228:363.

Esterbauer, H., 1985b, Lipid peroxidation products: formation, chemical properties and biological activities, in: "Free radicals in liver injury", G. Poli, K.H. Cheeseman, M.U. Dianzani, T.F.Slater, eds, IRL Press, Oxford, England.

Ferro, M., Marinari, U.M., Poli, G., Dianzani, M.U., Fauler, G., Zollner, H. and Esterbauer, H., 1988, Metabolism of 4-hydroxynonenal by the rat hepatoma cell line MH_1C_1, Cell Biochem. Funct., 6:245.

Gornall, A.G., Bardawill, C.J. and David, M., 1949, Determination of serum proteins by means of the biuret reaction, J. Biol. Chem., 177:751.

Hauptlorenz, S., Esterbauer, H., Moll, W., Pumpel, R., Schauenstein, E. and Puschendorf, B., 1985, Effects of the lipidperoxidation product 4-hydroxynonenal and related aldehydes on proliferation and viability of cultured Ehrlich ascites tumor cells, Biochem. Pharmacol., 34:3803.

Lin, K-H, Leach, M.F., Winters, A.L. and Lindahl, R., 1986, Characteristics and aldehyde dehydrogenase activity of four rat hepatoma cell lines produced by diethylnitrosamine-phenobarbital treatment, In Vitro Cell Develop. Biol., 22:263.

Lindahl, R. and Evces, S., 1984, Rat liver aldehyde dehydrogenase. I. Isolation and characterization of four high Km normal liver isoenzymes, J. Biol. Chem., 259:11986.

Lindahl, R. and Evces, S., 1987, Changes in aldehyde dehydrogenase activity during diethylnitrosamine-initiated rat hepatocarcinogenesis, Carcinogenesis, 8:785.

Mitchell, D.Y. and Petersen, D.R., 1987, The oxidation of   -unsaturated aldehydic products of lipid peroxidation by rat liver aldehyde dehydrogenase, Toxicol. Appl. Pharmacol., 87:403.

Poli, G., Gravela, E., albano, E. and Dianzani, M.U., 1979, Studies on fatty liver with isolated hepatocytes, Exp. Mol. Pathol., 30:116.

Poli, G., Dianzani, M.U., Cheeseman, K.H., Slater, T.F., Lang, J. and Esterbauer, H., 1985, Separation and characterization of the aldehydic products of lipid peroxidation stimulated by carbon tetrachloride or ADP-iron in isolated rat hepatocytes and rat liver microsomal suspensions, Biochem. J., 227:629.

Poli, G., Cecchini, G., Biasi, F., Chiarpotto, E., Canuto, R.A., Biocca, M.E., Muzio, G., Esterbauer, H. and Dianzani, M.U., 1986, Resistance to oxidative stress by hyperplastic and neoplastic rat liver tissue monitored in terms of production of unpolar and medium polar carbonyls, Biochim. Biophys. Acta, 883:207.

Schauenstein, E., Esterbauer, H. and Zollner, H., 1977, In: Aldehydes in

Biological Systems, their natural occurrence and biological actvities. Pion Limited, London.

Tottmar, O., Pettersson, H., Kiessling, K.H., 1973, The subcellular distribution and properties of aldehyde dehydrogenase in rat liver, Biochem. J., 135:577.

Weiner, H. and Wermuth, B., 1982, in: "Enzymology of carbonyl metabolism: aldehyde dehydrogenase and aldo-keto reductase", A.R. Liss, New York.

ROLE OF ALDEHYDE DEHYDROGENASE (ALDH) IN THE DETOXICATION OF
CYCLOPHOSPHAMIDE (CP) IN RAT EMBRYOS

Philip E. Mirkes*, Aaron Ellison and Sally A. Little

Division of Embryology, Teratology and Congenital
Defects, Department of Pediatrics, University of
Washington, Seattle, WA 98195
*Affiliate of Child Development and Mental Retardation
Center, University of Washington

INTRODUCTION

One of the fundamental questions in Teratology is why
teratogens kill some cells but not others in tissues
destined to develop abnormally. Our work with the teratogen
cyclophosphamide (CP) has revealed that although CP induces
elevated levels of cell death in exposed embryos, not all
cells or tissues are sensitive to the cell-killing effects
of CP. The most dramatic example of this is the complete
resistance of the cells of the developing heart to the
cytotoxicity of CP (Mirkes and Greenaway, 1985).

In order for CP to exert its cytotoxic and teratogenic
effects, the parent compound must be metabolically activated
(Fantel et al., 1979). The initial metabolite formed in
this activation process is 4-hydroxycyclophosphamide (4OHCP)
(Hill et al., 1973). Once formed 4OHCP equilibrates with
its ring-opened tautomer, aldophosphamide (AP). Subsequent
metabolism involves either spontaneous breakdown of 4OHCP/AP
to yield phosphoramide mustard (PM) and acrolein (AC) or
detoxication (Domeyer and Sladek, 1980) to yield
carboxyphosphamide (CaP) or 4-ketocyclophosphamide
(4-ketoCP). Detoxication of 4OHCP/AP is catalyzed primarily
by aldehyde dehydrogenase (ALDH) (Cox et al., 1975 and
Domeyer and Sladek, 1980).

Research presented here was initiated to examine the
hypothesis that heart cell resistance is related to the
level of ALDH in these cells. Precedent for this hypothesis
comes from the work of Hilton (1984) which has shown that a
CP-resistant L1210 cell line is resistance because it
contains elevated levels of ALDH compared to the
CP-sensitive cell line.

MATERIALS AND METHODS

Materials
CP was purchased from Mead Johnson (Evansville, IL) and
4CH₃CP and 4-hydroperoxycylophosphamide (4OOHCP) were gifts

Enzymology and Molecular Biology of Carbonyl Metabolism 3
Edited by H. Weiner *et al.*, Plenum Press, New York, 1990

from Dr. Michael Colvin, The Johns Hopkins University. Antibodies to various rat ALDH isozymes were generously provided by Dr. Ronald Lindahl, University of South Dakota.

In vitro Embryo Culture

A modification of the rat embryo explant system of New (1978) was used in some of these studies (Mirkes et al., 1981). Primigravida Sprague-Dawley rats were obtained from a local supplier. The morning following copulation was termed day 0 of gestation.

Determination of CP and 4CH₃CP Embryotoxicity

Because CP and 4CH₃CP require activation in order to be embryotoxic, culture medium contained 200 μl of phenobarbital-induced male rat liver supernatant fraction (S-9), NADPH at 5×10^{-4} M and glucose-6-phosphate at 5×1^{-3} M final concentrations. Embryos were exposed to drug continuously for 24-26 hours. At the end of the culture period, explants with active heartbeat and yolk sac circulation, referred to as viable embryos, were further characterized as previously described (Mirkes et al., 1981). Protein determinations were made according to the procedure of Bradford (1976), using bovine gamma globulin as a standard. Statistical analyses were performed by using one-way ANOVA with Dunnet's multiple comparisons (Zar, 1984). For histological analyses embryos were fixed, sectioned and stained as previously described (Mirkes et al., 1981).

ALDH Enzyme activity

ALDH activity was measured according to the method of Hilton (1984). The enzymatic production of NADH from NAD in the presence of substrate (either 1mM propionaldehyde or 100 μM aldophosphamide) was monitored spectrofluorometrically (excitation, 348 nm; emission, 460 nm). The assay was carried out in a 1 ml volume of assay buffer (0.1 M potassium phosphate, pH 7.4 1mM EDTA, 5 μM 2-mercaptoethanol, and 100 μM NAD). In assays using aldophosphamide as substrate, 5 mM 4OOHCP was reduced with 50 mM thiosulfate for 1 hr. Aliquots of this reduced 4OOHCP (i.e., aldophosphamide) were then used as substrate at a final concentration of 100 μM. In addition, 5 mM N-acetyl cysteine was added to the final assay mixture. In some assays the ALDH inhibitor, diethylaminobenzaldehyde (DEAB), was added to a final concentration of 10 uM).

Extracts of day 10 embryos were prepared by homogenization in assay buffer containing 0.5% Triton X-100. The homogenate was frozen and thawed three times and then centrifuged (48,000 x g) for 1 hr at 4°C. Aliquots (50-100 μl) of the resultant supernatant were used for ALDH assays. Supernatant extracts were also prepared from resistant and sensitive L1210 cell lines in the same manner except the homogenization step was omitted.

ALDH immunohistochemistry

Day 10-11 embryos were fixed in methyl Carnoys, embedded in paraffin, and sectioned at 6-8 microns. Following deparaffinization and rehydration, parasagittal sections were exposed for 18 hours to rabbit antibodies

specific for various isozymes of ALDH. Primary antibodies
were visualized with biotinylated secondary antibody (donkey
anti-rabbit 1gG, 1:60 dilution, 2 hrs) followed by Texas
Red-conjugated stretavidin (1:100, 30 minutes). Stained
sections were examined using a Nikon Microphot microscope
equipped with an epifluorescent attachment and a G-2A filter
block for Texas Red fluorescence.

RESULTS

In order to assess the role of ALDH in heart cell
resistance, the following experiments were performed. ALDH
activity in extracts prepared from day 10 embryos, isolated
hearts from day 10 embryos, and CP-resistant L1210 cells is
presented in Table 1. Using propionaldehyde as substrate,
detergent extractable ALDH activity in CP-resistant L1210
cells is 16.51 nmoles/min/mg protein, whereas activity in
embryos is 0.63 nmoles/min/mg protein. In both extracts
100% inhibition of ALDH activity is achieved by adding 10 μM
DEAB. In addition, total detergent extractable ALDH
activity in isolated hearts from day 10 embryos is 0.57
nmoles/min/mg protein. Using aldophosphamide as substrate,
detergent extractable ALDH activity in nmoles/min/mg protein
is 1.10 and 0.09 for CP resistant L1210 cells and embryos,
respectively. These data indicate that day 10 rat embryos
contain ALDH and that this enzyme can use CP
(aldophosphamide) as a substrate).

Because ALDH exists as a family of isozymes, we next
attempted to determine which ALDH isozyme was responsible
for resistance observed in CP-resistant L1210 cells and
whether this isozyme was present in rat embryo tissues,
particularly the heart. To do this we used a battery of
antibodies specific for mitochondrial, microsomal,
tumor-associated, and phenobarbital-inducible ALDHs to
screen CP-resistant or sensitive L1210 cell lines. Of the
available antibodies, only the one recognizing the
phenobarbital-inducible ALDH (Pb-ALDH) stained the

Table 1. ALDH activity (nmoles/min/mg protein) in extracts from day
10 rat embryos, isolated hearts from day 10 embryos and
cyclophosphamide resistant L1210 cells

Substrate	DEAB [a]	Embryo	Isolated hearts	CP resistant L1210 cells
Propionaldehyde	-	0.53 0.74 ------ $\bar{X}=0.63$	0.57	14.31 18.70 ------ $\bar{X}=16.51$
	+	0	0	*
Aldophosphamide	-	0.10 0.08 ----- $\bar{X}=0.09$	ND	1.02 1.17 ----- $\bar{X}=1.10$
	+	0	ND	0

[a] 10 μM DEAB added to reaction mixture either before or after
substrate.
* Inhibition was greater than 70% at 10 μM DEAB
ND Not determined

CP-resistant but not the sensitive L1210 cells (Figure 1). Using this antibody (Pb-ALDH), which apparently is specific for the ALDH isozyme that catalyzes the detoxification of CP, we next determined the presence and localization of this isozyme in the rat embryo. In the day 9 embryo (presomite stage), the antibodies to Pb-ALDH specifically stain the extraembryonic endoderm (Figure 2) while leaving the embryo proper unstained. On day 10, when heart cells are known to be resistant to CP, staining is observed in the extraembryonic membranes and now also in the developing gut epithelium (Figure 3). No obvious staining is observed in the heart, however, low levels of this ALDH in the heart cannot be rigorously ruled out. On day 11 intense staining is noted in gut epithelium, cloaca, and mesonephric tubules (Figure 4). In addition, significant staining is now also observed in the heart, primarily in the myocardium.

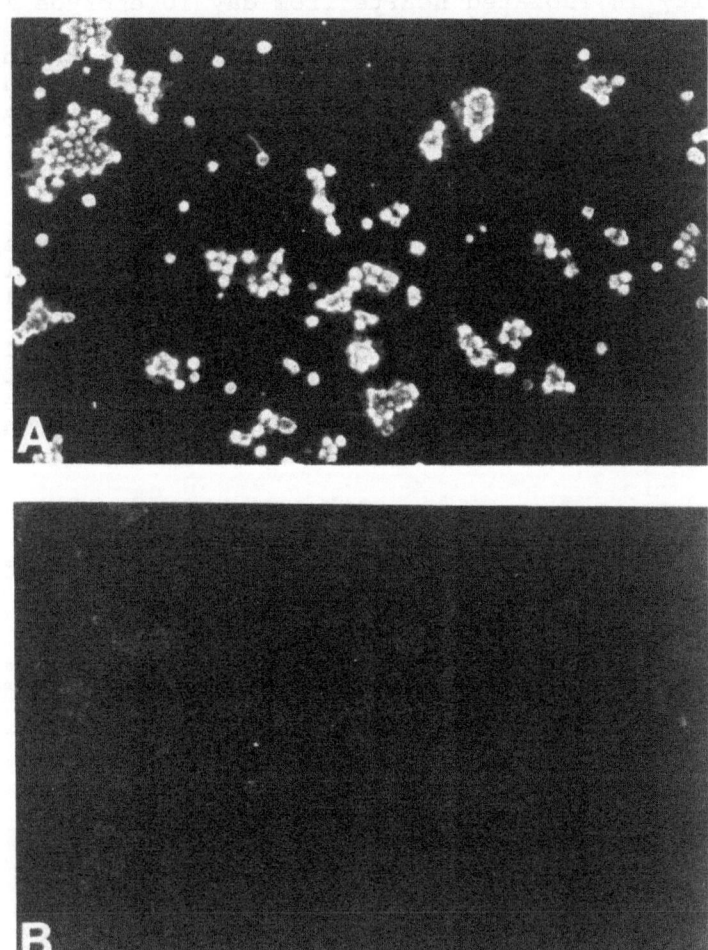

Figure 1. CP-resistant (A) and sensitive (B) L1210 cells stained with an antibody that recognizes a phenobarbital-inducible ALDH (anti Pb-ALDH).

Because the enzyme assays and immunohistochemistry indicated that day 10 rat embryos contain ALDH, some of which could potentially detoxify CP, we undertook the following experiments to determine the significance of this ALDH activity. Unlike CP, the 4-methyl analogue 4CH₃CP, cannot be detoxified by ALDH and therefore will be converted entirely to PM and methylvinylketone (Figure 5). If the ALDH present in day 10 embryos, particularly the heart, plays a significant role in protecting heart cells from the cytotoxic effects of CP, then 4CH₃CP should kill heart cells. Histological sections (Figure 6) from embryos exposed to 4CH₃CP show that a cytotoxic response has been induced in the central nervous system (Figure 6C) and in the mandibular arch mesenchyme (Figure 6B), but not in the heart (Figure 6D). Data in Figure 7 also indicate that on a molar basis the overall embryotoxicities of CP and 4CH₃CP, as measured by total embryo protein content, are similar.

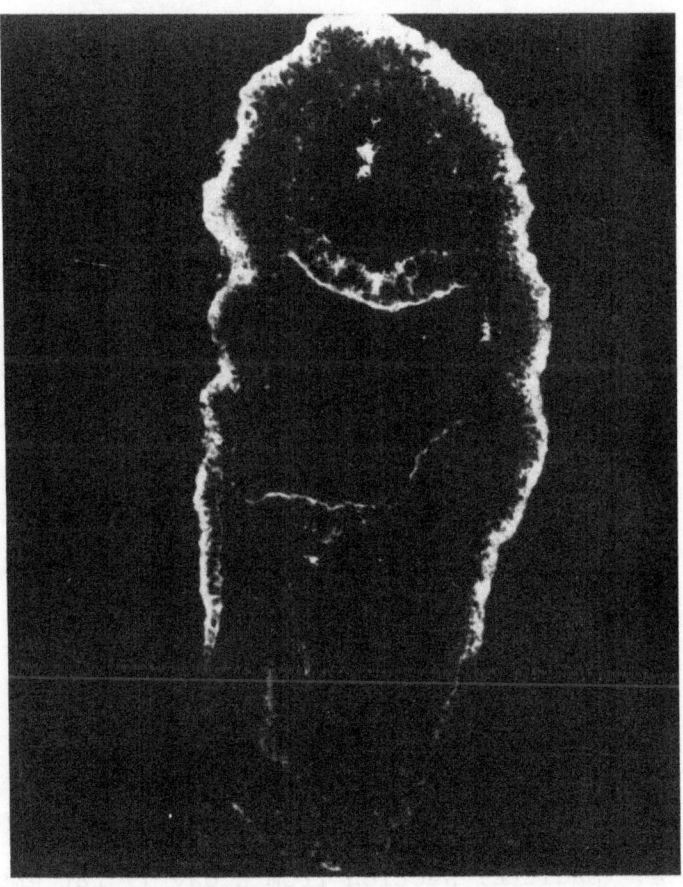

Figure 2. Sagittal section from a day 9 (presomite) rat embryo stained with anti Pb-ALDH.

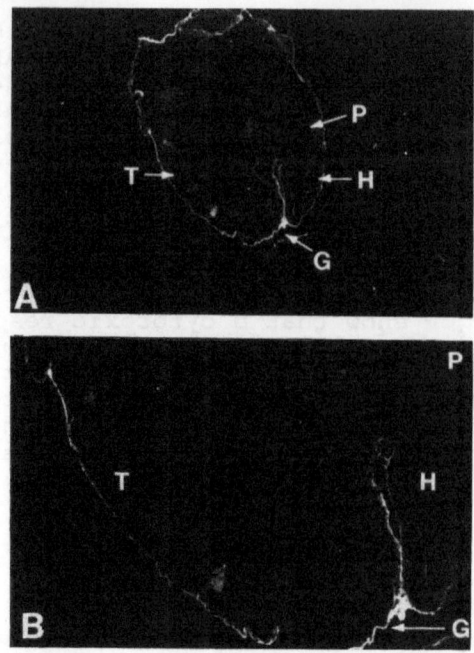

Figure 3. Sagittal section from a day 10 rat embryo (6-10 somites) stained with anti Pb-ALDH. Low power magnification (A) and high power magnification (B) howing staining localized to gut epithelium (G).

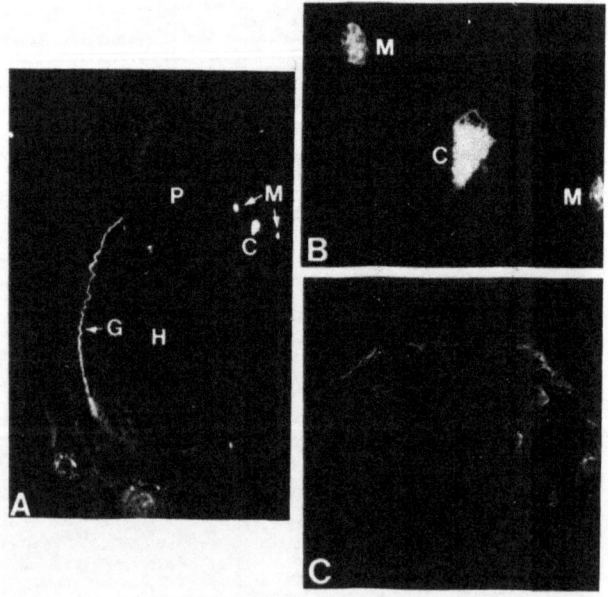

Figure 4. Sagittal section from a day 11 rat embryo stained with anti Pb-ALDH. Low power montage (A) showing staining localized to gut epithelium (G) mesonephros (M) and cloaca (C). B is higher magnification photo showing staining in mesonephros (M) and cloaca (C). C is a higher magnification showing staining in the heart (H).

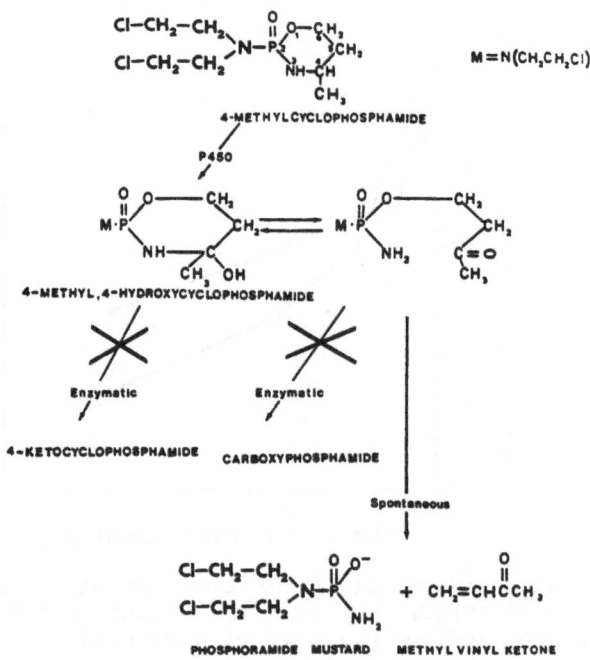

Figure 5.   Metabolic pathways for 4-methylcyclophosphamide based upon data from the literature.

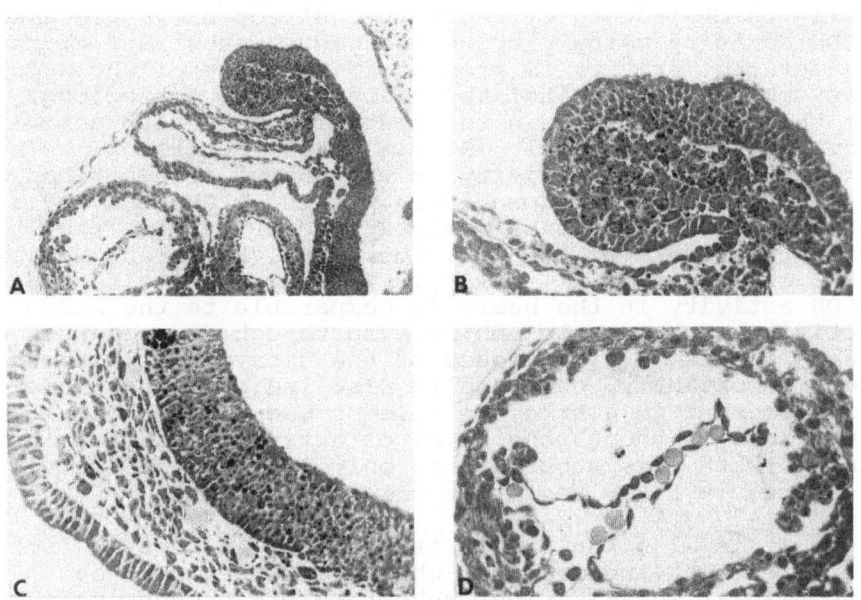

Figure 6.   Histology of an embryo exposed to 5.6 μM 4CH$_3$CP, rat liver S-9 and cofactors for 24 hours from day 10 to 11. Figure A is a low magnification micrograph depicting the mandibular arch and heart. Figure B is a high magnification micrograph depicting the mandibular arch, Figure C the prosencephalic neuroepithelium, and surrounding mesenchyme, and Figure D the heart. Note the necrosis in the mesenchyme of the arch and in the neuroepithelium, but not in the heart.

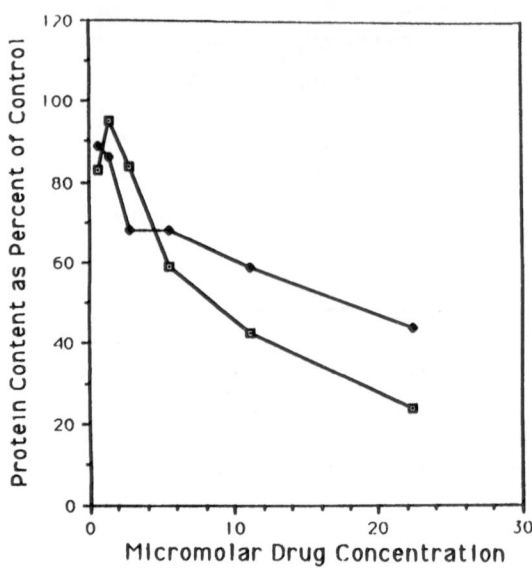

Figure 7. Comparison of molar concentration-response curves for CP (▣) and 4CH₃CP (◆) as a function of embryo protein content (expressed as percent of control).

DISCUSSION

The goal of the work presented here was to assess the role of ALDH in the resistance observed in rat embryo heart cells to CP-induced cytotoxicity. Direct enzymatic assay of ALDH activity using propionaldehyde as substrate shows that measurable activity is present in extracts of whole day 10 rat embryos and in isolated hearts. To our knowledge, this is the earliest time in rat gestation that ALDH activity has been detected (Lindahl, 1977). Total (detergent extractable) ALDH activity in the embryo or heart, however, is only 4% and 3% respectively, of the ALDH activity in L1210 cells known to be resistant to CP by virtue of elevated levels of ALDH (Hilton, 1984, Russo and Hilton, 1988 and Russo et al., 1988). These results also show that ALDH activity in the heart is comparable to the ALDH activity in the whole embryo even though drug-induced cell death occurs in all tissues of the embryo except the heart. More importantly, our results also indicate that some of the ALDH present in embryos is able to use aldophosphamide as substrate, although the level of activity measured in the embryo with this substrate is only 2.5% of the activity measured in L1210 resistant cell extracts.

It is also known that ALDH exists in several isozymic forms each having distinct characteristics such as subcellular localization, substrate preference, cofactor requirements and enzyme kinetics (Hempel and Lindahl, 1989). We were interested to know, therefore, which of these isozymes were present in the day 10 embryo, their spatial distribution in the embryo, and which isozyme(s) detoxifies CP. Our results and those of (Dunn et al., 1989) indicate that the Pb-ALDH is the ALDH specific for CP metabolism because it is this isozyme that is found in L1210 resistant cells. Although we could not detect the Pb-ALDH in heart cells from day 10 embryos, detectable Pb-ALDH was observed in hearts from day 11 embryos. Thus, at some time between day 10 (6-8 somite stage) and day 11(21-24 somite stage) heart ALDH either first makes it appearance or this specific ALDH reaches levels detectable by our immunohistochemical technique. In addition, we have demonstrated that this ALDH is expressed in a localized pattern that changes as development proceeds from day 9 to day 11, i.e., extraembryonic localization on day 9, embryonic gut epithelium on day 10, and embryonic gut epithelium, mesonephric tubules, cloaca and heart on day 11. Although we do not know the function of this ALDH in these tissues, the very localized distribution in tissues that in one way or another connect to the exterior of the embryo suggests that the enzyme may be involved in removing potentially toxic aldehydes from the embryonic compartment.

Both ALDH enzymology and immunohistochemistry, therefore, indicate that day 10-11 rat embryo heart cells, resistant to the cytotoxic effects of CP, contain not only ALDH activity, but also a specific ALDH isozyme associated with resistance to CP in a mouse leukemia cell line. Although these findings suggest that ALDH could play a role in heart cell resistance, the following evidence argues against ALDH as the mechanism of heart cell resistance. First, the levels of ALDH activity in heart cells (or embryos), whether based upon propionaldehyde or aldophosphamide as substrate, represents only 2.5 -4% of the activity present in CP-resistant L1210 cells. Second, heart cells are completely resistant to the 4-methyl analog of CP. If heart cell ALDH is detoxifying significant amounts of CP, thereby conferring resistance by reducing the PM concentration to a non-cytotoxic level, then heart cells exposed to $4CH_3CP$, which cannot be detoxified by ALDH, should die. This argument is independent of the fact that in addition to PM, methylvinylketone is concommittantly formed from $4CH_3CP$ whereas acrolein is formed from CP. Third, the level of CP-induced DNA crosslinking, thought to be the cytotoxic lesion (Mehta et al., 1980, Erickson and Zlotogorski, 1984, and Zwelling et al., 1980), is similar in heart cells and head cells when assayed immediately after a 5 hour drug exposure (data not shown). If heart cell ALDH had detoxified CP with the resultant reduction in the level of PM, the expected result would be a significant reduction in PM-induced DNA crosslinking. The fact that DNA crosslinks in the heart are not related to subsequent cytotoxicity in that tissue suggests that either other factors, eg. repair, modulate the ultimate effect of these crosslinks or that some other drug induced lesion is responsible for CP-induced cytotoxicity. Nonetheless, taken together the data presented indicate that ALDH does not appear to play a major role in conferring resistance to the cytotoxic effects of CP in the rat embryo.

The failure of ALDH present in the embryo and/or heart to detoxify significant amounts of CP could be related to several factors. First, the level of ALDH present may be insufficient to reduce the level of PM below a "cytotoxic" level. Second, ALDH and/or the substrate (aldophosphamide) may be sequestered such that enzyme-substrate interaction is precluded. This is a somewhat unlikely explanation given that both the mouse (Russo and Hilton, 1988) and rat (Dunn et al., 1989) ALDH isozyme, known to use CP as a substrate, are cytosolic (soluble) enzymes. Third, it may be that although the appropriate ALDH is present in the day 10 embryo, the enzyme is functionally inactive. Perhaps the rat embryo ALDH, whose normal function in the embryo is unrelated to the detoxification of CP, is present in day 10 embryos only to be functionally activated at a later stage of development. At present we are pursuing research aimed at understanding the regulation and function of this and other ALDH isozymes in the mammalian postimplantation embryo.

ACKNOWLEDGMENTS

This work was supported by National Institutes of Health grants HD 16287 and HD 22095 to Philip E. Mirkes and in part to grant HD 00836 to Thomas H. Shepard. We thank Dr. Ronald Lindahl, Department of Biochemistry and Molecular Biology, University of South Dakota for generously supplying the ALDH antibodies and Rajan John Puthenpurackal for secretarial assistance.

REFERENCES

Bradford, M. (1976) A rapid and sensitive method for the quantitation of microgram quantities of protein utilizing the principle of dye binding. Anal. Biochem. 72:248-254.

Cox, P.J., P.B. Farmer, and M. Jarman (1975) The microsomal metabolism of some analogues of cyclophosphamide:4-methyl-cyclophosphamide and 6-methylcyclophosphamide. Biochem. Pharmacol. 24:599-606.

Domeyer, B.E. and N.E. Sladek (1980) Metabolism of 4-hydroxycyclophosphamide/aldophosphamide in vitro. Biochem. Pharmacol. 29:2903-2912.

Dunn, T.J., A.J. Koleske, R. Lindahl, and H.C. Pitot (1989) Phenobarbital-inducible aldehyde dehydrogenase in the rat. cDNA sequence and regulation of the mRNA by phenobarbital in responsible rats. J. Biol. Chem. 264:13057-13065.

Erickson, L., and C. Zlotogorski (1984) Induction and repair of macromolecular damage by alkylating agents. Cancer Treat. Rep. II (supp. A):25-35.

Fantel, A.G., J.C. Greenaway, M.R. Juchau, and T.H. Shepard (1979) Teratogenic bioactivation of cyclophosphamide in vitro. Life Sci. 25:67-72.

Hempel, J., and R. Lindahl, M. (1989) Class III aldehyde dehydrogenase from rat liver: superfamily relationship to classes I and II and functional interpretations. In: Progress in Clinical and Biological Research: Enzymology and Molecular Biology of Carbonyl Metabolism 2, H. Weiner and T.G. Flynn, eds., Volume 290, pp. 3-17.

Hill, D.L., W.R. Laster, Jr., M.C. Kirk, S. ElDareer, and R.F. Struck (1973) Metabolism of phosphamide (2-(2-chloroethylamino)-3(2-chloroethyl)-tetrahydro 2H-1, 3,2,-oxazaphosphorine 2-oxide) and production of a toxic phosphamide metabolite. Cancer Res. 33:1016-1022.

Hilton, J. (1984) Role of aldehyde dehydrogenase ion cyclophosphamide-resistant L1210 leukemia. Cancer Res. 44:5156-5160.

Lindahl, R. (1977) Aldehyde dehydrogenase in 2-acetamidofluorene-induced rat hepatomas. Ontogeny and evidence that the new isoenzymes are not due to normal gene de-repression. Biochem. J. 164:119-123.

Mehta, J.M. Przybylski, and D. Ludlum (1980) Alkylation of guanosine and deoxyguanosine by phosphoramide mustard. Cancer Res. 40:4183-4186.

Mirkes, P.E., A.G. Fantel, J.C. Greenaway, and T.H. Shepard (1981) Teratogenicity of cyclophosphamide metabolites: phosphoramide mustard, acrolein, and 4-ketocyclophosphamide in rat embryos cultured in vitro. Toxicol. Appl. Pharmacol. 58:322-330.

Mirkes, P.E. and J.C. Greenaway (1985) Uptake and binding of tritium from (chloroethyl 3H) cyclophosphamide by rat embryos in vitro. Teratology 31:373-380.

New, D.A.T. (1978) Whole embryo culture and the study of mammalian embryos during organogenesis. Biol. Rev, 53:81-122.

Russo, J. and J. Hilton (1988) Characterization of cytosolic aldehyde dehydrogenase from cyclophosphamide resistant L1210 cells. Cancer Res. 48:2963-2968.

Russo, J., J. Hilton, and M. Colvin (1988) Aldehyde dehydrogenase (ALDH) activity confers resistance to cyclophosphamide (CP) in murine small intestine. Proc. Am. Assoc. Cancer. Res. 29:296.

Zar, J.H. (1984) In: Biostatistical analysis, Prentice Hall, Englewood Cliffs, New Jersey, pp. 401-403.

Zwelling, L., S. Michaels, H. Schwartz, P. Dobson and K. Kohn (1980) DNA crosslinking as an indicator of sensitivity and resistance of mouse L1210 Leukemia to cis-diamminedichloroplatinum (II) and L-phenyl-alanine mustard. Cancer Res. 40:640-649.

HUMAN AND MOUSE HEPATIC ALDEHYDE DEHYDROGENASES IMPORTANT IN

THE BIOTRANSFORMATION OF CYCLOPHOSPHAMIDE AND THE RETINOIDS

N. E. Sladek, P. A. Dockham, and M. O. Lee

Department of Pharmacology, University of Minnesota
3-249 Millard Hall, 435 Delaware Street S.E.
Minneapolis, Minnesota 55455

INTRODUCTION

The rate at which aldehyde dehydrogenase-catalyzed biotransformation occurs in various tissues can be of major importance with regard to the ultimate therapeutic efficacy of drugs and other xenobiotics that are aldehydes or that give rise to aldehydes (Sladek, *et al.*, 1989). Examples of such agents are cyclophosphamide and the retinoids, retinol and beta-carotene.

Cyclophosphamide is a widely used anticancer and immunosuppressive agent that, *per se*, is pharmacologically inert. Hepatic microsomal mixed-function oxidases catalyze its conversion to aldophosphamide, also without pharmacological activity. Aldophosphamide undergoes one of two metabolic fates: beta-elimination to phosphoramide mustard, the pharmacologically active species, and oxidation to carboxyphosphamide. The latter is without pharmacological activity. Thus, cyclophosphamide is bioinactivated (detoxified) when aldehyde dehydrogenase-catalyzed oxidation of aldophosphamide to carboxyphosphamide occurs (Sladek, 1988).

Retinoids such as retinol (vitamin A alcohol) and beta-carotene give rise to retinoic acid, presumably via the intermediate, retinaldehyde. Oxidation of retinaldehyde to retinoic acid can be catalyzed by aldehyde dehydrogenases (Futterman, 1962; Olson, 1967; Napoli, 1986). Retinoic acid exhibits a number of biological effects. For example, it promotes the differentiation of epithelial and other cells (Favennec and Cals, 1988). Depending upon the system under investigation, retinol, beta-carotene and retinaldehyde appear to be less/minimally potent in that regard (Williams and Napoli, 1985; Fontana, *et al.*, 1986; Amatruda and Koeffler, 1986; Connor, 1988). Thus, aldehyde dehydrogenase-catalyzed oxidation of retinaldehyde to retinoic acid can be viewed, at least on occasion, as a bioactivation.

Eleven aldehyde dehydrogenases have been identified in mouse liver (Manthey and Sladek, 1989; Manthey, *et al.*, 1990). One, *viz.*, AHD-2, the most abundant cytosolic enzyme, accounts for the bulk (>80%) of the total activity when the aldophosphamide concentration is pharmacological, Figure 1. This is not the hepatic aldehyde dehydrogenase that is most important in catalyzing the oxidation of acetaldehyde. That distinction goes to the most abundant mitochondrial enzyme, *viz.*, AHD-5.

Similarly, in mouse cyclophosphamide-resistant L1210/OAP leukemia cells, all of the aldehyde dehydrogenase-catalyzed oxidation of aldophosphamide to carboxyphosphamide is by AHD-2 (Manthey, 1988).

The situation is quite different in mouse stomach epithelium. AHD-2 is not present in this tissue (Manthey, *et al.* 1990). Conversely, AHD-4 is found in stomach, but not in liver.

Enzymology and Molecular Biology of Carbonyl Metabolism 3
Edited by H. Weiner *et al.*, Plenum Press, New York, 1990

The only other aldehyde dehydrogenases found in mouse stomach are AHD-5 and AHD-8. In this tissue, AHD-8 and AHD-4 contribute most to catalyzing the oxidation of aldophosphamide.

Of the eleven aldehyde dehydrogenases present in mouse liver, only two, *viz.*, AHD-2 and AHD-7, another cytosolic enzyme, catalyze the oxidation of retinaldehyde to retinoic acid (Lee and Sladek, 1990). In addition, the dehydrogenase form of xanthine oxidase also catalyzes it. Catalysis by AHD-2 accounts for more than 90% of the total NAD-dependent activity in mouse liver, Figure 1.

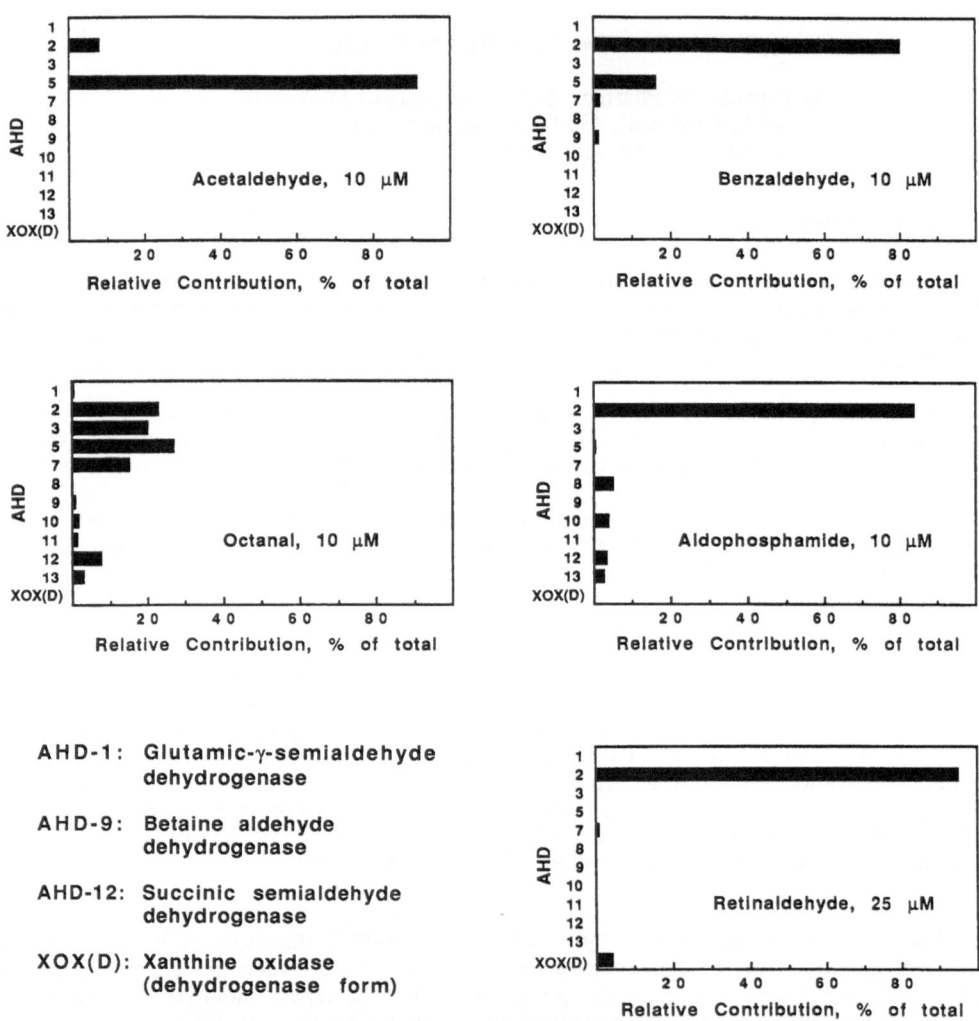

Figure 1.   Relative contribution (percent of total) of DBA/2 mouse hepatic aldehyde dehydrogenases to the catalytic oxidation of various aldehydes.

There are at least eight "aldehyde dehydrogenases", *viz.*, ALDH's 1-5, and succinic semialdehyde, betaine aldehyde and γ-aminobutryaldehyde dehydrogenases, in human liver. These are listed in Table 1 as are their putative mouse counterparts. Present efforts in our laboratory are directed towards ascertaining which of the human aldehyde dehydrogenases catalyze the oxidation of aldophosphamide and retinaldehyde to their respective acids.

Table 1. Human and mouse aldehyde dehydrogenases. Compiled from Koivula, 1975; Duley, *et al.*, 1985; Rout and Holmes, 1985; Holmes, 1987; Ryzlak and Pietruszko, 1988; Kurys, *et al.*, 1989; Manthey and Sladek, 1989; and Manthey, *et al.*, 1990.

Enzymes		Subcellular Locus	Km (acetaldehyde)
Human	Mouse		
ALDH-1	AHD-2	cytosolic	μM
ALDH-2	AHD-5	mitochondrial	nM/μM
ALDH-3	AHD-7 AHD-4	cytosolic	mM
ALDH-4	AHD-1	mitochondrial	mM
ALDH-5	AHD-3	microsomal	mM
Betaine aldehyde DH'ase	AHD-9	cytosolic	μM
Succinic semialdehyde DH'ase	AHD-12	mitochondrial	mM
γ-Amino-butyraldehyde DH'ase	?	cytosolic(?)	μM

MATERIALS AND METHODS

Human liver samples were obtained from a 51-year old Caucasian female. They were stored at -70°C prior to analysis. Aldophosphamide was generated from 4-hydroperoxycyclophosphamide by chemical reduction with methyl sulfide (Manthey, *et al.*, 1990). Aldehyde dehydrogenases present in the 105,000 x g supernatant fractions obtained from homogenized human liver samples treated with lubrol (0.3%) were separated from each other by ion exchange, affinity (5'-AMP-sepharose), and chromatofocusing column chromatography essentially as previously described (Manthey, *et al.*, 1990). Retinaldehyde, aldophosphamide, acetaldehyde, benzaldehyde, octanal, succinic semialdehyde, betaine aldehyde, glutamic-γ-semialdehyde and γ-aminobutyraldehyde were used as substrates to monitor enzyme activity. Routinely, aldehyde dehydrogenase activity was quantified by monitoring the appearance of

NAD(P)H (and retinoic acid when retinaldehyde was the substrate) at 340 nm with the aid of a Beckman DU-70 recording spectrophotometer as previously described (Manthey, *et al.*, 1990). In some cases, retinoids were extracted with an organic solvent mixture, the extract was passed through an HPLC column, and individual retinoids were quantified at 340 nm with the aid of a Beckman UV detector 160. Isoelectric focusing and kinetic analysis were also as previously described (Manthey, *et al.*, 1990). A schematic presentation of the experimental protocol is provided in Figure 2.

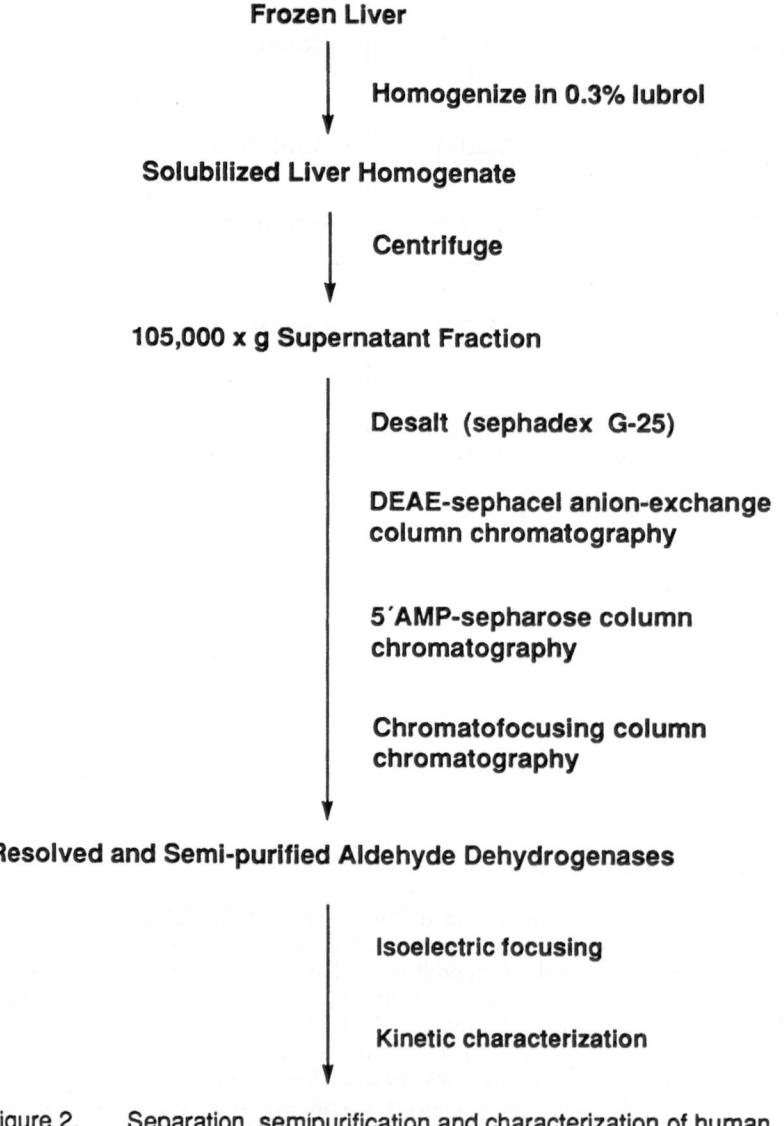

Figure 2.   Separation, semipurification and characterization of human liver aldehyde dehydrogenases: experimental scheme.

RESULTS AND DISCUSSION

In very early experiments, aldehyde dehydrogenase-catalyzed oxidation of aldophosphamide and retinaldehyde was quantified in 105,000 x g supernatant and particulate fractions prepared from a liver that had never been frozen. The donor was a 2.5-year-old male Caucasian. About two-thirds of the total dehydrogenase-catalyzed aldophosphamide oxidation, and all of the dehydrogenase-catalyzed retinaldehyde oxidation, took place in the supernatant fraction. The frozen human liver sample described in MATERIALS AND METHODS was used in all subsequent experiments because of the limited availability of fresh human liver samples. Preliminary experiments showed that enzymes known to be restricted to the particulate fraction appeared in quantity in the 105,000 x g supernatant fraction, and vice versa, when frozen liver was the sample source. Therefore, subcellular fractionation was not subsequently attempted; instead, the frozen liver sample was homogenized in a solution containing lubrol to solubilize the particulate fraction. This was then centrifuged at 105,000 x g and the supernatant saved for further analysis. At concentrations of 25 µM retinaldehyde and 160 µM aldophosphamide, the lubrol-solubilized hepatic preparation catalyzed the NAD-dependent oxidation of these substrates at rates of 325 and 2,834 nmoles/min/g liver, respectively. About 80% of the total enzyme-catalyzed oxidation of retinaldehyde was NAD-dependent. No attempt was made to determine whether NAD-independent enzyme catalysis of aldophosphamide also occurred.

DEAE-sephacel column chromatographic separation of hepatic 105,000 x g lubrol-solubilized aldehyde dehydrogenases is shown in Figure 3.

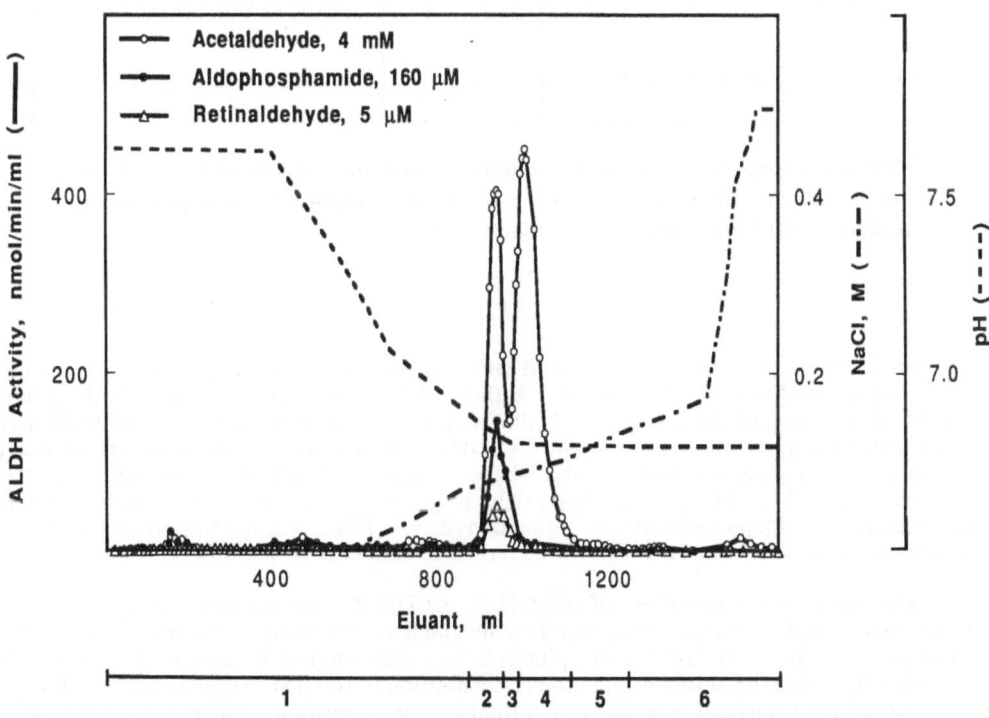

Figure 3.   DEAE-sephacel column chromatography of solubilized human hepatic aldehyde dehydrogenases.   The 6 pools that were collected are delineated at the bottom.

Dehydrogenase-catalyzed oxidation of retinaldehyde and aldophosphamide was largely by an enzyme in pool 2. After further purification by affinity and chromatofocusing column chromatography, isoelectric focusing and substrate preference studies identified this enzyme as ALDH-1, Table 2.

Table 2.   Basis for enzyme identification.

ENZYME	pI	Preferred Substrate(s)[a]
ALDH-1	5.1	Octanal, Acetaldehyde
ALDH-2	4.9	Octanal, Acetaldehyde
ALDH-4	6.9	Glutamic-γ-semialdehyde
ALDH-5	6.0 - 7.5	Octanal
Succinic semialdehyde DH'ase	6.0 - 7.5	Succinic semialdehyde
Betaine aldehyde DH'ase	6.5	Betaine aldehyde

[a]Acetaldehyde, benzaldehyde, octanal, glutamic-γ-semialdehyde, succinic semialdehyde, betaine aldehyde, γ-aminobutyraldehyde, aldophosphamide and retinaldehyde were the substrates utilized.

Found in pool 1 were ALDH-4 and succinic semialdehyde dehydrogenase. Neither catalyzed the oxidation of retinaldehyde. Either or both may account for the dehydrogenase-catalyzed oxidation of aldophosphamide that was observed in this pool. The major aldehyde dehydrogenase present in pool 4 was ALDH-2. It did not catalyze the oxidation of retinaldehyde and only poorly catalyzed the oxidation of aldophosphamide. ALDH-5 (present in pool 6) and betaine aldehyde dehydrogenase (present in pool 5) did not catalyze the oxidation of aldophosphamide or retinaldehyde. ALDH-3 was not present in the sample; a distinct γ-aminobutyraldehyde dehydrogenase also could not be identified.

The relative contribution of ALDH-1, ALDH-2, and ALDH-4 and/or succinic semialdehyde dehydrogenase to human hepatic aldehyde dehydrogenase-catalyzed oxidation of aldophosphamide to carboxyphosphamide was estimated to be about 70, 5, and 25%, respectively, when the aldophosphamide concentration is 160 μM. Human hepatic aldehyde dehydrogenase-catalyzed oxidation of retinaldehyde to retinoic acid is apparently all by ALDH-1. The Km values for ALDH-1-catalyzed oxidation of aldophosphamide and retinaldehyde were 37 and <1 μM, respectively.

At least four conclusions can tentatively be made from these observations. Firstly,

ALDH-1 is likely to be an important determinant of cellular sensitivity to a) the cytotoxic action of cyclophosphamide and other oxazaphosphorines, and b) retinoid-induced differentiation; certain mitochondrial aldehyde dehydrogenases may also be important determinants with regard to the cytotoxic action of the oxazaphosphorines, particularly in bone marrow pluripotent progenitor cells. Increased expression of ALDH-1, and perhaps certain mitochondrial aldehyde dehydrogenases, may account for acquired resistance to cyclophosphamide by neoplastic cells. Secondly, since ALDH-2 is likely to be unimportant with regard to catalyzing the detoxification of aldophosphamide, individuals expressing the nonfunctional ALDH-2 variant phenotype (Yoshida, et al., 1984) would not be vulnerable to increased risk of clinical toxicity following cyclophosphamide administration. Thirdly, human and mouse liver are quite similar in that the bulk of aldehyde dehydrogenase-catalyzed aldophosphamide and retinaldehyde oxidation is by the major low Km (acetaldehyde) enzyme present in the cytosol, viz., ALDH-1 in human liver and AHD-2 in mouse liver. Thus, the mouse would seem to be a good model when answers to questions about the pharmacologic/biologic relevance of human hepatic aldehyde dehydrogenase are being sought. Fourthly, the ability to catalyze NAD-dependent retinaldehyde oxidation may be a good marker for the presence of ALDH-1.

ACKNOWLEDGEMENTS

Technical assistance was provided by Matt Hedge and Theresa Sladek. Financial support was by USPHS Grant CA 21737.

REFERENCES

Amatruda, T. T. III and Koeffler, H. P., 1986, Retinoids and cells of the hematopoietic system, in: "Retinoids and Cell Differentiation," Sherman, M. I., ed., CRC Press, Boca Raton, pp. 79-104.

Connor, M. J., 1988, Oxidation of retinol to retinoic acid as a requirement for biological activity in mouse epidermis, Cancer Res., 48:7038-7040.

Duley, J. A., Harris, O. and Holmes, R. S., 1985, Analysis of human alcohol- and aldehyde-metabolizing isozymes by electrophoresis and isoelectric focusing, Alcoholism: Clin. Exp. Res., 9:263-271.

Favennec, L. and Cals, M.-J., 1988, The biological effects of retinoids on cell differentiation and proliferation, J. Clin. Chem. Clin. Biochem., 26:479-489.

Fontana, J. A., Reppucci, A., Durham, J. P. and Miranda, D., 1986, Correlation between the induction of leukemia cell differentiation by various retinoids and modulation of protein kinases, Cancer Res., 46:2468-2473.

Futterman, S., 1962, Enzymatic oxidation of vitamin A aldehyde to vitamin A acid, J. Biol. Chem., 237:677-680.

Holmes, R. S., 1987, Biochemistry and genetics of enzyme metabolism. The mouse model, Prog. Clin. Biol. Res., 241:141-157.

Koivula, T., 1975, Subcellular distribution and characterization of human liver aldehyde dehydrogenase fractions, Life Sciences, 16:1563-1570.

Kurys, G., Ambroziak, W. and Pietruszko, R., 1989, Human aldehyde dehydrogenase. Purification and characterization of a third isozyme with low Km for γ-aminobutyraldehyde, J. Biol. Chem., 264:4715-4721.

Lee, M.-O. and Sladek, N. E., 1990, Identification of mouse liver aldehyde dehydrogenases that catalyze the oxidation of retinaldehyde to retinoic acid, Fed. Proc., 4:A659.

Manthey, C. L., 1988, Resolution and characterization of the aldehyde dehydrogenases important in cyclophosphamide metabolism, Ph.D. Dissertation, University of Minnesota, Minneapolis.

Manthey, C. L. and Sladek, N. E., 1989, Aldehyde dehydrogenase-catalyzed bioinactivation of cyclophosphamide, Prog. Clin. Biol. Res., 290:49-63.

Manthey, C. L., Landkamer, G. J. and Sladek, N. E., 1990, Identification of the mouse aldehyde dehydrogenases important in aldophosphamide detoxification, Cancer Res., 50: 4991-5002.

Napoli, J. L., 1986, Retinol metabolism in LLC-PK$_1$ cells, J. Biol. Chem., 261:13592-13597.

Olson, J. A., 1967, The metabolism of vitamin A, *Pharmacol. Rev.*, **19**:559-596.

Rout, U. K. and Holmes, R. S., 1985, Isoelectric focusing studies of aldehyde dehydrogenases from mouse tissues: variant phenotypes of liver, stomach and testis isozymes, *Comp. Biochem. Physiol. B*, **81B**:647-651.

Ryzlak, M. T. and Pietruszko, R., 1988, Human brain "high Km" aldehyde dehydrogenase: purification, characterization, and identification as NAD^+-dependent succinic semialdehyde dehydrogenase, *Arch. Biochem. Biophys.*, **266**:386-396.

Sladek, N. E., 1988, Metabolism of oxazaphosphorines, *Pharmac. Ther.*, **37**:301-355.

Sladek, N. E., Manthey, C. L., Maki, P. A., Zhang, Z. and Landkamer, G. J., 1989, Xenobiotic oxidation catalyzed by aldehyde dehydrogenases, *Drug Metabol. Rev.*, **20**:697-720.

Williams, J. B. and Napoli, J. L., 1985, Metabolism of retinoic acid and retinol during differentiation of F9 embryonal carcinoma cells, *Proc. Natl. Acad. Sci., USA*, **82**: 4658-4662.

Yoshida, A., Huang, I.-Y. and Ikawa, M., 1984, Molecular abnormality of an inactive aldehyde dehydrogenase variant commonly found in Orientals, *Proc. Natl. Acad. Sci., USA*, **81**:258-261.

A PARADIGM FOR ALDEHYDE OXIDATION: HISTIDINOL DEHYDROGENASE

Charles Grubmeyer

Department of Biology
New York University
100 Washington Square
New York, NY 10003

INTRODUCTION

Histidinol dehydrogenase (HDH, EC 1.1.1.23) catalyzes the oxidation of histidinol to histidine, using two moles of NAD. The reaction is the final step in the biosynthesis of histidine in bacteria, plants, and fungi. The enzyme is of particular interest in what it can tell us about dehydrogenase action: the reaction contains both alcohol and aldehyde dehydrogenase steps, apparently occurring at a single active site. Two other enzymes, UDP-glucose dehydrogenase (UDPGDH, EC 1.1.1.22) and hydroxymethyl glutaryl CoA reductase (HMGR, EC 1.1.1.34) catalyze conceptually similar 4-electron oxidations. Although the latter two have important roles in health and disease, for the enzymologist HDH offers the advantage of a long and interesting genetic history, and is particularly well suited to molecular approaches.

Pioneering work by Adams (1955) showed that the reaction proceeded at a single active site, and that although externally added histidinaldehyde was competent as a substrate for either oxidation or reduction by the enzyme, no histidinaldehyde was released to the medium in the overall oxidation of histidinol. More recent work by Gorisch and Holke (1985) has shown that externally added aldehyde is bound very tightly to the enzyme with an off-rate of $2.5 \times 10^{-5} s^{-1}$, and K_D of $10^{-11}M$. For both the overall reaction and HDH-aldehyde complexes, the bound intermediate was protected from base-catalyzed degradation or aldehyde trapping reagents (Adams, 1955; Gorisch and Holke, 1985). These observations led to a model in which the intermediate aldehyde is captured as an active site thiohemiacetal. Thiohemiacetals have been proposed for many of the aldehyde dehydrogenases. Although this model can explain most catalytic data for HDH, it does not yet relate catalysis to the structure of the enzyme.

Salmonella typhimurium HDH is the product of the *hisD* gene, one of the 8 genes of the extensively studied histidine operon (Carlomagno et al., 1988). The *hisD* gene product has been overproduced and purified (Yourno and Ino, 1968), and sequenced (Kohno and Gray, 1981). It is 433 amino acid residues in length, containing 7 Cys and 5 His residues. Physical studies indicate a stable dimeric structure (Burger et al., 1979). The *S. typhimurium* enzyme is known to contain 1 Zn(II) per monomer (Grubmeyer et al., 1989b). The *hisD* genes of both *E. coli* and *S. typhimurium* have been cloned and sequenced (Carlomagno et al., 1988). Kinetically HDH is bi-uni-uni-bi ping-pong, with binding of histidinol preceding NAD (Burger and Gorisch, 1981a).

Enzymology and Molecular Biology of Carbonyl Metabolism 3
Edited by H. Weiner *et al.*, Plenum Press, New York, 1990

Table 1. Stereochemistry of hydride transfers for 4e⁻ dehydrogenases.

Enzyme	NAD	Stereochemistry Alcohol	Aldehyde
HDH	R	S	R
HMGR	R	S	R
UDPGDH	S	R	S

References: HDH, Grubmeyer et al., (1989); HMGR, Rogers et al., 1983; UDPGDH, Ridley and Kirkwood, 1973.

Two types of study suggest that the dimeric structure of the enzyme is important in the chemistry of the reaction. First is the interesting complementation behavior of the *S. typhimurium hisD* gene. Genetic studies (Greeb et al., 1971) showed that mutants in *hisD* fell into three major complementation groups. Two of these, *hisDa* and *hisDb*, when present as separate alleles in the same cell, could complement to produce active enzyme, as judged by histidine prototrophy. Mapping showed that *hisDa* and *hisDb* mutants fell into separate regions, *hisDa* mapping to a small portion of the carboxy terminal, with *hisDb* mapping throughout the middle of the gene. Since the enzyme catalyzes a two-step reaction, it is tempting to think that perhaps the two complementation groups correspond to reaction steps. Experiments by Loper and Adams (1965) however, showed that of 23 mutant HDH types tested, none performed either partial reaction. Complementation provides an excellent system for the study of structure-function relations in HDH, and we have developed a system to study the phenomenon in vitro (Lee and Grubmeyer, 1987). The second type of evidence for subunit interactions in catalysis is from work by Kirkwood (Eccleston et al., 1979) who showed that low levels of urea caused subunit dissociation and activity loss in the overall reaction, but that either of the two half-reactions was relatively unaffected. Thus we might want to propose a role for both subunits in every reaction cycle.

Analogy with other enzymes, especially the two other 4-electron dehydrogenases, can provide a guide for our thinking about HDH mechanism. In the case of UDPGDH, work by Ordman and Kirkwood (1977) has clearly delineated a two-step mechanism in which both imine and thiohemiacetal intermediates take part. In the case of HMGR, the normal reaction is the reduction of a thiolester to an alcohol, occurring by well-defined steps (Sherban et al., 1985). We can quite reasonably ask if the 4-electron dehydrogenases form a "family" in the sense of being more closely related to one another evolutionarily than to other enzymes. Dr. Barbara Stitt and I undertook sequence comparisons of an active site peptide of UDPGDH (Franzen et al., 1981), the 537 carboxy-terminal residues of human HMGR (Luskey & Stevens, 1985), and HDH, and were unable to find any significant areas of similarity. Another way to look at the reactions is biochemically. The complete stereochemistry has been identified for all three enzymes, and is shown in Table 1. There is no suggestion of a close relationship. More indirect evidence for the independence of the individual 4e⁻ dehydrogenases is the fact that LADH can perform the HDH reaction, albeit slowly (Dutler et al., 1986). What we gather from other enzymes therefore are paradigms which allow us to construct models for HDH mechanism.

A model combining information from UDPGDH, HMGR, and observations on HDH was constructed in which the initial aldehyde intermediate, formed at one subsite of the active site, is transferred, as a lysine imine, to a second site where formation of a thiohemiacetal takes place. Oxidation is followed by hydrolytic cleavage and release of product acid (Eccleston et al., 1979). We have tested several aspects of this model.

TESTS FOR IMINES

The group of Gorisch has shown that HDH reacts with pyridoxal phosphate to produce an inactive enzyme (Burger and Gorisch, 1981b). Interpretation of these results is made difficult by the very high PLP levels required to inactivate HDH, and the consequent lack of reaction specificity. Our own studies have not yet located a lysine-reactive reagent that provides a substrate-protected inactivation of HDH.

An alternative approach is to use $NaBH_4$ or $NaBCNH_3$ to trap an imine or carbinolamine intermediate from the steady state of histidinol oxidation. These reagents are generally not inactivators, but by effectively reducing imines or carbinolamines to amines, provide irreversible modification of the active site. Such an approach provided convincing evidence for a lysine adduct in aldolases (Lai et al., 1965). A problem in applying this approach to HDH is that generation of the imine requires NAD, which is also readily reduced by hydrides. To carry out this test, we tried several combinations of NAD, hydride, histidinol and enzyme, but failed to observe any enzyme inactivation. Although the group of Gorisch has developed a method for producing HDH-histidinaldehyde complexes, they have not reported on reduction of the complexes with $NaBH_4$.

Imine formation from carbonyl groups, when it occurs, excludes the original carbonyl oxygen. In cases such as aldolase, it was shown that the enzyme catalyzed the rapid exchange of ^{18}O between the 2 position of fructose-1,6-bisphosphate and HOH, demonstrating the catalytic relevance of the Schiff base. We have done experiments to follow the oxygen atoms in the HDH reaction (Grubmeyer and Insinga, 1990). In the HDH reaction, product histidine contains two oxygens, one clearly arising from solvent. An imine intermediate would result in the loss of the original histidinol oxygen, and a product containing two solvent oxygens. To determine the origin of the two oxygens, we employed a simple and accessible technique developed at Purdue by Robert Van Etten, the ^{18}O isotope shift in ^{13}C NMR (Risley and Van Etten, 1979). Van Etten showed that the ^{13}C resonance of oxygen-linked carbon is shifted when ^{16}O is replaced by ^{18}O. These shifts are small, about 0.02 ppm for each oxygen in carboxyl groups, but are readily detected by modern NMR instruments. We prepared highly enriched [hydroxymethyl-^{13}C]histidinol by reduction of [carboxy-^{13}C]histidine (Merck Isotopes) via its methyl ester. The substrate was then oxidized by HDH in 50% enriched $H^{18}OH$, containing 50 mM Na-glycine, 50 mM histidinol, 10 mM NAD, and 0.1 mM $MnCl_2$. FMN and O_2 were used as an NAD regeneration system (Jones and Taylor, 1976). After reaction (about 4 h), EDTA (50mM) was added to complex Mn(II), the samples were dried, dissolved in 99.8% D_2O, dried, and redissolved in 99.998% D_2O. In the NMR at 75 MHz, one can readily discern the shift caused by ^{18}O labelling. This is shown in Fig 1A, which is the spectrum of ^{13}C acetate allowed to equilibrate with 20% enriched $H^{18}OH$. Three peaks are seen, corresponding to acetate with 0, 1 or 2 ^{18}O atoms, in the ratio 64:32:4, as expected for equilibration. Fig 1B shows that histidine itself does not exchange its oxygens, even when exposed to $H^{18}OH$ and subjected to the complete workup procedure. Fig 1C is the ^{13}C NMR spectrum of histidine produced by the HDH reaction. The spectrum shows two peaks, a result that can only be accomodated by incorporation of a single solvent oxygen, leading to histidine with 0 or 1 ^{18}O. In contrast, incorporation of two solvent oxygen atoms would lead to product showing three peaks in a 25:50:25 ratio. Thus, the histidine product of the HDH reaction retains the original histidinol oxygen and an imine cannot be an intermediate.

TESTING THE THIOHEMIACETAL MODEL

A thiohemiacetal intermediate demands an active site cysteine residue. HDH from S. typhimurium has 7 cysteine residues. Fink's laboratory (Donahue et al., 1982) has provided a sequence of the yeast hisD homolog, HIS4, which codes for a multifunctional protein, and examination of the region of homology shows that only two of the 7 cysteines, Cys-116 and Cys-153, are conserved. Modification studies have proven confusing. The non-conserved Cys-159 was modified with bulky maleimides (Bitar et al., 1977). However, in a situation familar to AlDH workers, not all activity was lost, even with exhaustive modification. Our own work with NEM modification showed only about 50% inactivation, and no substrate protection was observed. My group has also found that iodoacetate does not inactivate or modify the protein, suggesting that most of the thiol groups are not accessible to solvent. However, when HDH is reduced and denatured and then modified with IAA the modified enzyme fails to regain activity when exposed to renaturing conditions. To achieve more specificity, we explored the use of NBD-Cl, a modifier of sulfhydryl and to a lesser extent, tyrosine and lysine groups, whose structure resembles that of histidinol (Grubmeyer and Gray, 1986). The compound inactivated HDH in a pseudo-first order fashion, and excellent protection against inactivation was afforded by histidinol. The inactive HDH-NBD adduct was readily reactivated by incubation with DTT, behavior indicative of sulfhydryl or tyrosine modification, and the spectrum was that of a sulfhydryl adduct with a peak at 420 nm. Stoichiometry studies showed that 3 sulfhydryl groups were modified per dimer at full inactivation. Using a two-step labelling procedure, it was possible to demonstrate that 1 residue/dimer was not protected by histidinol, and its modification had no effect on activity, while the other 2 NBD-reactive residues were protected and their modification led to inactivation.

The base-catalyzed migration of NBD adducts with cysteine has hampered past studies using this modifier. Other researchers have employed Zn dust to reduce and thus stabilize the adducts. With William Gray, we were able to digest modified HDH in acid with pepsin, and purify labelled peptides with reverse-phase HPLC. Analysis showed that the substrate-protected cysteine was largely Cys-116, with some contribution by Cys-377. Cys-116 and Cys-153 are conserved in the yeast HDH homolog. Although the identification of Cys-116 as an active site residue appears clear, we can still question whether it functions in catalysis as a thiohemiacetal.

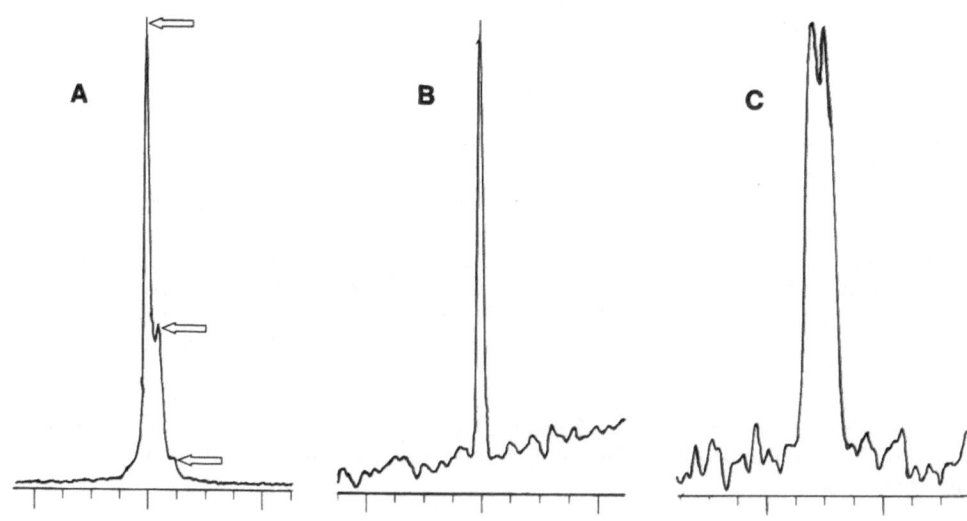

Figure 1. 75.5 MHz ^{13}C NMR spectra of carboxyl groups (from Grumeyer and Insinga, 1990). A. [1-^{13}C]acetate, equilibrated with 20% H^{18}OH. B. [carboxy-^{13}C]histidine. C. [carboxy-^{13}C]histidine produced by oxidation of ^{13}C-histidinol with HDH in 50% H^{18}OH. In each case one scale division is equivalent to 0.05 ppm, and the peaks are centered on 181.6 ppm (A) and 172.6 (B and C), relative to a standard of TMS.

Table 2. Kinetic isotope effects with deuterated histidinols

Substrate	^{D}V	$^{D}V/K$
(D,D)histidinol	2.55	1.30
(H,D)histidinol	1.18	1.13
(D,H)histidinol	2.39	1.19

OTHER MODELS

A thiohemiacetal, which is our choice, does not yet offer a convincing explanation for all data. One of the more attractive features of a thiohemiacetal is its redox poise, which is similar to that of alcohol/aldehyde. A thiohemiacetal/thiolester oxidation linked to NAD should be readily reversible (as demonstrated for HMGR (Sherban et al., 1985) and UDPGDH (Ridley et al., 1975)). To examine the reversibility of the two hydride transfers my group used exchange reactions catalyzed by HDH (Grubmeyer et al., 1987). Under conditions of high NADH concentration, HDH catalyzes two exchanges: of the ^{3}H from [$4R$-^{3}H]NADH into histidinol, and of [$4S$-^{3}H]NADH into NAD. The former exchange allowed us to look for reversibility of the second half-reaction, which would result in histidinol containing label from 2 molecules of NADH. In fact results clearly showed incorporation only of one ^{3}H which must be that resulting from the reversal of the first half-reaction. The result thus appears to indicate irreversibility of the second half-reaction. However, such apparent irreversibility could have a kinetic origin: if either the hydrolysis of the thiolester is rapid, or precedes NADH release, no reversal will be observed.

Another approach to determining the mechanism of the second step is to use isotope effects. Cleland and his school have largely been responsible for the development of this technique as a tool for determination of the kinetic mechanism and transition state structure for enzymes. Work on NAD-linked dehydrogenases has recently been reviewed by Cook and Bertagnolli (1986).

Dideuterated histidinol is readily synthesized using NaBH$_{4}$ reduction of histidine methyl ester. The compound shows a 2.5-fold ^{D}V isotope effect (Table 2). Thus, it appears that one of the two hydride transfer steps is is partially rate-limiting for the overall reaction, but it is not clear which. Second, it is not clear to what extent the intrinsic isotope effect for that step is manifested in the observed ^{D}V. The first question is readily answered using chiral monodeuterated histidinols. The substrates are synthesized using HDH-catalyzed exchange reactions between appropriately labelled histidinol and NAD, and are described by the step in which hydride is removed. Thus (D,H)histidinol (which is [hydroxymethyl-S^{2}H]histidinol, Grubmeyer et al., 1989) transfers deuterium in the first step; this compound shows very slight ^{D}V effects, although surprisingly, a $^{D}V/K$ effect is also observed. In contrast, with (H,D)histidinol, in which deuterium is removed in the aldehyde dehydrogenase step, a ^{D}V of 2.4 is observed. Thus, the second hydride transfer accounts for all, or nearly all, of the ^{D}V effect of (D,D)histidinol.

Determining the extent to which the second transfer limits rates is more difficult. The use of the alternative nucleotide acetyl-NAD, which is reduced with 3-fold slower Vmax, provided similar ^{D}V effects with (D,D)histidinol, however experiments with monodeuterated histidinols indicated some participation from the first step in the rate limitation. Conclusive evidence for full rate limitation by the second step is thus still lacking. To look for other proton transfers that might be involved in a rate limitation, we have performed the HDH reaction in 75% D$_{2}$O, and observed no change in rate. Thus, our provisional conclusion is that the second hydride transfer in HDH occurs with an intrinsic isotope effect of about 2.5. For many NAD-linked hydride transfers intrinsic effects are 5-6, although yeast AlDH shows a ^Dk of about 3 (Scharschmidt et al., 1984), and formate dehydrogenase demonstrates a ^Dk of 2.2 (Hermes et al., 1984). These deviant values have been attributed to transition states that are early or late, in other words for which the substrate is particularly reducing or oxidizing. Neither is expected for a thiohemiacetal, although it might be that unusual binding or deprotonation alters the postion of the transition state. Thus for HDH the chemistry of the second hydride transfer may be unusual.

Alternatives to the thiohemiacetal include the possible participation of an aldehyde hydrate as an intermediate. This mechanism has two potential drawbacks. First, the hydrate is far more reducing than a thiohemiacetal, violating the principle that intermediates should be isoenergetic. However, the high reducing potential would explain both the irreversibility of the second step and a deviant value for ^Dk. Second, past work has indicated a very tight binding of intermediate aldehyde, and this was interpreted to indicate covalency. However, tight noncovalent binding of a hydrate is not incompatible with past observations.

We are left with an interesting enzyme whose mechanism is not definitively established. Our future work is directed toward identification of the aldehyde-level intermediate, and its relationship to the subunit interactions of the enzyme.

Acknowledgements: I would like to thank the many lab associates who performed the studies described here, including technician Salvatore Insinga, undergraduates Marios Skiadopoulos, Shuk Yi Lee, Kim Wai Chu and Nader Moazami. Our work was supported by grants and REU fellowships from the National Science Foundation (DMB84-09256, DMB87-05583).

REFERENCES

Adams, E. (1955) L-Histidinal, a biosynthetic precursor of histidine. J. Biol. Chem. 217, 325-344.

Bitar, K. G., Firca, J. R., and Loper, J. C. (1977) Histidinol dehydrogenase from Salmonella typhimurium and Escherichia coli. Purification, some characteristics and the amino acid sequence around a reactive thiol group. Biochim. Biophys. Acta 493, 429-440.

Burger, E., and Gorisch, H. (1981a) Patterns of product inhibition of a bifunctional dehydrogenase; L-histidinol:NAD+ oxidoreductase. Eur. J. Biochem. 116, 137-142.

Burger, E., and Gorisch, H. (1981b) Evidence for an essential lysine at the active site of L-histidinol:NAD+ oxidoreductase; a bifunctional dehydrogenase. Eur. J. Biochem. 118, 125-130.

Burger, E., Gorisch, H., and Lingens, F. (1979) The catalytically active form of histidinol dehydrogenase from Salmonella typhimurium. Biochem. J. 181, 771-774.

Carlomagno, M. S., Chiariotti, L., Alifano, P., Nappo, A. G., & Bruni, C. (1988) Structure and function of the Salmonella typhimurium and Escherichia coli K-12 histidine operons. J. Mol. Biol. 203, 585-606.

Cook, P. F., & Bertagnolli, B. L. (1986) Kinetics of pyridine nucleotide-utilizing enzymes, in: Pyridine Nucleotide Coenzymes (Dolphin, D., Avramovic, O, & Poulsen, R., eds.) Part A, pp 405-447. Wiley, New York.

Donahue, T. F., Farabaugh, P. J., & Fink, G. R. (1982) The nucleotide sequence of the HIS4 region of yeast. Gene 18, 47-59.

Dutler, H., Ambar, A., & Donatsch, J. (1986) Function of zinc in liver alcohol dehydrogenase, in: Zinc Enzymes (Bertini, I., Luchinat, C., Maret, W., & Zeppezauer, M., eds.) Birkhauser, Boston, pp 471-483.

Eccleston, E. D., Thayer, M. L., & Kirkwood, S. (1979) Mechanisms of action of histidinol dehydrogenase and UDP-Glc dehydrogenase: evidence that the half-reactions proceed on separate subunits. J. Biol. Chem. 254, 11399-11404.

Franzen, B., Carrubba, C., Feingold, D. S., Ashcom, J., & Franzen, J. S. (1981) Amino acid sequence of the tryptic peptide containing the catalytic-site thiol group of bovine liver uridine diphosphate glucose dehydrogenase. Biochem. J. 199, 599-602.

Gorisch, H., & Holke, W. (1985) Binding of histidinal to histidinol dehydrogenase. Eur. J. Biochem. 150, 305-308.

Greeb, J., Atkins, J. F., & Loper, J. C. (1971) Histidinol dehydrogenase (hisD) mutants of Salmonella typhimurium. J. Bacteriol. 106, 421-431.

Grubmeyer, C., Chu, K. W., & Insinga, S. (1987) Kinetic mechanism of histidinol dehydrogenase: histidinol binding and exchange reactions. Biochemistry, 26, 3369-3373.

Grubmeyer, C. T., and Gray, W. R. (1986) A cysteine residue (cysteine-116) in the histidinol binding site of histidinol dehydrogenase. Biochemistry 25, 4778-4784.

Grubmeyer, C. T., and Insinga, S. (1990) Histidinol dehydrogenase: ^{18}O isotope shift in ^{13}C NMR reveals origin of histidine oxygens. J. Am. Chem. Soc. (in press).

Grubmeyer, C., Insinga, S., Bhatia, M., & Moazami, N. (1989a) Salmonella typhimurium histidinol dehydrogenase: complete reaction stereochemistry and active site mapping. Biochemistry 28, 8174-8180.

Grubmeyer, C., Skiadopoulos, M., and Senior, A. E. (1989b) L-Histidinol dehydrogenase, a Zn^{2+}-metalloenzyme. Arch. Biochem. Biophys. 272, 311-317.

Hermes, J. D., Morrical, S. W., O'Leary, M. H., and Cleland, W. W. (1984) Variation of transition-state structure as a function of the nucleotide in reactions catalyzed by dehydrogenases. 2. Formate dehydrogenase. Biochemistry 23, 5479-5488.

Jones, J. B., and Taylor, K. E. (1976) Nicotinamide coenzyme regeneration. Flavin mononucleotide (riboflavin phosphate) as an efficient, economical, and enzyme compatible recycling agent. Can J. Chem. 54, 2969-2973.

Kohno, T., & Gray, W. R. (1981) Chemical and genetic studies on L-histidinol dehydrogenase of Salmonella typhimurium: isolation and structure of the tryptic peptides. J. Mol. Biol. 147, 451-464.

Lai, C. Y., Tchola, O., Cheng, T., and Horecker, B. L. (1965) The mechanism of action of aldolases. VIII. The number of combining sites in fructose diphosphate aldolase. J. Biol. Chem. 240, 1347-1355.

Lee, S.-Y, and Grubmeyer, C. T. (1987) Purification and in vitro complementation of mutant histidinol dehydrogenases. J. Bacteriol. 169, 3938-3944.

Loper, J. C., & Adams, E. (1965) Purification and properties of histidinol dehydrogenase from Salmonella typhimurium. J. Biol. Chem. 240, 788-795.

Luskey, K. L. & Stevens, B. (1985) Human 3-hydroxy-3-methylglutaryl coenzyme A reductase: conserved domains responsible for catalytic activity and sterol regulated degradation. J. Biol. Chem. 260, 10271-10277.

Model, P., Ponticorvo, L., and Rittenberg, D. (1968) Catalysis of an oxygen-exchange reaction of fructose 1,6-diphosphate and fructose 1-phosphate with water by rabbit muscle aldolase. Biochemistry 7, 1339-1347.

Ordman, A. B., and Kirkwood, S. (1977) Mechanism of action of uridine diphosphoglucose dehydrogenase: evidence for an essential lysine residue at the active site. J. Biol. Chem. 252, 1320-1326.

Ridley, W. P., and Kirkwood, S. (1975) The stereospecificity of hydrogen abstraction by uridine diphosphoglucose dehydrogenase. Biochem. Biophys. Res. Commun. 54, 955-960.

Ridley, W. P., Houchins, J. P., and Kirkwood, S. (1975) Mechanism of action of uridine diphosphoglucose dehydrogenase: evidence for a second reversible dehydrogenation step involving an essential thiol group. J. Biol. Chem. 250, 8761-8767.

Rogers, D. H., Panini, S. R., and Rudney, H. (1983) Properties of HMGCoA reductase and its mechanism of action, in: 3-Hydroxy-3-methylglutaryl CoA Reductase (Sabine, J. R., Ed.) pp 57-75, CRC PRess Cleveland.

Scharschmidt, M., Fisher, M. A., and Cleland, W. W. (1984) Variation of transition state structure as a function of the nucleotide in reactions catalyzed by dehydrogenases. 1. Liver alcohol dehydrogenase with benzyl alcohol and yeast aldehyde dehydrogenase with benzaldehyde. Biochemistry 23, 5471-5478.

Sherban, D. G., Kennelly, P. J., Brandt, K. G., & Rodwell, V. W. (1985) Rat liver 3 hydroxy-3-methylglutaryl-CoA reductase: catalysis of the reverse reaction and two half-reactions. J. Biol. Chem. 260, 12579-12585.

Yourno, J., and Ino, I. (1968) Purification and crystallization of histidinol dehydrogenase from *Salmonella typhimurium* LT-2. J. Biol. Chem. 242, 3273-3276.

STRUCTURAL STUDIES OF PIG LENS ALDOSE REDUCTASE: REVERSIBLE DIMERIZATION OF THE ENZYME

Jean-Michel Rondeau[1], Dino Moras[2], Frédérique Tête[2], Alberto Podjarny[2], Alain Van Dorsselaer[3], Jean-Marc Reymann[4], Patrick Barth[4], Jean-François Biellmann[4]

[1]Biostructure, 8 rue Gustave Hirn, 67000 Strasbourg
[2]Laboratoire de Cristallographie Biologique, I.B.M.C., C.N.R.S., 15 rue René Descartes, 67084 Strasbourg [3]Laboratoire de Chimie Organique des Substances Naturelles, URA 31, Département de Chimie, Université Louis Pasteur, 1 rue Blaise Pascal, 67008 Strasbourg [4]Laboratoire de Chimie Organique Biologique, URA 31, Département de Chimie, Université Louis Pasteur, 1 rue Blaise Pascal, 67008 Strasbourg (France)

Drugs inhibiting Aldose Reductase (AR) may have a profound therapeutic impact for patients suffering for diabetes Mellitus. However, a prerequisite for such a long term treatment is the absence of side effects. For the drug design, the knowledge of the structure of AR will be a most useful contribution and will answer a number of puzzling facts such as the surprisingly large number of aldehydes which are substrates for AR. In this account, we present preliminary results of a structural study of pig lens Aldose Reductase.

The purification procedure of Aldose Reductase from pig lenses has been improved and it can now be performed in four-five days. The procedure and the results are presented in Table 1.

<u>Table 1</u> . Pig lens Aldose Reductase purification from fresh frozen tissue (110g of lenses). All steps were carried out at 4°C.

Step	Protein (mg)	Specific Activity (xylose)U/mg	Purification factor	Yield %
Crude extract	43 000	0.0004	1	100
2O%(w/v)PEG-3000 fractionation	360	0.039	100	80
Ion exchange QAE$_{250}$ZETA Prep	21	0.39	1000	50
Chromatofocusing Mono P HR 5/20	6	1.24	3100	43

The enzyme was homogeneous according to SDS gel electrophoresis and ·Isoelectric focusing experiments (Fig. 1).

Purity criteria of AR

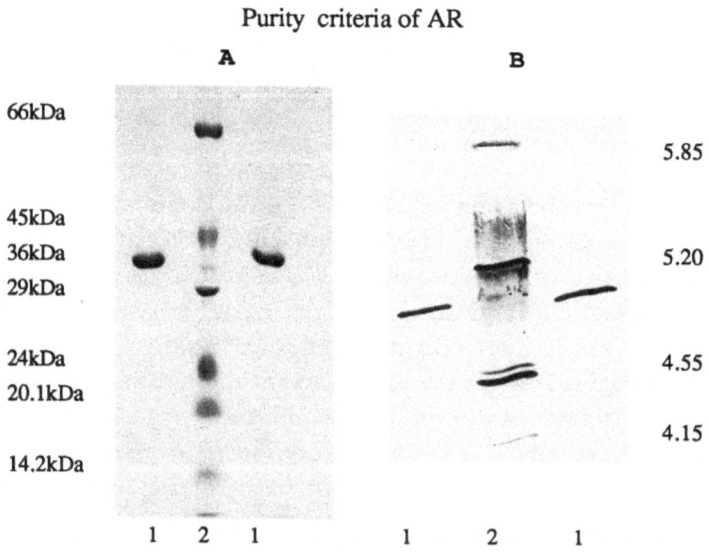

<u>Figure 1</u> . **A**. SDS-PAGE of purified Aldose Reductase (Coomassie blue G-250 staining). Lanes 1: purified Aldose Reductase ; Lane 2: molecular weight standards (indicated in kilodaltons). The molecular weight of Aldose Reductase is approximately 36,500 daltons. **B**. Isoelectric focusing on Phastgel (IEF 4.0-6.5; Pharmacia) of purified Aldose Reductase (Silver nitrate staining). Lanes 1: purified Aldose Reductase; Lane 2: pI standards. This technique confirmed the pI of 4.9-5.0 for Aldose Reductase estimated by chromatofocusing using a mono P HR 5/20 column.

High performance size exclusion chromatography of purified AR (TSK PW 3000) gave a single protein peak and no further increase in specific activity. The mass spectrum of Aldose Reductase by electrospray ionisation reviewed by Fenn et al.[1] gave additional evidence for the homogeneity of our preparation and indicated a molecular weight of 35,775±3Da for the pig lens AR (Fig. 2).

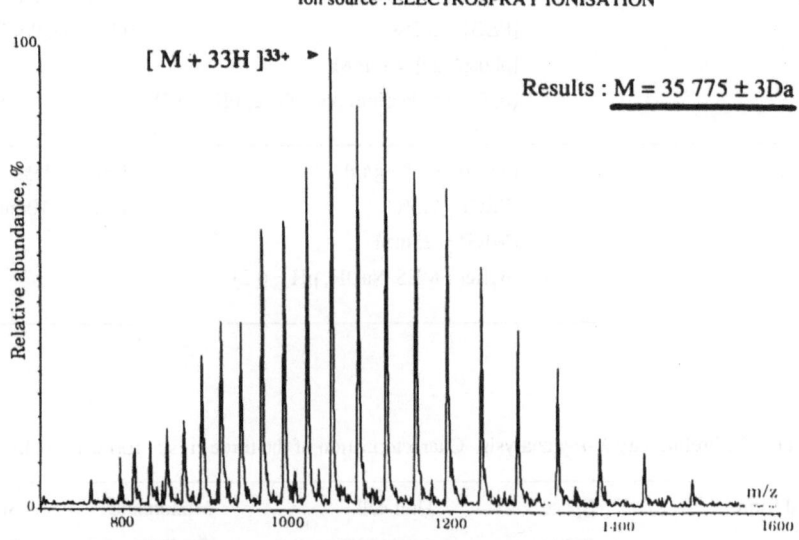

Figure 2. Electrospray mass spectrometer (ESMS) spectrum of purified Aldose Reductase. An average mass of 35,775±3 Da was calculated from four separated experiments.

The crystallization of AR at +4°C under different conditions using the vapor diffusion in hanging drops method gave three crystalline forms : triclinic, hexagonal and tetragonal. The conditions are indicated in Table 2 and the cell parameters in Table 3.

The triclinic and tetragonal crystals diffract to high resolution and are suitable for X ray structure determination. In presence of NADPH, the triclinic crystals dissolve, whereas the tetragonal crystals are stable.

Triclinic crystals were dissolved after 9-11 months storage at 4°C and the specific activity was measured and compared to that of the freshly prepared AR. We found that crystallization and storage did not alter the AR enzymatic activity.

Table 2. Experimental conditions of crystallization of purified Aldose Reductase by the vapor diffusion in hanging drop technique at +4°C.

Crystal form	Droplet	Well
Triclinic	[enzyme] = 5mg/ml [PEG] = 2.5% [citrate] = 15mM (buffer : ammonium citrate, pH = 6.2)	[PEG] = 20% [citrate] = 75mM
Hexagonal	[enzyme] = 5mg/ml [PEG] = 2.5% [phosphate] = 15mM (buffer : potassium phosphate, pH = 6.2)	[PEG] = 20% [phosphate] = 75mM
Tetragonal	[enzyme] = 5mg/ml [PEG] = 2.5% [MES] = 25mM (buffer : MES-NaOH, pH = 6.2)	[PEG] = 25% [MES] = 100mM

Table 3. Preliminary X-ray analysis. Characterization of the three crystal forms of Aldose Reductase.

Crystal form	Space group	Unit cell parameters	Monomers/ Asymm. unit	Specific volume (Å^3/Dalton)
Triclinic	P1	a=81.3 Å b=85.9 Å c=56.6 Å α=102.3(1)° β=103.3(1)° γ=79.0(1)° V=371 000 Å^3	4	2,65
Hexagonal	$P6_222$ $P6_422$	a=b=101.0Å c=257.0Å V=2 270 000 Å^3	2	2,70
Tetragonal	$P4_12_12$ $P4_32_12$	a=b=68.9 Å c=155.4 Å V=737 700 Å^3	1	2,59

From the preliminary X-ray analysis, it became clear that the AR was present as a dimer in one crystal form. Indeed AR is present as a non-crystallographic dimer in the triclinic crystals. In the tetragonal crystals too, the existence of dimers cannot be ruled out because of the crystallographic diad. However, AR has been found until now as a monomer in solution. Therefore, the question arose whether dimers of AR could be detected at high concentration.

The molecular weight determination by gel filtration on a Sephadex G-100 column gave a M.W. of 31,300 kDa ; this value has to be compared to 35,775±3 Da as determined by Mass Spectrometry. This result suggests that AR was retarded on the column through specific interaction with the solid phase. Moreover, it should be pointed out that this molecular weight determination was carried out at a low protein concentration so that in case of a reversible dimerization, only the monomer may be detected.

Ultracentrifugation experiments on AR in the concentration range 0.5-5mg/ml showed that the sedimentation coefficient (Fig. 3) varied with the protein concentration with a maximum at 1mg/ml. Such a profile is consistent with a monomer-dimer system in reversible equilibrium as discussed by Gilbert and Gilbert[2].

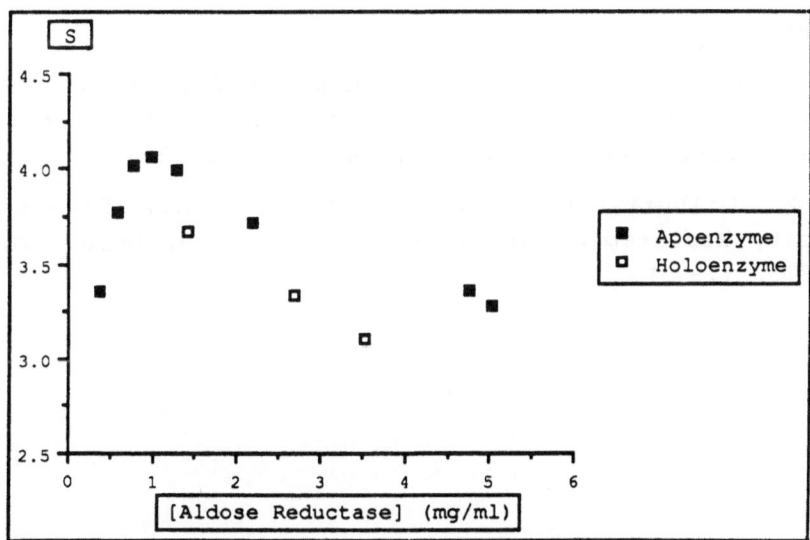

Figure 3 . Dependance of the sedimentation coefficient $S_{20,W}$ of Aldose Reductase in function of its concentration. The enzyme solution was prepared in a 50mM potassium phosphate buffer pH 6.2, 1 mM dithiothreitol. For the holoenzyme, the concentration of NADPH was 10mM.

Small angle neutron scattering experiments of AR were then undertaken at Institut Laue Langevin in Grenoble with the assistance of Dr. G. ZACCAI. In the concentration range 1-23mg/ml of AR in 0.05 M phosphate buffer pH 6.2, the average molecular weight was 50±2 kDa and seemed to depend on the enzyme preparation. An increase in the buffer concentration or an addition of sodium chloride resulted in a decrease in the apparent M.W.. The addition of the coenzyme in molar excess decreased the M.W. to 32-33 kDa. Furthermore, sorbinil had no effect on the apparent-MW of the apoenzyme, while tolrestat decreased the M.W. to 35 kDa. These results strongly support the existence of a monomer/dimer equilibrium. However, the value of the M.W. in the protein concentration range 1-23 mg/ml did not vary to any significant extend and did not correspond to the M.W. of about 70 kDa which is expected for a dimer. One explanation could be that AR is a mixture of two species : one undergoing dimerization at high concentration and one monomeric. Crosslinking experiments in progress in our group agree with the presence in solution of dimers of the apoenzyme, which dissociate in presence of the coenzyme.

Electron density maps have been obtained and their interpretation is in progress. We hope to present in a near future the structure of Aldose Reductase and interpret a number of its enzymological features in structural terms.

REFERENCES

1. Fenn, J.B., Mann, M., Meng, C.K., Wong, S.F., and Whitehouse,C.M., Electrospray Ionization for Mass Spectrometry of Large Biomolecules, Science, 246: 64, (1989).

2. Gilbert, L.M., and Gilbert, G.A., Sedimentation Velocity Measurement of Protein Association, Meth. Enzymol., 27: 273, (1973).

cDNA CLONING AND EXPRESSION OF HUMAN ALDOSE REDUCTASE

Chihiro Nishimura, Yoshiharu Matsuura#, Tsuyoshi Tanimoto*,
Takashi Yamaoka, Tai Akera, and T. Geoffrey Flynn+

Dept. of Pediatric Pharmacol., Natl. Children's Med. Res.
Ctr.; #Dept. of Vet. Sci., Natl. Institute of Health
*Div. of Biol. Chem., Natl. Institute of Hygienic Sci.
Tokyo, Japan; +Dept. of Biochem., Queen's University
Kingston, Canada

INTRODUCTION

Aldose reductase (ALR2; EC 1.1.1.21), a member of the monomeric
NADPH-dependent aldoketoreductase family, has been strongly suggested to
play a key role in the development of various diabetic complications
(Gabbay et al., 1966; Engerman and Kern, 1984; Nishimura et al., 1987;
Kinoshita and Nishimura, 1988).

A number of ALR2 inhibitors have already been developed to serve as a
new means to prevent and treat these complications. The efficacy of these
inhibitors, however, is usually examined using enzyme preparations
obtained from animal sources. Because several biochemical properties of
this enzyme protein are known to differ considerably among species (Kador
et al., 1985), the experimental data on the efficacy of inhibitors against
ALR2 obtained from animal sources have a limited value in predicting their
efficacy against human enzyme.

It thus appeared essential to study the protein structure of human
enzyme for the development of more potent, as well as selective, ALR2
inhibitors. To this end, we have attempted to determine the primary
structure of human ALR2 by nucleotide sequence analysis of cloned cDNA for
human ALR2. The complete amino acid sequence of ALR2 expressed in retina
and muscle was obtained from clones isolated using synthetic
oligonucleotide probes, which were designed based on partial amino acid
sequences of purified human muscle ALR2. This approach also enabled us to
construct recombinant baculovirus containing one of the cDNA clones and
overproduce a large amount of human ALR2 in a powerful baculovirus
expression system.

MATERIALS AND METHODS

Human ALR2 was purified from psoas muscle and subjected to amino acid
sequencing after endoproteinase Lys-C digestion as described by Morjana et
al.(1989). Based on these partial amino acid sequence data, five
oligonucleotide probes were designed using codon usage tables by Lathe

(1985). Initial screening of human retina cDNA library (Nathans et al., 1986) yielded eight clones with inserts ranging in length from 1.0 to 1.3 kb. Using isolated insert of one of these retina clones, fetal muscle library (Koenig and Kunkel, 1987) was subsequently screened and an additional eight clones were isolated from this library. The nucleotide sequence of these clones was determined following subcloning into pGEM-4Z (Promega) by the dideoxy chain termination method (Chen and Seeburg, 1985) using synthetic oligonucleotide primers.

Secondary structure prediction was made according to Garnier's method (Garnier et al., 1978). Decision constants were set at DCH = -75, and DCE = 50, based on the estimation of 49 % alpha helical content in bovine kidney ALR2 from circular dichroism spectra recently reported by Grimshaw et. al.(1989). Hydrophilicity was calculated according to the algorithm of Hopp and Woods (1981) using the University of Wisconsin Genetics Computer Group programs.

A baculovirus transfer vector pAcYM1/AR was constructed by introducing isolated cDNA fragment from one of the muscle clones (L-1) into pAcYM1 (Matsuura et al., 1987) as previously described (Nishimura et al., in press). The nucleotide sequence of this construct was confirmed to contain the upstream sequence of the polyhedrin gene, the first A of the initiating ATG codon for the polyhedrin coding sequence, 11 nucleotides derived from DNA linkers, and 8 nucleotides of 5'-untranslated leader sequence prior to the initiation codon for ALR2.

The insect cell line *Spodoptera frugiperda* (SF9) was transfected with mixtures of purified wild-type AcNPV DNA and pAcYM1/AR as described previously (Matsuura et al., 1987; Smith et al., 1983). One of the isolated recombinant baculovirus AcAR was subsequently used to infect monolayers of SF9 cells for the production of recombinant human ALR2.

SDS-polyacrylamide gel electrophoresis of the supernatant fraction of the SF9 cell homogenates or culture medium was carried out according to the method of Laemmli (1970). Enzyme activity was determined in a reaction mixture containing 0.1 M sodium phosphate buffer (pH 6.2), 120 µM NADPH, 10 mM DL-glyceraldehyde, and the enzyme preparation in a total volume of 1 ml. The reaction was started by the addition of enzyme, and activity was estimated by measuring NADPH oxidation from the decrease in absorbance of 340 nm at 25 °C.

RESULTS AND DISCUSSION

Screening of 5×10^5 plaques from a λgt10 human retina library yielded eight clones and one of the clones (UU-1) had the putative initiation codon with 32 nucleotides of 5'-untranslated sequence. This clone, however, contained an unrelated sequence 332 nucleotide downstream, which seemed to be a cloning artifact. Using a 1256-base pair cDNA insert of the retina clone OO-1, 6×10^5 plaques of human fetal muscle library in λgt10 were subsequently screened. Among eight muscle clones obtained, four had a single open reading frame encoding 316 amino acids with the initiation methionine. These retina and muscle clones demonstrated the identical nucleotide sequence, shown in Fig. 1 as the complete ALR2 cDNA sequence compiled from UU-1 (dotted line) and the rest of the clones obtained from both libraries. UU-1 includes nucleotides 1-332, OO-1 nucleotides 71-1326, and the muscle clone L-1 includes 25-1371 plus 2 more A residues at the 3'-end. Lines between amino acids indicate regions matched with the peptide sequence data. Arrowheads indicate a potential polyadenylation signal. Comparing with cDNA sequences of human placental ALR2 reported by Bohren et. al.(1989) and Chung and LaMendola (1989), our cDNA sequence was essentially identical to that of Bohren et al. except for a difference in the 5'-noncoding region.

```
          10                      30                      50
CGGTACGTGCGGCCTTGGGGAGCGCAGCAGCCATGGCAAGCCGTCTCCTGCTCAACAACG
••••••••••••••••••••••••••••••••••  M   A   S   R   L   L   L   N   N   G
          70                      90                      110
GCGCCAAGATGCCCATCCTGGGGTTGGGTACCTGGAAGTCCCCTCCAGGGCAGGTGACTG
 A   K   M→P   I   L   G→L→G→T→W→K→S→P→P→G→Q→V→T→E
          130                     150                     170
AGGCCGTGAAGGTGGCCATTGACGTCGGGTACCGCCACATCGACTGTGCCCATGTGTACC
→A→V→K   V   A→I→D→V→G→Y→R→H→I→D→C→A   H   V→Y   Q
          190                     210                     230
AGAATGAGAATGAGGTGGGGGTGGCCATTCAGGAGAAGCTCAGGGAGCAGGTGGTGAAGC
 N→E   N   E   V   G   V   A   I   Q   E   K   L   R   E→Q→V→V→K→R
          250                     270                     290
GTGAGGAGCTCTTCATCGTCAGCAAGCTGTGGTGCACGTACCATGAGAAGGGCCTGGTGA
→E→E→L→F→I→V→S→K→L→W→C→T→Y→H→E→K   G   L   V   K
          310                     330                     350
AAGGAGCCTGCCAGAAGACACTCAGCGACCTGAAGCTGGACTACCTGGACCTCTACCTTA
 G   A   C   Q   K   T→L→S→D→L→K→L→D→Y→L→D→L→Y→L→I
          370                     390                     410
TTCACTGGCCGACTGGCTTTAAGCCTGGGAAGGAATTTTTTCCCATTGGATGAGTCGGGCA
→H→W→P   T   G   F→K   P   G   K   E→F→F→P→L→D→E→S→G→N
          430                     450                     470
ATGTGGTTCCCAGTGACACCAACATTCTGGACACGTGGGCGGCCATGGAAGAGCTGGTGG
→V→V→P→S→D   T   N   I   L   D   T   W   A   A   M   E   E   L   V   D
          490                     510                     530
ATGAAGGGCTGGTGAAAGCTATTGGCATCTCCAACTTCAACCATCTCCAGGTGGAGATGA
 E   G   L   V   K   A→I→G→I→S→N→F→N→H→L→Q→V→E→M→I
          550                     570                     590
TCTTAAACAAACCTGGCTTGAAGTATAAGCCTGCAGTTAACCAGATTGAGTGCCACCCAT
→L→N→K   P   G   L   K   Y→K→P→A→V→N→Q→I→E→C   H   P→Y
          610                     630                     650
ATCTCACTCAGGAGAAGTTAATCCAGTACTGCCAGTCCAAAGGCATCGTGGTGACCGCCT
→L→T→Q→E→K   L   I→Q→Y   C   Q→S→K→G→I→V→V→T→A→Y
          670                     690                     710
ACAGCCCCCTCGGCTCTCCTGACAGGCCCTGGGCCAAGCCCGAGGACCCTTCTCTCCTGG
→S→P→L→G→S→P→D→R→P→W→A→K   P   E   D   P   S   L   L   E
          730                     750                     770
AGGATCCCAGGATCAAGGCGATCGCAGCCAAGCACAATAAAAACTACAGCCCAGGTCCTGA
 D   P   R   I   K   A   I   A   A   K   H   N   K   T→T→A→Q→V→L→I
          790                     810                     830
TCCGGTTCCCCATGCAGAGGAACTTGGTGGTGATCCCCAAGTCTGTGACACCAGAACGCA
→R→F→P→M→Q→R→N→L→V→V→I→P→K→S→V→T→P→E→R   I
          850                     870                     890
TTGCTGAGAACTTTAAGGTCTTTGACTTTGAACTGAGCAGCCAGGATATGACCACCTTAC
 A   E   N→F→K   V   F   D   F   E   L   S   S   Q   D   M   T   T   L   L
          910                     930                     950
TCAGCTACAACAGGAACTGGAGGGTCTGTGCCTTGTTGAGCTGTACCTCCCACAAGGATT
 S   Y   N   R   N   W   R   V   C   A   L   L   S   C   T   S   H   K   D→Y
          970                     990                     1010
ACCCCTTCCATGAAGAGTTTTGAAGCTGTGGTTGCCTGCTCGTCCCCAAGTGACCTATAC
→P→F   H   E   E   F
          1030                    1050                    1070
CTGTGTTTCTTGCCTCATTTTTTTCCTTGCAAATGTAGTATGGCCTGTGTCACTCAGCAG
          1090                    1110                    1130
TGGGACAGCAACCTGTAGAGTGGCCAGCGAGGGCGTGTCTAGCTTGATGTTGGATCTCAA
          1150                    1170                    1190
GAGCCCTGTCAGTAGAGTAGAAGTCTCTTCCAGTTTGCTTTGCCCTTCTTTCTACCCTGC
          1210                    1230                    1250
TGGGGAAAGTACAACCTGAATACCCTTTTCTGACCAAAGAGAAGCAAAATCTACCAGGTC
          1270                    1290                    1310
AAAATAGTGCCACTAACGGTTGAGTTTTGACTGCTTGGAACTGGAATCCTTTCAGCAAGA
          1330        ▼▼▼▼▼▼       1350                    1370
CTTCTCTTTGCCTCAAATAAAAAGTGCTTTTGTGAAAAAAAAAAAAAAAAAA
```

Fig. 1   Nucleotide sequence of human retina and muscle ALR2 cDNA and deduced amino acid sequence

Fig. 2a represents secondary structure predictions based on the estimate of high alpha helical content recently reported by Grimshaw et al.(1989). H indicates alpha helix; E, beta strand; A, antiparallel beta strand; P, parallel beta strand; T, turn; and C, coil. A strong hydrophobic region was found near the carboxyl-terminal end (Fig. 2b). The examination of amino acid sequence indicated a possible N-glycosylation site at position 242. When the recombinant human ALR2 expressed in the baculovirus system was treated with N-glycosidase F and analyzed on SDS-polyacrylamide gel electrophoresis, however, no significant change in mobility was found comparing with untreated ALR2. In addition to this finding, sequence analysis of the acid cleaved peptide fragment including Asn-242 demonstrated a clear peak of phenylthiohydantoin-Asn, suggesting that this amino acid is scarcely modified (data not shown). This finding is consistent with those recently reported by Schade et al.(1990) with purified bovine lens ALR2.

a

```
        10        20        30        40        50        60        70
MASRLLLNNGAKMPILGLGTWKSPPGQVTEAVKVAIDVGYRHIDCAHVYQNENEVGVAIQEKLREQVVKR
HHHHHHHHH    PPPPP    CCCC  HHHHHHHHHHH                HHHHHHHHHHHHHHHHHH

        80        90       100       110       120       130       140
EELFIVSKLWCTYHEKGLVKGACQKTLSDLKLDYLDLYLIHWPTGFKPGKEFFPLDESGNVVPSDTNILD
HHHHHHHHHHHHHHHHH   HHHHHHHHHHHHHHHHHHHHH      CCC    HHH              HH

       150       160       170       180       190       200       210
TWAAMEELVDEGLVKAIGISNFNHLQVEMILNKPGLKYKPAVNQIECHPYLTQEKLIQYCQSKGIVVTAY
HHHHHHHHHHHHHHHHHHHHHCCC   HHHHHHHH       HHH       HHHHHHHHHTTTTEEEEEE

       220       230       240       250       260       270       280
SPLGSPDRPWAKPEDPSLLEDPRIKAIAAKHNKTTAQVLIRFPMQRNLVVIPKSVTPERIAENFKVFDFE
EETTTCC C  CCCCCCHHHHHHHHHHHHHHHHHHHHHHHHHHEEEEE     PPPP  CCCHHHHHHHHHHHHHH

       290       300       310
LSSQDMTTLLSYNRNWRVCALLSCTSHKDYPFHELF
HHHHHHHAAAA  TTTT        TTTCCCHHHH
```

b

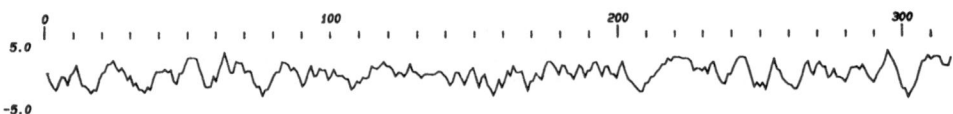

Fig. 2   Secondary structure prediction (a) and hydrophilicity profile (b) of human ALR2

A cDNA insert of one of the muscle clones L-1, containing the complete ALR2 coding region (shaded box, Fig. 3) was introduced into the transfer vector pAcYM1 downstream of its polyhedrin promotor (black box) in a 5'- to 3'-orientation. This vector pAcYM1/AR contained no initiation codons upstream of the ALR2 translation initiation site (ATG). SF9 cells were cotransfected with a mixture of pAcYM1/AR and DNA of wild-type baculovirus AcNPV. Recombinant baculovirus AcAR containing ALR2 cDNA in its original viral genome was produced by homologous recombination between the transfer vector and AcNPV DNA.

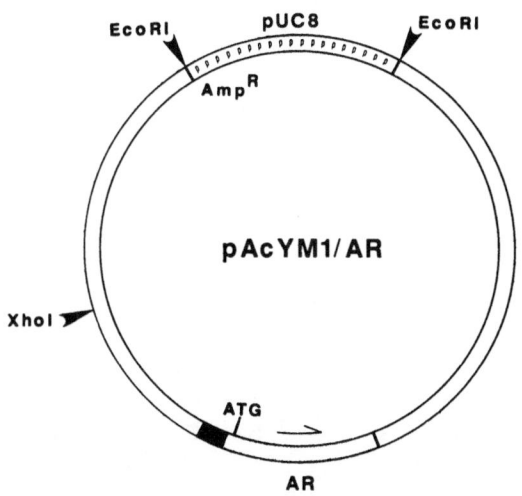

Fig. 3 Schematic diagram of pAcYM1/AR

When SF9 cells were infected with purified recombinant AcAR, expression of ALR2 protein was detected not only in the total cell lysate but also in the culture medium (Fig. 4). Lane 1 represents molecular size markers; lanes 2-5, total cellular proteins of SF9 cells harvested 2 days (lanes 2 and 4) or 3 days (lanes 3 and 5) after infection with wild-type virus AcNPV (lanes 2 and 3) or with recombinant virus AcAR (lanes 4 and 5); lanes 6-9, culture medium of SF9 cells collected 2 days (lanes 6 and 8) or 3 days (lanes 7 and 9) after infection with AcNPV (lanes 6 and 7) or with AcAR (lanes 8 and 9); lane 10, purified human ALR2 from psoas muscle. The arrowhead indicates polyhedrin protein expressed in AcNPV-infected cells. An abundant protein with an apparent molecular size of 36 kd in AcAR-infected cells were clearly observed in contrast with the exclusive expression of polyhedrin protein in the cell lysate of wild-type baculovirus-infected cells.

This recombinant ALR2 protein exhibited high enzymatic activity possessing all the characteristics reported for human ALR2 (Tables 1 and 2). These characteristics are; preference of DL-glyceraldehyde over glucuronate as the substrate; utilization of NADPH or NADH as cofactor; significant activation by sulfate ion; inhibition with chloride ion; inhibition with low concentrations of commercially developed ALR2 inhibitors such as Sorbinil (Pfizer), Tolrestat (Ayerst), and Statil (ICI); and the lack of inhibition with sodium valproate, a potent inhibitor of aldehyde reductase (EC 1.1.1.2).

Thus it is evident that we have cloned functional cDNA for human retina and muscle ALR2 and developed a practical expression system to overproduce active enzyme protein. Since our cDNA sequence data turned out essentially identical not only in coding but in 3'-noncoding region to that of Bohren et al.(1989), it appears obvious that this is the form of ALR2 commonly expressed in placenta, muscle and retina, one of the target organs of diabetic complications. The recombinant ALR2 expressed in SF9 cells demonstrated all the characteristics known for purified human ALR2.

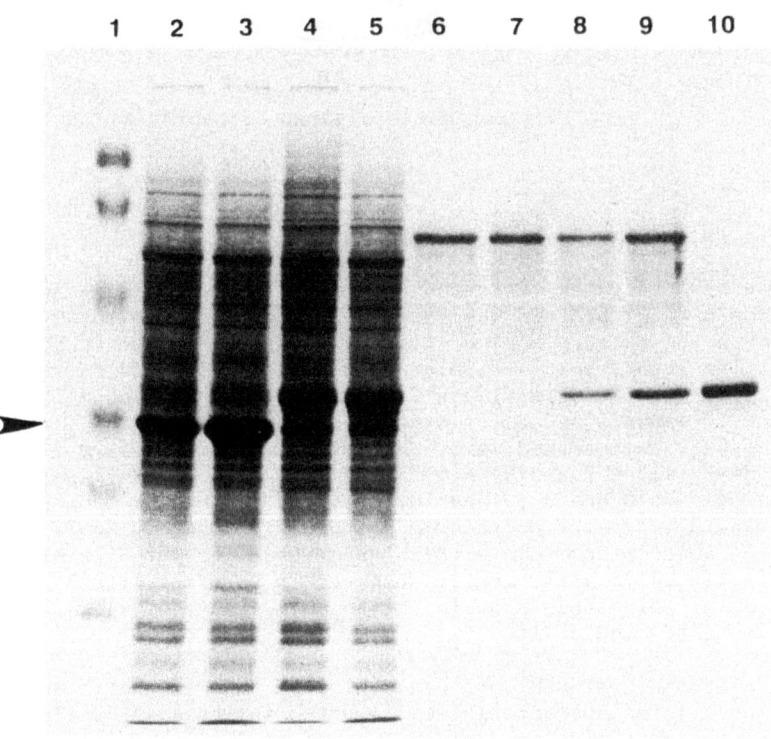

Fig. 4. SDS-polyacrylamide gel electrophoresis demonstrating expression of human aldose reductase.

Table 1   Substrate specificity of recombinant
         human aldose reductase

Substrate	Relative V_{max} (%)
DL-Glyceraldehyde	100
D-Xylose	100
D-Glucose	77
D-Galactose	83
D-Glucuronate	66
NADPH	100
NADH	90

Table 2   Effect of salts and inhibitors on
         recombinant human aldose reductase

Effector	Relative Activity (%)
None	100
0.3 M $(NH_4)_2SO_4$	235
0.3 M Li_2SO_4	257
0.3 M Na_2SO_4	230
0.3 M NH_4Cl	58
0.3 M NaCl	53
10 µM Sorbinil	35
10 µM Tolrestat	2
10 µM Statil	14
0.1 mM Valproate	99

Although sequence analysis of deduced amino acids pointed out a possible glycosylation site at position 242, no definite indication of actual modification on the Asn residue was obtained through N-glycosidase F treatment or sequence analysis of the acid-cleaved peptide fragment containing this site.

The fact that infinite amounts of the recombinant human enzyme can be supplied by this insect cell expression system should markedly advance more refined structural analysis of this protein, especially with regard to its inhibitor associating site(s). Although ALR2 inhibitors of diverse chemical structure are known to be equivalently effective on the enzyme, sites where inhibitors bind or associate with the protein molecule may possibly be not identical depending on the type of inhibitors. It may therefore be necessary to define these sites on the enzyme protein to design the new generation of ALR2 inhibitors, more effective and selective for ALR2 of human origin. In addition, this recombinant human ALR2 may serve as authentic preparations for screening the potency and efficacy in the process of developing new inhibitors.

REFERENCES

Bohren,K.M.,Bullock,B.,Wermuth,B.and Gabbay,K.H.,1989,The aldo-keto reductase superfamily,J.Biol.Chem.,264:9547.
Chen,E.Y.and Seeburg,P.H.,1985,Supercoil sequencing:A fast and simple method for sequencing plasmid DNA,DNA,4:165.
Chung,S.and LaMendola,J.,1989,Cloning and sequence determination of human placental aldose reductase gene,J.Biol.Chem.,264:14775.
Engerman,R.L. and Kern,T.S.,1984,Experimental galactosemia produces diabetic-like retinopathy,Diabetes,33:97.
Gabbay,K.H.,Merola,L.O.and Field,R.A.,1966,Sorbitol pathway: Presence in nerve and cord with substrate accumulation in diabetes,Science, 151:209.
Garnier,J.,Osguthorpe,D.J.and Robson,B.,1978,Analysis of the accuracy and implications of simple methods for predicting the secondary structure of globular proteins,J.Mol.Biol.,120:97.
Grimshaw,C.E.,Shahbaz,M.,Jahangiri,G.,Putney,C.G.,McKercher,S.R.and Mathur, E.J.,1989,Kinetic and structural effects of activation of bovine kidney aldose reductase,Biochemistry,28:5343.
Hopp,T.P.and Woods,K.R.,1981,Prediction of protein antigenic determinants from amino acid sequences,Proc.Natl.Acad.Sci.U.S.A., 78:3824.
Kador,P.F.,Robison,W.G.and Kinoshita,J.H.,1985,The pharmacology of aldose reductase inhibitors,Ann.Rev.Pharmacol.Toxicol.,25:691.
Kinoshita,J.H.and Nishimura,C.,1988,The involvement of aldose reductase in diabetic complications,Diabetes Metab.Rev.,4:323.
Koenig,M.,Hoffman,E.P.,Bertelson,C.J.,Monaco,A.P.,Feener,C.and Kunkel, L.M.,1987,Complete cloning of the Duchenne muscular dystrophy (DMD) cDNA and preliminary genomic organization of the DMD gene in normal and affected individuals,Cell,50:509.
Laemmli,U.K.,1970,Cleavage of structural proteins during the assembly of the head of bacteriophage T4,Nature,227:680.
Lathe,R.,1985,Synthetic oligonucleotide probes deduced from amino acid sequence data. Theoretical and practical considerations,J.Mol.Biol., 183:1.
Matsuura,Y.,Possee,R.D.,Overton,H.A.and Bishop,D.H.L.,1987,Baculovirus expression vectors:the requirements for high level expression of proteins, including glycoproteins,J.gen.Virol.,68:1233.

Morjana,N.A.,Lyons,C.and Flynn,T.G.,1989,Aldose reductase from human psoas muscle,J.Biol.Chem.,264:2912.

Nathans,J.,Thomas,D.and Hogness,D.S.,1986,Molecular genetics of human color vision:The genes encoding blue, greeen, and red pigments, Science,232:193.

Nishimura,C.,Lou,M.F.and Kinoshita,J.H.,1987,Depletion of myo-inositol and amino acids in galactosemic neuropathy,J.Neurochem.,49:290

Nishimura,C.,Matsuura,Y.,Kokai,Y.,Akera,T.,Carper,D.,Morjana,N.,Lyons,C. and Flynn,T.G.,Cloning and expression of human aldose reductase, J.Biol.Chem.,in press.

Schade,S.Z.,Early,S.L.,Williams,T.R.,Kezdy,F.J.,Heinrikson,R.L.,Grimshaw, C.E.,Doughty,C.C.,1990,Sequence analysis of bovine lens aldose reductase,J.Biol.Chem.,265:3628.

Smith,G.E.,Summers,M.D.and Fraser,M.J.,1983,Production of human beta interferon in insect cells infected with a baculovirus expression vector,Mol.Cell.Biol.,3:2156.

IN VITRO EXPRESSION OF HUMAN PLACENTAL ALDOSE REDUCTASE

IN *ESCHERICHIA COLI*

Deborah Carper, Sanai Sato, Susan Old,
Stephen Chung[+], and Peter F. Kador

National Eye Institute, NIH, 9000 Rockville Pike
Bethesda, MD 20892 and [+]Eugene Tech International
4 Pearl Ct., Allendale, NJ 07401

INTRODUCTION

Aldose reductase (alditol: NADP$^+$ oxidoreductase, EC 1.1.1.21, AR) reduces glucose to sorbitol. The accumulation of sorbitol has been implicated in the secondary complication of diabetes, including retinopathy, cataracts, neuropathy and nephropathy (Kinoshita, et al., 1990; Kador, 1988; Dvornik, 1987; Kinoshita, 1974). Blocking the accumulation of sorbitol with a number of aldose reductase inhibitors has proven to be effective in preventing or retarding several clinical complications associated with diabetes in galactosemic and diabetic animal models (Kinoshita, et al., 1990; Robison, et al., 1989; Kador, et al., 1988; Fukushi, Merola and Kinoshita, 1983; Yue, et al., 1982). Aldose reductase inhibitors have also shown promise in clinical studies (Sakamoto, et al., 1986; Judzewitsch, et al., 1983).

A better understanding of the structure of aldose reductase with emphasis on the physico-chemical interactions between the enzyme and aldose reductase inhibitors is required for the logical development of more specific, potent inhibitors. For such studies large quantities of enzyme are required. A number of studies have utilized human placental AR (Sawada, et al., 1989; Vander Jagt, et al., 1988; Herrmann, Kador and Kinoshita, 1983). One method for obtaining a safe and readily available supply of human placental AR is through molecular biology. Here, we report, the construction of a human placental aldose reductase in a bacterial expression vector, and the subsequent production, purification, and characterization of this recombinant enzyme.

MATERIALS AND METHODS

Plasmid Construction

A full length DNA sequence corresponding to the coding region and 3' untranslated sequence of human placental aldose reductase was constructed in the expression vector pKK233-2 (Pharmacia) (Fig. 1). The cDNA clone B1, previously cloned in the Bluescript vector M13 (Stratagene) (Chung and LaMendola, 1989), illustrated in Fig 1. as M13-B1, was partially digested with Nco I and Hind III. The digests were run on a 5% polyacrylamide gel and the 1338-bp fragment was isolated and purified by electroelution

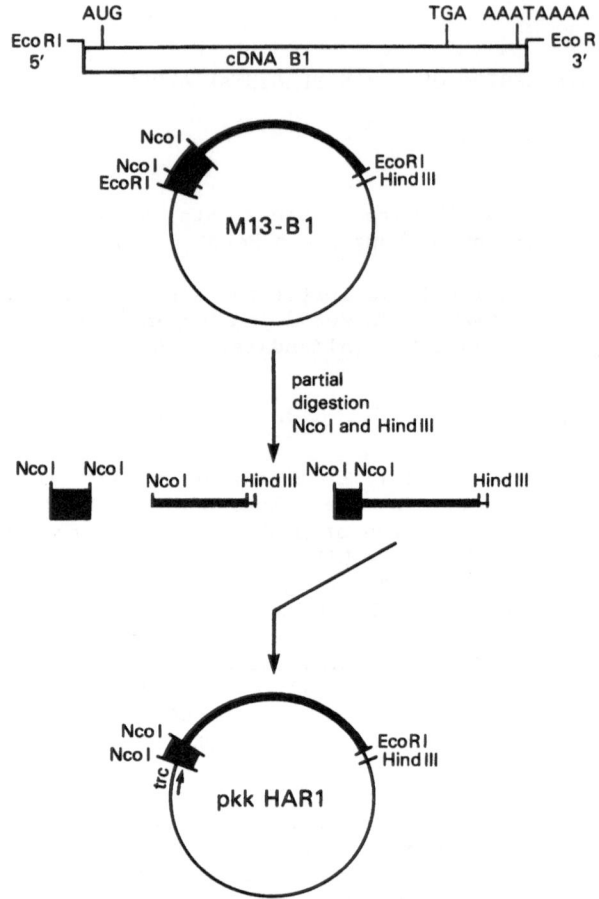

Fig. 1. Schematic diagram of the construction of pKK HAR1.
Details of the construction are found in the text.
AUG, start codon; TGA, stop codon; AAATAAAA, poly(A)
sequence; trc, expression vector promoter. Restriction
enzyme sites labeled: EcoR I, Hind III, Nco I.
Sequence data revealed that the fifth amino acid at the
amino-terminal end (Chung and La Mendola, 1989) should
be leucine (CTC) instead of isoleucine (ATC).

(International Biotechnologies, Inc.). This Nco I—Hind III fragment was ligated into the expression vector pKK233-2 (Pharmacia), which contains a trc promoter. All cloning was performed in *Escherichia coli* DH5α (Bethesda Research Laboratory). Nine clones were analyzed with eight of the nine clones containing the correct size aldose reductase insert as judged by Nco I—Hind III and Bam HI digestions. Construct pKK HAR1 was sequenced at the 5' ligation site to ensure that the most 5' Nco I site, and thus the AUG initiation codon, of the original clone was present (Sequenase Kit; United States Biochemical).

Antibody Reactions

The *E. coli* strains harboring the aldose reductase constructs were grown overnight in 500 ml of Luria broth containing 50 μg of ampicillin per ml. The cells were centrifuged (Beckman; 5,000 x g; 15 min) and washed twice in 20 mM phosphate buffer, pH 6.2 containing 7 mM 2—mercaptoethanol and 0.5 mM EDTA. The cell pellet was resuspended in 30 ml of the same buffer and sonicated (five bursts, 3 to 5 sec pulses at 50% power) to break the bacterial cell walls. The sonicate was centrifuged and the protein concentration of the supernatant was determined by the Bradford protein assay (Bio—Rad). The supernatant proteins were separated on the Phastgel system (Pharmacia) according to manufacturer's instructions. The protein gels were either stained with Coomassie blue or transferred to nitrocellulose for immunoblot analysis. The nitrocellulose filters were incubated with a human placenta aldose reductase antibody (1:250 dilution) (Kador, Carper and Kinoshita, 1981) or a rat lens aldose reductase antibody (1:250 dilution) (Shiono, et al., 1987) overnight at 4^{0}C after 30 min in a 0.2% milk blocking solution. A Vectastain ABC kit (Vector Laboratories) was used to visualize the antigen—antibody complex. Recombinant aldose reductase protein was compared to human placental aldose reductase that was partially purified by Sephadex G—100 column chromatography.

Protein Purification

Cells containing construct pKK HAR1 were grown overnight in 500 ml to 4 liter of Luria broth containing 50 μg of ampicillin per ml. The bacterial cell supernatant was obtained essentially as described above for antibody reactions, but with an increase in the final volume of the sonicate (40 ml maximum for the 4 liter preparations). The recombinant protein was purified by applying the bacterial cell supernatant onto a Sephadex G—100 size exclusion column (Pharmacia) followed by a Matrex gel orange A affinity column (Amicon) and lastly, a Mono P chromatofocusing column (Pharmacia) (Sato and Kador, 1990; Old, et al., 1990). Fractions from each column were monitored for NADPH reducing activity using 10 mM DL—glyceraldehyde, 10 mM D—xylose and 10 mM D—glucuronate as substrates. DH5α cells were used as controls in the purification process.

Aldose Reductase Activity

Aldose reductase activity was photometrically determined by monitoring the decrease in absorbance of NADPH at 340 nm using 10 mM DL—glyceraldehyde, 10 mM D—xylose, 10 mM D—glucuronate, 10 mM D—galactose, or 10 mM D—glucose as substrates. One unit of activity was defined as the activity consuming 1 μmol of NADPH per min at 22^{0}C. Recombinant aldose reductase activity was compared to activity of purified human placental aldose reductase and purified human kidney aldehyde reductase, both purified in three steps as outlined above.

RESULTS AND DISCUSSION

    An immunologically and enzymatically active protein has been obtained
from the construction of a recombinant human placental aldose reductase
protein as outlined in Fig. 1 and *Materials and Methods*.   Immunoblot
analyses indicated that pKK HAR1 and pKK HAR5 contained a protein that
reacted strongly with the human placental aldose reductase antibody (Fig.
2b, lanes 3 and 5).   In fact, eight clones containing the aldose reductase
DNA construct had an additional protein which was immunologically reactive
to the antibody, while one clone with no apparent aldose reductase construct
showed no additional protein (unpublished data).   The reactive protein from
clones pKK HAR 1 and pKK HAR5 had an approximate M_r of 35,000 (Fig. 2b,
lanes 3 and 5), which is similar to partially purified human placental
aldose reductase (Fig. 2b, lane 2).   The DH5α control cells (Fig. 2b, lane
4) demonstrated trace reactivity to the aldose reductase antibody.

    An antibody against rat lens aldose reductase reacted weakly to the
recombinant human placental aldose reductase (Fig. 3).   Immunological
differences have already been observed between rat lens and human placental

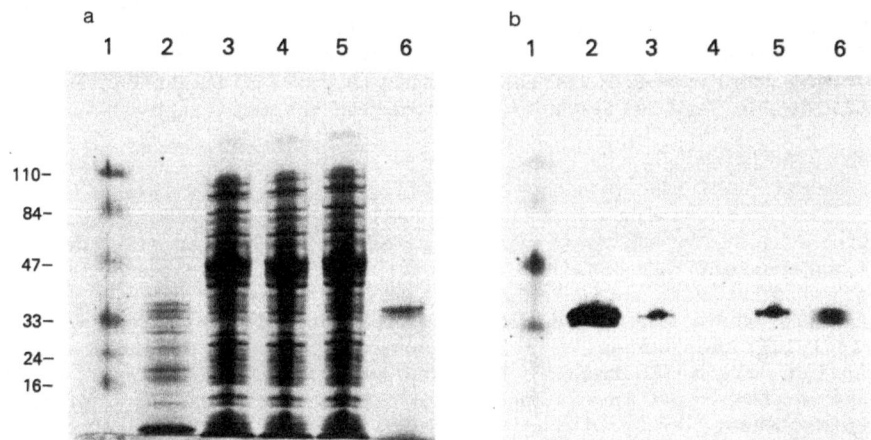

Fig. 2.   SDS—polyacrylamide gel (panel a) and immunoblot (panel b)
          of the human placental aldose reductase constructs using
          an antibody against human placental aldose reductase.
          Lanes: lane 1, apparent molecular weight markers (Bio—
          Rad prestained markers), molecular weight in daltons
          listed from top to bottom of gel: phosphorylase B,
          110,000; bovine serum albumin, 84,000; ovalbumin, 47,000;
          carbonic anhydrase, 33,000; soybean trypsin inhibitor,
          24,000; lysozyme, 16,000; lane 2, human placental aldose
          reductase partially purified from a Sephadex G—100 column,
          0.5 μg; lanes 3 and 5, constructs pKK HAR1 and pKK HAR5,
          respectively, 2.4 μg; lane 4, DH5α cells with no plasmid,
          2.4 μg; lane 6, purified recombinant aldose reductase from
          pKK HAR1 eluted from a Mono P column, 0.2 μg loaded on
          SDS/PAGE (a) and 0.02 μg loaded onto gel for immunoblot (b).

aldose reductase, indicating that the proteins are not identical (Hermann, Kador and Kinoshita, 1983). As seen in figure 3b, lane 2, the rat lens antibody reacted strongly to rat aldose reductase, with an apparent molecular weight of 35,000. The antibody showed a weak reaction to the recombinant human placental aldose reductase, which also has a molecular weight of 35,000 (Fig. 3, lanes 4, 5 and 7). These lanes contain the cells harboring the cDNA construct. The DH5α control cells have no AR construct and exhibited no detectable aldose reductase reactivity at an M_r of 35,000 (Fig. 3b, lane 6). Thus, both antibodies (placental and rat lens) recognize the recombinant aldose reductase protein. Their reactivities, however, are different. A strong reaction was observed in all bacterial cells to a protein with an apparent molecular weight of 41,000. This protein was also present in studies using a rat lens aldose reductase construct (Old, et al., 1990). One explanation previously proposed was that the 41,000 molecular weight protein is a bacterial aldose reductase. Another likely explanation is that the goat used in antibody production elicited an antibody response to a bacterial infection, so that, the final serum contained antibodies against aldose reductase and to a bacterial protein. The rat lens antibody, as did the placental antibody (not shown), reacted to higher molecular weight proteins in the total placental protein preparation (Fig 3, lane 3). These proteins have not been identified, however placental aldose reductase is known to bind to other proteins (Kador, Carper and Kinoshita, 1981).

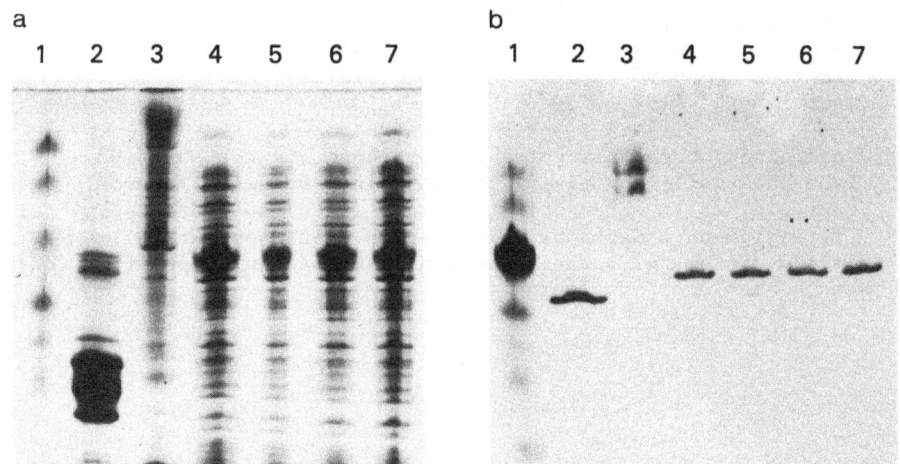

Fig. 3. SDS—polyacrylamide gel (panel a) and immunoblot (panel b) of the human placental aldose reductase contructs using an antibody against rat lens aldose reductase. Lanes: lane 1, apparent molecular weight markers as described in Fig. 2; lane 2, soluble rat lens proteins, 2.4 μg; lane 3, total human placental proteins, 2.4 μg; lane 4, construct pKK HAR1, 2.4 μg; lane 5, construct pKK HAR 5, 2.4 μg; lane 6, DH5α control cells with no plasmid, 2.4 μg; lane 7, a second preparation of pKK HAR1, 2.4 μg. The same sample preparations were used in figure 2 and in figure 3 for the lanes containing proteins from pKK HAR1, DH5α and pKK HAR5.

Purification of the recombinant protein resulted in a single peak of enzymatically active human placental aldose reductase (Fig 4., Tables 1 and 2). Table 1 shows the stepwise purification of the recombinant placental aldose reductase. Chromatofocusing on a Mono P column, the third and final step in the purification, yielded a major and a minor protein peak (Fig. 4). The major peak (peak 1) was identified as the recombinant human placental aldose reductase by using the substrates listed in Table 2 and by using aldose reductase inhibitors (Sato, et al., in preparation). The relative activity of the recombinant protein was very similar to aldose reductase purified from human placental tissue (Table 2). Purified human placental aldehyde reductase (ALR) had a much different substrate preference profile. The second minor peak from the Mono P column had a strong affinity for glyceraldehyde, but little, if any, reaction to other substrates (Table 2). Further analysis is needed to determine whether this protein is the immunologically reactive protein observed in Fig. 2 in the control DH5α cells (Fig. 2b, lane 4).

Table 1. Purification of aldose reductase from pKK HAR1

Steps	Total protein mg	Specific activity milliunits/mg	Total activity milliunits	Purification fold	Yield %
Homogenate	512	14.0	7316	1	100
G–100	138	39.4	5426	3.3	42.1
Orange A	0.90	2365	2121	197	29.0
Mono P	0.36	2076	743	148	10.2

Recombinant human placental aldose reductase was purified from DH5α cells containing pKK HAR1 in a stepwise fashion. The bacterial cells were sonicated (homogenate) and purified by gel filtration on a Sephadex G–100 column (size separation), an Amicon Matrex gel orange A column (NADPH affinity purification), and a Mono P column (chromatofocusing).

The recombinant human placental protein purified from the Mono P column had the same M_r as partially purified human placental aldose reductase (Fig 2., lanes 6 and 2, respectively). Although the purified recombinant aldose reductase protein was essentially a single band on SDS/PAGE, the immunoblot consistently stained a larger area, which appeared as a doublet. Doublet formation can also be observed with the partially purified aldose reductase from placental tissue (Fig. 2b, lane 2). This phenomenum has been observed before with aldose reductase from several sources, including rat lens and recombinant rat lens aldose reductase (Old, et al., 1990) and human kidney aldose reductase (Sato, unpublished). It has been suggested that this doublet results from the oxidation of aldose reductase and that the addition of dithiotheitol could reverse this effect.

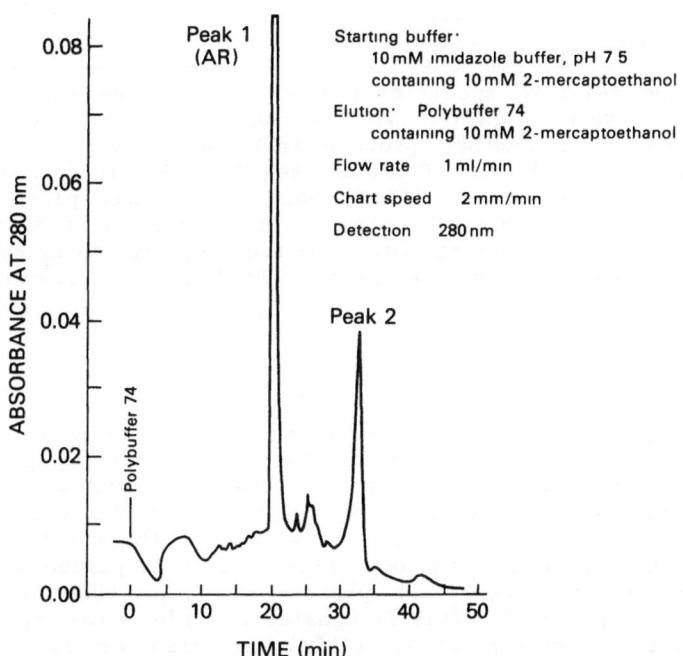

Fig. 4. Profile of the recombinant human placental aldose
reductase protein from clone pKK HAR1 purified on
a Mono P column. One peak (peak 1) appears at
19–20 min and a second peak (peak 2) appears at
30–31 min after Polybuffer 74 is started. The flow
rate is 1 ml/min. Peak 1 was identified immunologically
and enzymatically as the recombinant human placental
aldose reductase (see Fig. 2 and Table 2). Peak 2,
which coeluted by molecular weight and NADPH binding
with peak 1, was identified as a glyceraldehyde reductase.

Table 2. Comparison of relative activity of aldose reductase with different substrates.

Substrate 10 mM	Relative Activity (%)			
	pKK HAR1 peak 1	pKK HAR1 peak 2	Human Placental Tissue AR	ALR
DL—Glyceraldehyde	100	100	100	100
D—Glucuronate	37.8	0.3	35.1	143.9
D—Xylose	49.4	3.9	44.7	1.6
D—Glucose	4.7	0.2	4.3	0.3
D—Galactose	12.0	0.2	10.6	0.3

The relative specificities of various substrates for aldose reductase were compared: the recombinant aldose reductase (pKK HAR1, peak 1), another protein (pKK HAR1, peak 2) that eluted separately from the Mono P column (see Fig. 4), aldose reductase (AR) and aldehyde reductase (ALR), both from human placental tissue purified from a Mono P column. The specific activity of DL—glyceraldehyde was set at 100%. The four proteins being compared were purified in three steps: Sephadex G—100, Matrex gel orange A, Mono P.

The present enzymatic and immunological data strongly suggest that the recombinant protein produced in the DH5α cells is human placental aldose reductase. This study demonstrates that a recombinant human aldose reductase protein can be easily obtained from bacterial cells, reducing the need for human tissue and eliminating some of the safety considerations associated with working with human tissues such as placenta. The level of expression of the recombinant protein was less than 1% of the total bacterial cell proteins. This is considered to be a low expression for the trc promoter. Grundmann, et al. (1990) have also very recently reported a low expression of human aldose reductase with the trc promoter. In contrast, a rat lens recombinant aldose reductase protein constructed in the same vector comprised approximately 4% of the total protein (Old, et al., 1990). The explanation for the low yield of recombinant human placental AR is not fully understood, however, it was observed that another new protein of approximately equal expression was eluted in the first purification step (Sephadex G—100 column, Fig. 5). This higher molecular weight peak was observed in two different bacterial clones, pKK HAR1 and pKK HAR5, but not in the control cells and the protein had enzymatic activity indicative of aldose reductase. This higher molecular weight protein could represent an altered or aggregated form of the recombinant aldose reductase. These data also point out the possible differences between human and rat aldose reductase, since purification of recombinant rat lens aldose reductase did not reveal a higher molecular form. Recently, a DNA construct expressed in a eukaryote baculovirus system has shown higher yields of a human muscle aldose reductase (Nishimura, et al., 1990). Although the expression of human placental aldose reductase is low, for most purposes the amount produced is sufficient and easily obtained. In addition, the protein is from a single coding sequence, which eliminates the question of isozymes or other aldo—keto reductases. The production of a recombinant human placental aldose reductase should facilitate continued inhibitor structure—activity and affinity labelling studies being carried out on human placental AR. In addition, characterization of the key elements of the substrate, NADPH, and inhibitor binding sites can be examined through the use of site—directed mutagenesis studies.

(A) Control E. coli

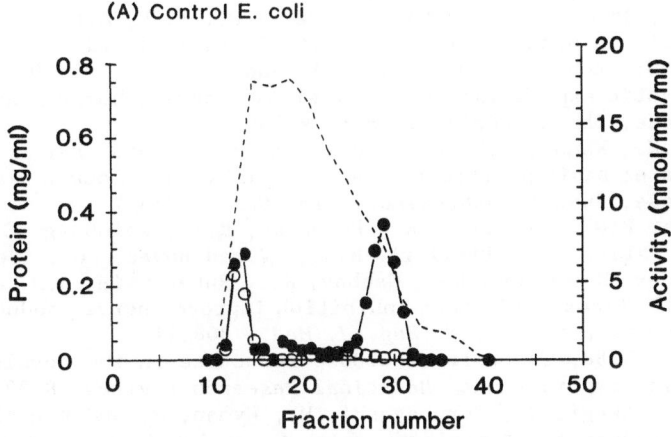

(B) HPAR clone

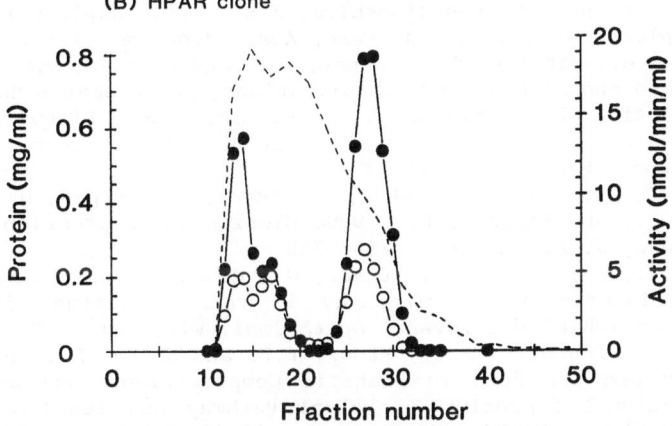

Fig. 5. Elution profiles from a Sephadex G–100 column of recombinant human placental pKK HAR1 aldose reductase protein (bottom panel, B HPAR clone) and of DH5α control bacterial cells (top panel, A Control E. coli). Recombinant aldose reductase was eluted in fraction numbers 25–30 (panel B, HPAR) and further purified. The shoulder in HPAR (bottom panel, fraction numbers 16–19) contained protein which reduced DL–glyceraldehyde and D–xylose. This shoulder was not present in the control cells (top panel). The activity in the higher molecular weight fraction could represent an altered form of aldose reductase. ––– protein; –•– DL–Glyceraldehyde; –o– D–Xylose

REFERENCES

Chung, S. and La Mendola, J., 1989, Cloning and sequencing determination of human placental aldose reductase gene, *J. Biol. Chem.*, **264**:14775.

Dvornik, D., 1987, Complications of chronic diabetes in man and their relation to hyperglycemia, in: An Approach to the Prevention of Diabetic Complications, D. Porte, ed., McGraw-Hill, New York.

Fukushi, S., Merola, L.O. and Kinoshita, J.H., 1980, Altering the course of cataracts in diabetic rats, *Invest. Ophthalmol. Vis. Sci.* **19**:313.

Grundmann, U., Bohn, H., Obermeier, R. and Amann, E., 1990, Cloning and prokaryotic expression of a biologically active human placental aldose reductase, *DNA and Cell Biology*, **9**:149.

Hermann, R.K., Kador, P.F. and Kinoshita, J.H., 1983, Rat lens aldose reductase: rapid purification and comparison with human placental aldose reductase, *Invest. Ophthalmol. Vis. Sci.*, **37**:467.

Judzewitsch, R.G., Jaspan, J.B., Polonsky, K.S., Weinberg, C.R., Halter, J.B., Halar, E., Pfeifer, M.A., Vukadinovac, C., Bernstein, L., Schneider, M., Liang, K-Y, Gabbay, K.., Rubenstein, A.H. and Porte, D., Jr, 1983, Aldose reductase inhibition improves nerve conduction velocity in diabetic patients, *N. Eng. J. Med.*, **308**:119.

Kador, P.F., 1988, The role of aldose reductase in the development of diabetic complications, *Medicinal Research Reviews*, **8**:325.

Kador, P.F., Akagi, Y., Terubayashi, H., Wyman, M. and Kinoshita, J.H., 1988, Prevention of pericyte ghost formation in retinal capillaries of galactose-fed dogs by aldose reductase inhibitors, *Arch. Ophthalmol.*, **106**:1099.

Kador, P.F., Carper, D. and Kinoshita, J.H., 1981, Rapid purification of human placental aldose reductase, *Anal. Biochem.*, **114**:53.

Kinoshita, J.H., Datiles, M.B., Kador, P.F. and Robison, W.G., 1990, Aldose reductase and diabetic eye complications, in: Diabetes Mellitus:Theory and Practice, H. Rifkin and D. Porte, Jr., eds., Elsevier, New York.

Kinoshita, J.H., 1974, Mechanisms initiating cataract formation, *Invest. Ophthalmol. Vis. Sci.*, **13**:713.

Nishimura, C., Matsuura, Y., Kokai, Y., Akera, T., Carper, D., Morjana, N., Lyons, C., and Flynn, T.G., 1990, Cloning and expression human aldose reductase, *J. Biol. Chem.*, **265**:9788.

Robison, G.W.,Jr., Nagata, M., Laver, N., Hohman, T and Kinoshita, J.H., 1989, Diabetic-like retinopathy in rats prevented with an aldose reductase inhibitor, *Invest. Ophthalmol. Vis. Sci.*, **30**:2285.

Sakamoto, N., Kinoshita, J.H., Kador, P.F. and Hotta, N., eds., Polyol Pathway and Its Role in Diabetic Complications: Proceedings of the International Symposium on Polyol Pathway and Its Role in Diabetic Complications, Kashikojima, Japan, 28-30 October, 1986, pp. 1-584, Excerpta Medica, Amsterdam.

Sato, S. and Kador, P.F., 1990, NADPH-dependent reductases of the dog lens, *Exp. Eye Res.*, **50**:629.

Sawada, H., Hamatake, M., Hara, A., Nakagawa, M. and Nakayama, T., 1989, Inhibition of human placenta aldose reductase by tannic acid, *Chem. Pharm. Bull.*, **37**:1662.

Shiono, T., Sato, S., Reddy, V.N., Kador, P.F. and Kinoshita, J.H., 1987, Purification and properties of rat lens aldose reductase, *Prog. Clin. Biol. Res.*, **232**:317.

Old, S.E., Sato, S., Kador, P.F. and Carper, D.A., 1990, *In vitro* expression of rat lens aldose reductase in *Escherichia coli*, *Proc. Natl. Acad. Sci.*, USA., in press.

Vander Jagt, D.L., Stangebye, L.A., Hunsaker, L.A., Eaton, R.P. and Sibbitt, W.L., 1988, Purification of aldose reductase from human placenta and stabilization of the inhibitor binding site, *Biochem. Pharm.*, **37**:1051.

Yue, D.K., Hanwell, M.A., Satchell, P.M. and Turtle, J.R., 1982, The effect of aldose reductase inhibition on motor nerve conduction velocity in diabetic rats, *Diabetes* **31**:789.

OSMOTIC STRESS INDUCES ALDOSE REDUCTASE IN GLOMERULAR

ENDOTHELIAL CELLS

Thomas C. Hohman, Deborah Carper*, Sarmila Dasgupta*
and Masayuki Kaneko*
Wyeth-Ayerst Laboratory, CN 8000, Princeton, NJ 08540
*National Eye Institute, National Institutes of Health
Bethesda, MD

INTRODUCTION

Compelling evidence from animal studies (Robison et al, 1989, Notvest and Inserra, 1987; McCaleb et al, 1988, Kador et al 1988; c.f. Dvornik, 1987) and recent results from human clinical trials (Sima et al, 1988; Boulton, 1989) have implicated aldose reductase (AR), the first enzyme in the polyol pathway, in the initiation of a number of the long-term complications of diabetes. These complications develop in tissues where glucose uptake is independent of insulin. In these tissues during diabetes-induced hyperglycemia there is an intense flux of glucose through the polyol pathway which initiates a cascade of biochemical alterations that slowly progress to cell dysfunction and structural damage. While the role of AR in the pathogenesis of diabetic complications is well documented, its physiologic role, if any, under normal glycemic conditions is still unknown.

AR has a low affinity for glucose, and at physiological glucose concentrations the polyol pathway is virtually inactive, as reflected by the barely detectable levels of sorbitol in most tissues from non-diabetic animals and man. These results indicate that AR assumes a role only during diabetic hyperglycemia. However, results from some recent studies in which AR induction and sorbitol accumulation occurred in a continuous line of rabbit medullary cells grown under hypertonic conditions (Burg, et al, 1988) together with other reports identifying a pool of sorbitol in the normal renal medulla (Wirthensohn et al, 1989) suggest that AR may play a role in cellular osmoregulation in the kidney medulla. The accumulation of sorbitol is thought to protect medullary cells from the steep osmotic gradients generated during the urine concentrating process. Our results show that osmotic stress also induces AR in primary cultures of kidney glomerular endothelial cells. In these glomerular cells sorbitol accumulation does not appear to play an essential role in cellular osmoregulation; cellular viability and proliferation were unaffected when sorbitol accumulation was inhibited in osmotically stressed cells. In contrast to the rabbit renal cell line where sorbitol was the only osmotically regulated organic solute, both myo-inositol and sorbitol accumulate in hypertonically stressed endothelial cells, suggesting that glomerular endothelial cells may have several mechanisms for adjusting to osmotic stress.

MATERIALS AND METHODS

Cultured Cells. Glomeruli were isolated from the kidneys of adult beagle dogs using a graded seiving technique and were plated for explant growth in RPMI-1640 with 20% fetal bovine serum (FBS) diluted (1:1) with conditioned media from Swiss 3T3 cells (Harper et al, 1984). After several weeks the cells which grew out from the glomeruli were released from the culture dish by a gentle trypsin digestion (0.25% in phosphate buffered saline) and were transferred to gelatin coated dishes (2.0%, Sigma, St. Louis, MO) where they were cultured in RPMI-1640 with 10% FBS and 10% NuSerum Type IV (Collaborative Research, Bedford, MA). Colonies of endothelial cells, identified by their distinctive morphology and their ability to internalize acetylated low density lipoprotein (AcLDL) were transferred to separate dishes and expanded (Vota et al, 1984). Internalization of AcLDL was visualized with fluorescence microscopy following a 3 h incubation in growth medium containing 10 µg/ml 1,1'-dioctadecyl-3,3,3',3'-tetramethyl-indocarbocyanine perchlorate (Biomedical Tech, Stoughton, MA). Endothelial cells in the sixth through the sixteenth passage were used.

Enzymatic Activity. Aldose reductase activity was determined spectrophotometrically by monitoring the oxidation of NADPH (340 nm, 37°C) in 30 mM sodium phosphate buffer, pH 6.2, with 0.2 M $LiSO_4$, 0.2 mM NADPH and 10 mM DL-glyceraldehye. Lactate dehydrogenase activity was similarly determined by monitoring the oxidation of NADH (340 nm, 37°C) in 100 mM potassium phosphate buffer, pH 7.5 with 0.3 mM NADH and 1 mM sodium pyruvate. Protein concentrations were quantitated with a modified Lowry procedure using bovine serum albumin as a standard (Peterson, 1977).

Carbohydrate Analysis. Sorbitol, fructose and myo-inositol (MI) levels were quantitated in cell homogenates by gas liquid chromatography with flame ionization detection using an HP-5890A chromatograph (Hewlett-Packard, Palo Alto, CA) equipped with an SP-2100 fused silica capillary column (Supelco, Bellefonte, PA). For these analyses, methylmannoside was added as an internal standard to each sample and the carbohydrates in aliquots of cell homogenates were converted to aldononitrile acetate derivatives following protein precipitation and lyophilization (Guerrant and Moss, 1984). The instrument was calibrated using samples of control cells with known concentrations of sorbitol, fructose and MI. Results are the means of triplicated measurements (±SD) expressed as nmol/mg protein.

Cell Proliferation. A spectrophotometric assay quantitating the reduction of 3-(4,5-dimethylthiazol-2-yl)-2,5-diphenyltetrazolium bromide (MTT) in intact cells was used to measure cell survival and proliferation (Mosmann, 1983). For this assay, cells in 24-well plates were first equilibrated for 30 min in 0.5 ml of serum-free medium. Following the addition of 50 µl of a sterile stock solution of MTT (5 mg/ml) the cells were incubated for 45 min at 37°C. The formazan product was solubilized in acetic isoproponol (1.6 ml of 1.0 N HCL in 38.4 ml of isoproponol) and aliquots from each well were transferred to a 96 well plate. The optical densities (reference wavelength 630 nm, test wavelength 570 nm) were measured on a Microplate Autoreader EL-309 (Bio Tek Instruments, Winooski, VT). Preliminary experiments demonstrated that the reduction of MTT was linear with time and cell number.

SDS-Polyacrylamide Gel Electrophoresis and Immunoblotting. Cells were homogenized in 50 mM Tris buffer, pH 7.5, containing 0.1 M KCl, 1 mM ethylenediaminetetraacetic acid, 0.02% NaN_3 and 0.07% 2-mercaptoethanol. Proteins in aliquots of the homogenates were denatured in 0.1 M Tris buffer, pH 6.8, containing 0.3% SDS, 0.8% 2-mercaptoethanol and 4.2% glycerol at 95°C for 2 min and electrophoretically separated on a 10% polyacrylamide gel using the Phastsystem (Pharmacia, Piscataway, NJ). Gels fixed in 30% ethanol with 10% acetic acid were stained with Coomassie Brilliant Blue (Pharmacia). Immunoblots were prepared as previously described (Kaneko et al, 1990). Briefly, proteins separated by electrophoresis were transferred to nitrocellulose membranes and incubated, first, for 2 h at room temperature in a 0.2% milk solution and then overnight at 4°C with the primary antibody. Two separate primary antibodies, gifts from Drs. P.F. Kador and S. Sato (National Eye Institute, NIH, Bethesda, MD), were used. The first was an antibody to rat lens AR and the second, an antibody to rat kidney aldehyde reductase. Both primary antibodies, prepared in goats, were used at a 1:400 dilution. After the incubation with the primary antibodies, the membranes were washed with two changes of 20 mM Tris buffer, pH 7.5, containing 0.2 M NaCl and 0.05% Tween 20 and incubated with a biotinylated anti-goat IgG (1:1000 dilution) at 37°C for 1 h. Following an additional wash, membranes were incubated with an avidin-biotin complex for 1 h at 37°C (Vector lab., Inc., Burlingame, CA), and reacted with a 4-chloro-1-naphthol solution containing hydrogen peroxide.

<u>Analysis of RNA</u>. Total RNA was isolated according to the method of Chirgwin et al (1986). For Northern blots RNA was denatured (95°C, 2 min) in 2.1 M formaldehyde and 50% formamide, separated by electrophoresis on a 1% agarose/2.2 M formaldehyde gel (Davis et al, 1986), and transferred to nylon membranes (ICN Biomedicals, Cleveland, OH). Slot blots were prepared according to the method of Bowes et al (1988). Membranes were prehybridized in 50 mM sodium phosphate buffer, pH 6.4, containing 50% formamide, 5 X SSC, 0.1% SDS, 5 X Denhardt's solution and 0.25 mg/ml denatured salmon sperm DNA at 42°C, overnight. Membranes were then hybridized at 42°C in a fresh preparation of the same buffer containing ^{32}P labeled DNA probes: AR (Carper et al, 1987) or actin (Oncor, Inc., Gaithersburg, MD). The probes were labelled to a specific activity of 8 to 300 X 10^7 cpm/μg DNA with the use of a random primer DNA labelling kit (Bethesda Research Laboratory, Gaithersburg, MD). Following hybridization, membranes were washed according to manufacturers protocols (ICN, Oncor) and were exposed to Kodak XAR-5 film with intensifying screens (Dupont Cronex Lighting Plus) at -70°C.

RESULTS

<u>Enzymatic Activity</u>. AR activity increased in kidney endothelial cells when they were incubated in hypertonic media. Increasing the medium tonicity to 500 mosm/kg with the addition of 100 mM NaCl resulted in a 6-fold increase in AR activity within 2 days and a maximal 34-fold increase, after 6 days (Figure 1A). This level of AR activity was sustained for as long as the cells were maintained in hypertonic media. When endothelial cells were returned to normal media (300 mosm/kg), AR activity remained unchanged for the first 4 days and then rapidly declined to near control levels by day 10.

The change in AR activity observed with increased tonicity could have resulted from either AR activation or increased synthesis of AR. To distinguish between these alternatives, AR activity was quantitated in cells when protein synthesis was inhibited with 5 mM cycloheximide. In these studies AR activity did not increase above basal levels when cycloheximide was added for the last 48 h of a 4 d incubation in hypertonic medium. These incubations were limited to 48 h, since with longer exposure to cycloheximide, cells rounded up and detached from the culture dish.

The induction of AR appeared to be a response to an increase in medium tonicity rather than a response to a specific osmolyte. The addition of glucose, cellobiose, sorbitol, myo-inositol or raffinose, but not glycerol, to the growth medium resulted in an increase in AR activity. The di- and trisaccharides, cellobiose and raffinose, were as effective as NaCl, while glucose, sorbitol and MI were less effective.

Differences between these solutes in the level of AR induction probably reflect differences in their ability to penetrate cell membranes. However, for each solute the increase in AR activity was proportional to the concentration of that solute in the medium. The results with NaCl, cellobiose, glucose and glycerol are shown in figure 1B.

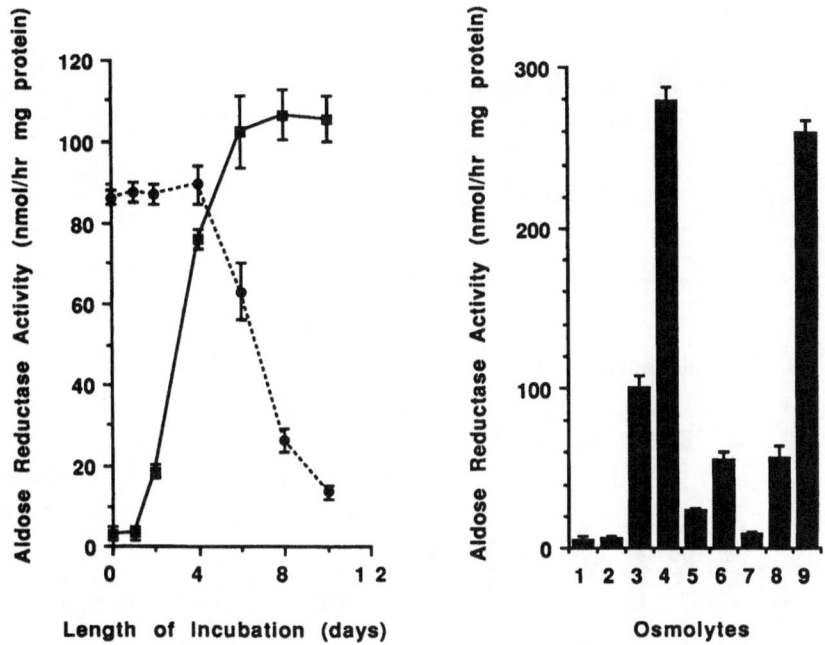

Figure 1(A). AR activity was quantitated in endothelial cells incubated in hypertonic medium (500 mosm/kg) for 12 h to 10 d (■——■) and in cells incubated in hypertonic media for 8 d and then returned to normal media (•——•). (B). AR activity was quantitated in cells incubated for 6 d in normal medium (1) or for 6 days in media containing 50 (2), 100 (3), 150 (4) mM NaCl; 150 (5), 250 (6) mM glucose; 250 mM glycerol (7); 150 (8), or 250 (9) mM cellobiose. Data point are the mean of triplicate samples ± 1 SD in a representative experiment.

In our studies AR activity in cell homogenates was quantitated by monitoring the oxidation of NADPH in the presence of DL-glyceraldehyde. To determine whether the increased consumption of NADPH by homogenates of osmotically stressed endothelial cells was due to an induction of AR or the induction of other NADPH consuming enzymes, three structurally distinct AR inhibitors (c.f. Dvornik, 1987), tolrestat (Wyeth-Ayerst), ponalrestat (ICI) and sorbinil (Pfizer) were tested for their ability to inhibit the enzymatic reaction.

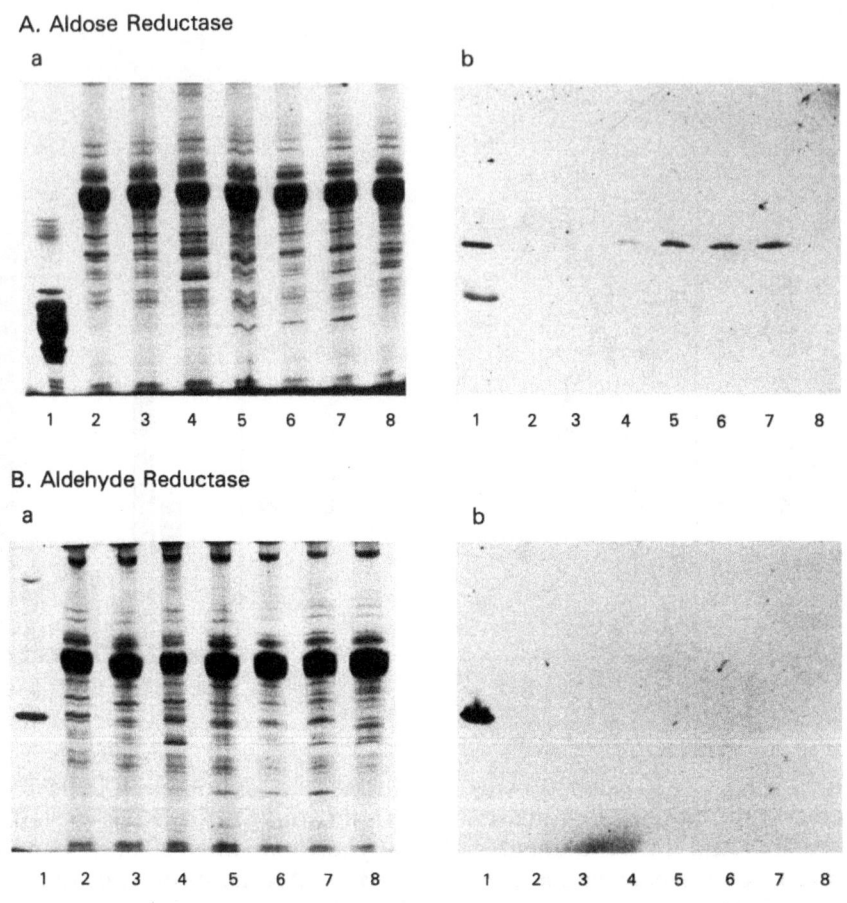

Figure 2. SDS-PAGE and Western blots of endothelial cells grown in hypertonic or isotonic medium. Lanes: 1, soluble lens proteins (A) or purified ALR (B); 2 and 8 soluble proteins from endothelial cells in isotonic media; 3-7 soluble proteins from endothelial cells in hypertonic media (500 mosm/kg) for 1, 2, 4, 6 and 8 days. 2 μg of protein were applied to each line. (Panel A) Western blot probed with an antibody to rat lens AR. (Panel B) Western blot probed with an antibody to rat kidney ALR.

The IC_{50} values, the concentrations which inhibited the enzymatic activity by 50%, were 0.52, 2.5, and 23.0 X 10^{-8}M, respectively. The IC_{50} values for these inhibitors in control endothelial cell preparations were not different from those of cells incubated in hypertonic medium, and were similar to the IC_{50} values observed with partially purified bovine lens AR, (3.6, 3.7 and 112 X 10^{-8}M for tolrestat, ponalrestat and sorbinil, respectively; Dvornik, 1987).

SDS Polyacrylamide Gel Electrophoresis and Immunoblotting.
Additional evidence that AR was specifically induced in hypertonically stressed endothelial cells was obtained from Western blots probed with a polyclonal antibody to rat lens AR. In endothelial cell preparations this antibody recognized a single protein band which co-migrated with AR from rat lens. This 35 kDa protein band was not observe in normal endothelial cells but appeared in cells incubated in hypertonic media (Figure 2A). The staining intensity of this band increased as cells were incubated for longer periods in hypertonic media. A polycolonal antibody prepared against rat kidney aldehyde reductase (ALR) reacted strongly with purified rat ALR but did not react with any of the bands in the endothelial cells (Figure 2B).

RNA Analysis. A cDNA probe for AR (Carper et al, 1987) was used to identify and to calculate relative concentrations of AR mRNA in control and hypertonically stressed endothelial cells. By Northern blot analysis the size of the mRNA was about 1.4 Kb, which corresponds to that previously reported for AR mRNA in other cell types (Nishimura et al, 1988; Kaneko et al, 1990). Slot blots (Figure 3) showed that the levels of AR mRNA in endothelial cells rapidly increased in osmotically stressed cells. Within 48 h the message level increased between 30 and 60-fold in cells incubated in media made hypertonic with either NaCl or cellobiose. To confirm that similar amounts of mRNA from control and osmotically stressed cells were applied, the AR cDNA probe was stripped from the slot blots and they were then reacted with a cDNA probe for actin.

Carbohydrate Analysis. Cellular levels of sorbitol and fructose were quantitated in hypertonically stressed endothelial cells to determine whether the induction of AR led to an increased flux of glucose through the polyol pathway. The results (Figure 4A) demonstrate the formation of a large intracellular pool of sorbitol which increased with the length of the cellular incubation in hypertonic medium and reached a plateau level only after 8 to 10 days. The cellular level of fructose in osmotically stressed cells was not different from that in control cells

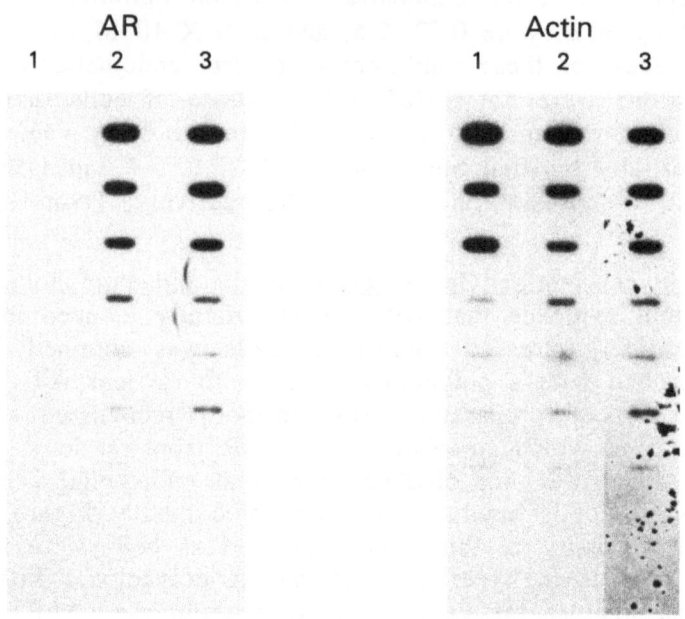

Figure 3. Slot blots of endothelial cell RNA were probed with AR cDNA, stripped, and then probed with actin cDNA. Serial dilutions (1:1) were made and applied to each slot column beginning with 3 µg at the top. Cells were incubated under the experimental conditions for 48 h. Lanes: 1, control cells; 2, cells incubated in media with 150 mM NaCl; 3, cells incubated in media with 250 mM cellobiose.

Carbohydrate analysis also revealed a second polyol, myo-inositol (MI), which accumulated in osmotically stressed cells more rapidly and to higher levels than sorbitol. MI accumulation appeared to be modulated in response to sorbitol levels. When endothelial cells were incubated in medium with 150 mM NaCl (600 mosm/kg) MI rapidly increased reaching a plateau, 12-fold increase, within 4 days. As the cellular level of sorbitol approached its steady-state level, MI concentration decreased to a new plateau (Figure 4A). When sorbitol accumulation was blocked by inhibiting AR with tolrestat, MI levels increased to an even higher level (Figure 4B).

When hypertonically stressed cells were returned to isotonic medium both sorbitol and MI rapidly returned to near control levels (Figure 5), while fructose levels were unchanged. The rapid loss of sorbitol and MI did not appear to be the result of cell lysis since the level of LDH in the culture medium was not increased (data not shown).

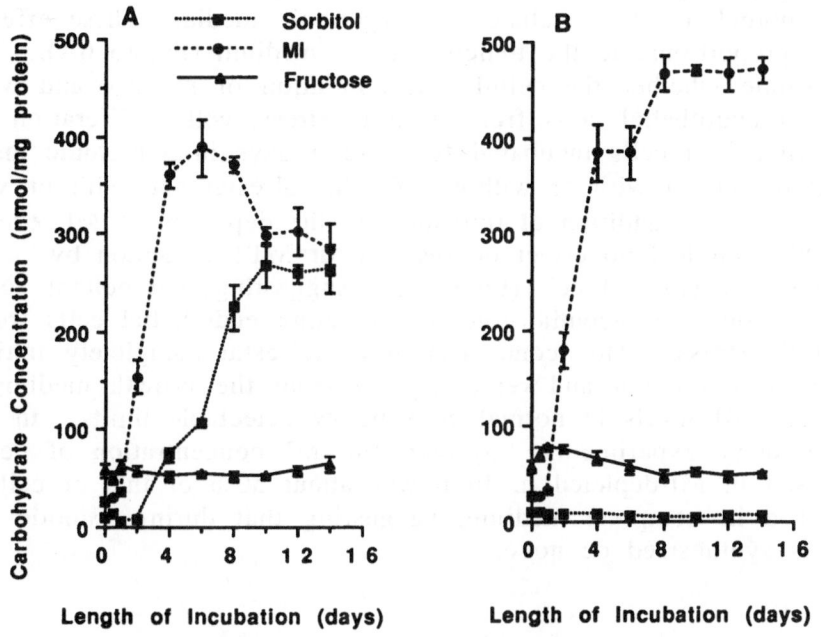

Figure 4. Sorbitol, fructose and MI were quantitated in endothelial cells incubated in hypertonic media (600 mosm/kg) without (A) or with (B) 20 µM tolrestat for 0 to 14 days.

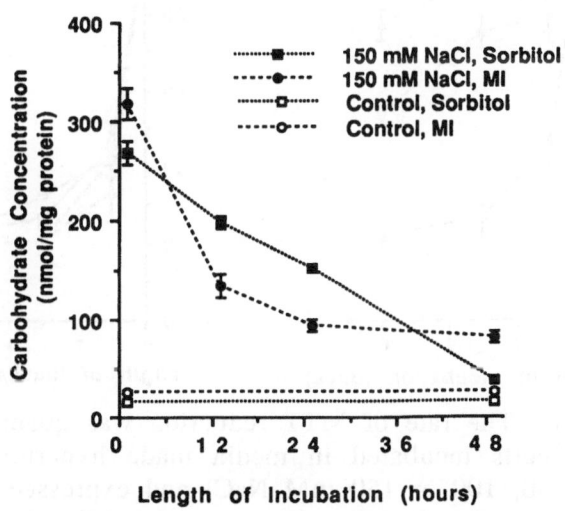

Figure 5. Sorbitol and MI were quantitated in endothelial cells incubated for 8 days in hypertonic medium (600 mosm/kg) and then returned to isotonic medium for 12 to 48 hours.

Cell Proliferation. Cellular viability and proliferation were decreased in endothelial cells incubated in hypertonic media. These effects were proportional to the tonicity of the medium (Figure 6A). To determine whether the cellular accumulation of sorbitol and MI help protect endothelial cells from osmotic stress, cell proliferation was quantitated in cells incubated for 1 to 6 days in hypertonic media (600 mosm/kg) with or without 20 μM tolrestat and with or without free MI. The addition of tolrestat and the depletion of MI, even in combination had no effect on the rate of MTT reduction by osmotically stressed cell (Figure 6) suggesting that neither sorbitol nor MI play an essential role in protecting endothelial cells from osmotic stress. This concentration of tolrestat completely inhibited sorbitol production and removing MI from the growth medium reduced MI levels in normal cells below detectable limits. In preliminary experiments, however, the MI concentration of cells stressed in MI-depleted medium was about 50% of that of cells stressed in complete medium, suggesting that during osmotic stress MI is synthesized de novo.

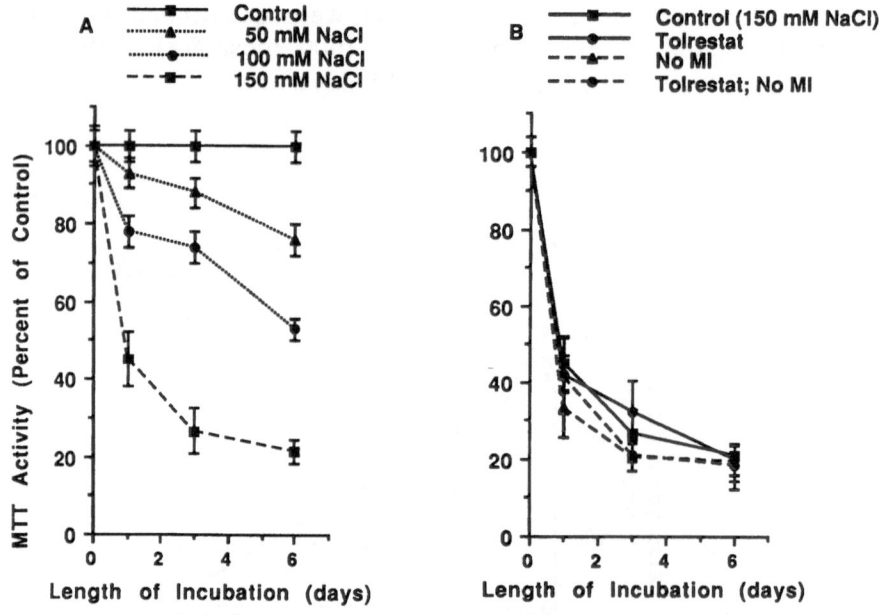

Figure 6(A). The rate of MTT reduction was quantitated in endothelial cells incubated in media made hypertonic with the addition of 50, 100 or 150 mM NaCl and expressed as a percent of the rate in control cells (B). The rate of MTT reduction was also quantitated in endothelial cells incubated in hypertonic media (600 mosm/kg) with or without MI and 20 μM tolrestat and expressed as a percent of the rate in cells incubated in normal isotonic media. Data points are the mean of triplicate samples ± 1 SD.

DISCUSSION

The changes in AR mRNA, AR synthesis, AR activity and sorbitol accumulation induced by hypertonicity in primary cultures of kidney endothelial cells were similar to those reported for a continuous line of rabbit renal papillary cells (Bedford et al, 1987; Bagnasco et al, 1987; Moriyama et al, 1989; Garcia-Perez et al, 1989) and appear to be to be part of an intracellular adaptation to chronic changes in extracellular tonicity. In a variety of cell types chronic hypertonic stress leads to the cellular accumulation of organic solutes, including amino acids, methlamines and polyols (c.f Yancey et al, 1982). The cellular accumulation of these organic solutes increases intracellular osmolarity and appears to help protect the cells from water loss and maintain a normal cellular level of sodium and potassium ions. Maintaining normal levels of these inorganic ions appears to be fundamentally important since sodium and potassium ion concentrations exist within very narrow ranges in the cytoplasm of most animals, plants and bacterial cells, and results from in vitro experiments (Yancey et al, 1982) suggest that concentrations of inorganic ions outside of this range may adversely affect the macromolecular structure and function of proteins. Osmotically stressed cells counterbalance the altered extracellular tonicity by accumulating neutral organic solutes which even at high concentration do not seem to perturb enzyme function.

Unlike the rabbit renal cell line (Yancey et al, 1990), the viability and proliferation of hypertonically stressed kidney endothelial cells were unaffected when sorbitol accumulation was inhibited. This result suggests that canine kidney endothelial cells may have several mechanisms for responding to osmotic stress. A second organic solute which these cells may use to regulate their intracellular tonicity is MI. This cyclic polyol rapidly increased in osmotically stressed endothelial cells, reaching concentrations greater than those of sorbitol, and rapidly decreased to near basal levels when cells were returned to isotonic media. MI is thought to function as an osmolyte in a number of other cell types (c.f. Yancey et al, 1982) including those of the renal medulla where it has been identified as one of five regulated osmolytes (Wirthensohn et al, 1989, Law and Turner, 1987; Grunewald and Kinne, 1989).

MI accumulation and sorbitol accumulation do not appear to be redundant mechanisms for responding to osmotic stress. In endothelial cells MI accumulation is a more acute response than sorbitol accumulation. Not only does MI accumulate more rapidly than sorbitol but its accumulation is stimulated by osmotic stresses which are insufficient to induce AR and sorbitol accumulation. However, these two mechanisms are not entirely independent. Experiments in which MI concentrations decreased once AR was induced and sorbitol accumulation reached its maximal level suggest that these two mechanisms act in concert. Experiments are now in progress to more fully characterize the integration of these two adaptive osmotic responses.

REFERENCES

Bagnasco, S.M., Uchida, S., Balaban, R.S., Kador, P.F. and Burg, M.B., 1987, Induction of aldose reductase and sorbitol in renal inner medullary cells by elevated extracellular NaCl. Proc. Natl. Acad. Sci., 84:1718.

Bedford, J.J., Bagnasco, S.M., Kador, P.F., Harris, H.W., Jr. and Burg, M.B., 1987, Characterization and purification of a mammalian osmoregulatory protein, aldose reductase, induced in renal medullary cells by high extracellular NaCl. J. Biol. Chem., 262:14255.

Boulton, A.J.M., Atiea, J., De Leeuw, I.H., Lemkes, H., MacLeod, A.F., MacRury, S., Scarpello, J.H.B., Somers, G., Tattersall, R.B. and Van Der Veen, E.A., 1989, The efficacy and safety of the aldose reductase inhibitor tolrestat in the treatment of chronic sensorimotor diabetic neuropathy. Diabetologia, 32:469A.

Bowes, C., van Veen, T. and Farber, D.B., 1988, Opsin, G-protein and 48-kDa protein in normal and rd mouse retinas: developmental expression of mRNAs and proteins and light/dark cycling of mRNAs. Exp. Eye Res., 47:369.

Burg, M., 1988, Role of aldose reductase and sorbitol in maintaining the medullary intracellular milieu. Kidney Inter., 33:635.

Carper, D., Nishimura, C., Shinohara, T., Dietzchold, B., Wistow, G., Craft, C., Kador, P., and Kinoshita, J.H., 1987, Aldose reductase and p-crystalline belong to the same superfamily as aldehyde reductase. FEBS Lett., 220:209.

Chirgwin, J.M., Prybyla, A.E., MacDonald, R.J. and Rutter, W.J., 1986, Isolation of biologically active ribonucleic acid from sources enriched in ribonuclease. Biochem., 18:5294.

Davis, L.G., Dibner, M.D. and Battey, J.F., 1986, Formaldehyde gel for electrophoretic separation of RNA and Northern blot, in: Basic Methods in Molecular Biology. Elsevier Science Publ. Co., Inc., New York, 143-146.

Dvornik, D., 1987, Aldose Reductase Inhibition: an Approach to the prevention of diabetic Complications. McGraw Hill, New York, pp368.

Garcia-Perez, A., Martin, B., Murphy, H.R., Uchida, S., Murer, H., Cowley, B.D., Handler, J.S. and Burg, M.B., 1989, Molecular cloning of cDNA coding for kidney aldose reductase. J. Bio. Chem., 264:16815.

Grunewald, R.W. and Kinne, R.K.H., 1989, Intracellular sorbitol content in isolated rat inner medullary collecting duct cells. Pflugers Arch., 414:178.

Guerrant, G.O. and Moss, C.W., 1984, Determination of monosaccharides as aldononitrile o-methyloxime, alditol and cyclitol acetate derivatives by gas chromatography. Anal. Chem., 56:633.

Harper, P.A., Robinson, J.M., Hoover, R.L., Wright, T.C. and Karnovsky, M.J., 1984, Improved methods for culturing rat glomerular cells. Kidney Inter., 26:875.

Kador, P.F., Akagi, Y., Terubayashi, H., Wymen, M., Kinoshita, J.H., 1988, Prevention of pericyte ghost formation in retinal capillaries of galactose fed dogs by aldose reductase inhibitors. Arch. Ophthalmol., 106:1099.

Kaneko, M., Carper, D., Nishimura, C., Millen, J., Bock, M., and Hohman, T.C., 1990, Induction of aldose reductase expression in rat kidney mesangial cells and Chinese hamster ovary cells under hypertonic conditions. Exp. Cell Res., 188:135.

Law, R.O. and Turner, D.P.J., 1987, Are ninhydrin-positive substances volume-regulatory osmolytes in rat renal papillary cells. J. Physiol., 386:45.

McCaleb, M.L., Sredy, J. Millen, J., Ackerman, D.M., Dvornik, D., 1988, Prevention of urinary albumin excretion in 6 month streptozocin-diabetic rats with the aldose reductase inhibitor tolrestat. J. Diab. Compl., 2:16.

Moriyama, T., Garcia-Perez, A. and Burg, M.B., 1989, Osmotic regulation of aldose reductase protein synthesis in renal medullary cells. J. Biol. Chem., 264:16810.

Mosmann, T., 1983, Rapid colorimetric assay for cellular growth and survival: Application to proliferation and cytotoxicity assays. J. Immunol. Methods, 65:55.

Nishimura, C., Graham, C., Hohman, T.C., Nagata, M., Robison, W.G., Jr. and Carper, D., 1988, Characterization of mRNA and genes for aldose reductase in rat. Biochem. Biophys. Res. Commun., 153:1051.

Notvest, R.R. and Inserra, J.J., 1987, Tolrestat, an aldose reductase inhibitor, prevents nerve dysfunction in conscious rats. Diabetes, 36:500.

Peterson, G.L., 1977, A simplification of the protein assay method of Lowry et al. which is more generally applicable. Anal. Biochem., 83:346.

Robison, W.G., Jr., Nagata, M., Laver, N., Hohman, T.C. and Kinoshita, J.H., 1989, Diabetic-like retinopathy in rats prevented with an aldose reductase inhibitor. Invest. Ophthalmol. Vis. Sci., 30:2285.

Sima, A.A.F., Bril, V., Nathaniel, V., McEwen, T.A.J., Brown, M.B., Lattimer, S.A., and Greene, D.A., 1988, Regeneration and repair of myelinated fibers in sural-nerve biopsy specimens from patients with diabetic neuropathy treated with sorbinil. New Engl. J. Med., 319:548.

Vota, J.C., Via, D.P., Butterfield, C.E. and Zetter, B.R., 1984, Identification and isolation of endothelial cells based on their increased uptake of acetylated low-density lipoprotein. J. Cell Biol., 99:2034.

Wirthensohn, G., Lefrank, S., Schmolke, M. and Guder, W.G., 1989, Regulation of organic osmolyte concentrations in tubules from rat renal inner medulla. American J. Physiol., 256:F128.

Yancey, P.H., Clark, M.E., Hand, S.C., Bowlus, R.D. and Somero, G.N., 1982, Living with water stress: evolution of osmolyte systems, Science, 217:1214.

Yancey, P.H., Burg, M.B. and Bagnasco, S.M., 1990, Effects of NaCl, glucose, and aldose reductase inhibitors on cloning efficiency of renal medullary cells. American J. Physiol., 258:C156.

PURIFICATION OF ALDOSE AND ALDEHYDE REDUCTASES FROM DOG KIDNEY

Sanai Sato

National Eye Institute
National Institutes of Health
Bethesda, MD 20892

INTRODUCTION

A broad family of NADPH-dependent oxidoreductases catalyze the reduction of various aldehydes and ketones to their corresponding alcohols. Members include the enzymes aldose reductase (EC 1.1.1.21), aldehyde reductase (EC 1.1.1.2) and carbonyl reductase (EC 1.1.1.184). Aldose reductase and aldehyde reductase, discovered by Hers (1960) in the late 1950's, were named for their ability to reduce aldonic sugars versus aliphatic and aromatic aldehydes (Felsted and Bachur, 1980; Turner and Flynn, 1982). Aldehyde reductase, however, is also known as L-hexonate dehydrogenase (EC 1.1.1.19) because of its ability to utilize $NADP^+$ to oxidize aldonic acids such as L-gulonate to D-glucuronate (Mano et al., 1961). Carbonyl reductase was first described in the 1970's as "aromatic aldehyde and ketone reductase" by Culp and McMahon (1968). More recently, it has been referred to as carbonyl reductase (Wermuth, 1981).

The physiological roles of these reductases remain unknown. Experimental evidence links the aldose reductase catalyzed production of sugar alcohols to the onset of diabetic complications such as cataract, keratopathy, retinopathy, neuropathy and possibly, nephropathy (Kinoshita, 1974; Kador, 1988; Kador et al., 1988). These studies have led to the development of potent aldose reductase inhibitors as potential therapeutic agents for the treatment of diabetic complications. Observations that aldose reductase inhibitors also inhibit aldehyde reductase has recently sparked interest in defining the biological role of aldose reductase and the effects of its inhibition (O'Brien et al., 1982; Srivastava et al., 1982; Sato and Kador, 1990).

In the kidney, studies suggest that the formation of polyols, catalyzed by aldose reductase, has both physiological and pathological importance. In the collecting tubules of the medulla, sorbitol has been suggested to serve as an osmolyte to counterbalance the high sodium environment surrounding these cells (Burg and Kador, 1988). In diabetes or galactosemia, the abnormal accumulation of sugar alcohols in the glomeruli and proximal collecting tubules of the cortex is associated with proteinuria and microalbuminuria (Beyer-Mears, 1986; Beyer-Mears et al., 1986; Terubayashi et al., 1988). Defining the relationship between aldose reductase and aldehyde reductase in the kidney is critical for understanding the pathogenesis of diabetic nephropathy and the effects of aldose

Enzymology and Molecular Biology of Carbonyl Metabolism 3
Edited by H. Weiner *et al.*, Plenum Press, New York, 1990

reductase inhibitors. Because of the growing importance of the dog as a model for diabetic complications, the presence of NADPH-dependent reductases have been investigated in the dog kidney and their properties have been compared to those in rat kidney and lens.

MATERIALS AND METHODS

Chemicals

All chemicals employed here were of reagent grade or as previously reported (Sato et al., 1988; Kador et al., 1981). Matrex Gel Orange A was a product of Amicon Corporation, Danvers, MA. Sephadex G-100, Mono P, polybuffer 74, and standard proteins for molecular weight determination and isoelectric focusing were obtained from Pharmacia Fine Chemicals, Piscataway, NJ. The aldose reductase inhibitors Al 1576, sorbinil, tolrestat, and Statil were gifts from Alcon Laboratories (Ft. Worth, TX), Pfizer Central Research (Groton, CT), Ayerst Research Inc. (Princeton, NJ) and ICI Americas (Wilmington, DE), respectively.

Enzyme assay

Reductase and dehydrogenase activities were spectrophoto-metrically assayed on a Gilford Response spectrophotometer by following the decrease or increase of NADPH at 340 nm as previously described (Sato et al., 1988). One unit of activity was defined as the activity consuming or producing 1 micromole of NADPH per min at 22°C. Kinetic calculations and IC_{50} calculations were conducted on the NIH Prophet computer system as previously described (Kador et al., 1981). Protein concentrations were determined according to the method of Bradford (1976).

Purification of Dog Kidney Aldose Reductase and Aldehyde Reductase

Kidneys were obtained from male beagle dogs immediately after euthanasia with a mixture of sodium pentobarbital and sodium phenytoin (Beuthanasia-D) and stored at -70°C. The kidneys were dissected into medullary and cortical regions while still partially frozen. The medullary region was utilized predominantly for aldose reductase purification while aldehyde reductase was purified from the cortex.

Five grams of either dog kidney medulla or cortex were homogenized with 15 ml of 20 mM phosphate buffer, pH 7.5, containing 0.5 mM EDTA, 7 mM β-mercaptoethanol and 0.3 M sodium chloride and then centrifuged at 10,000 x g for 15 min. The obtained supernatant was applied to a column (2.5 x 100 cm) of Sephadex G-100 which was equilibrated with the same phosphate buffer. The column was developed with the same phosphate buffer and the eluent was collected into 150-drop fractions (about 10 ml). This step yielded two activity peaks with DL-glyceraldehyde as substrate. The first peak, subsequently identified as carbonyl reductase, was not further purified because of its instability. The second peak, containing both aldose reductase and aldehyde reductase, was pooled, dialyzed against 20 mM phosphate buffer, pH 7.5, containing 0.5 mM EDTA and 7 mM mercaptoethanol, and applied to a Matrex Gel Orange A column (2.5 x 15 cm) equilibrated with the same 20 mM phosphate buffer. After the column was washed with a minimum of 500 ml of the phosphate buffer and then 250 ml of 10 mM imidazole buffer, pH 7.6, containing 1 mM β-mercaptoethanol, both aldose reductase and aldehyde reductase were eluted with the same imidazole buffer containing 0.1 mM NADPH. This enzyme fraction was applied to a Mono P column (0.5 x 20 cm) equilibrated with 20 mM imidazole buffer, pH 7.6. The elution was performed with Polybuffer 74 which was diluted 10 times with 1 mM β-mercaptoethanol at the flow rate of 1.5 ml/min. The protein was monitored at 280 nm and the aldose reductase and aldehyde reductase peaks were collected.

Purification of Rat lens Aldose Reductase and Rat Kidney Aldehyde Reductase

Both rat lens aldose reductase and rat kidney aldehyde reductase were purified as previously described (Sato et al., 1988; Shiono et al., 1987).

Polyacrylamide gel electrophoresis

Sodium dodecylsulfate-polyacrylamide gel electrophoresis (SDS-PAGE), performed according to the method of Laemmli (1970), and 10-15% gradient polyacrylamide gel electrophoresis were performed on the Pharmacia PhastSystem. Protein was stained with Coomassie blue. The molecular weight of enzymes were calculated from the mobilities of standard proteins; phosphorylase B (94K), bovine serum albumin (67K), ovalbumin (43K), carbonic anhydrase (30K), soy bean trypsin inhibitor (20K), and alpha-lactalbumin (14.4K).

Isoelectric focusing

Isoelectric focusing was conducted on PhastGel IEF 3-9 (Pharmacia) using the Pharmacia PhastSystem. Protein bands were visualized by Coomassie blue stain.

Antibodies

Antibodies against purified enzymes were produced in goats by multiple injections at 2-week intervals of 250 micrograms of purified enzyme mixed with equal volumes of complete Freud's adjuvant. Double immunodiffusion was performed on Ouchterlony plate.

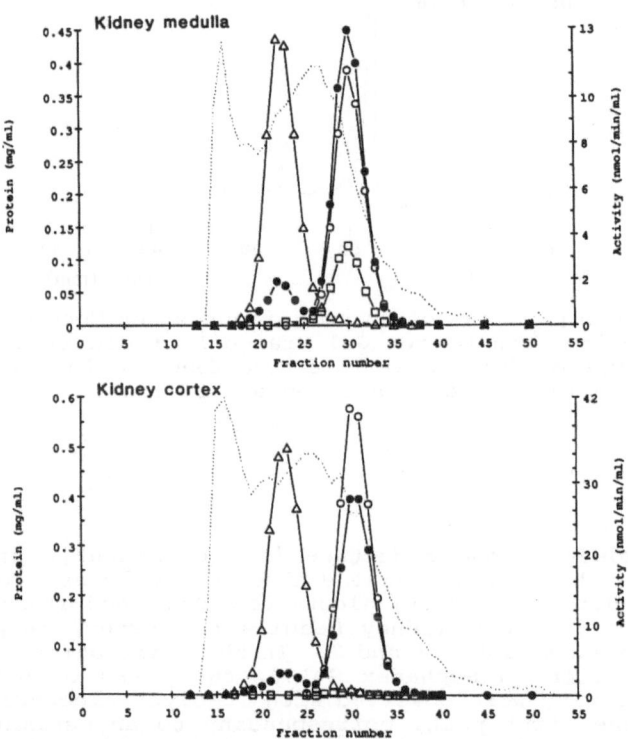

Fig. 1. Elution profiles of the crude extracts of dog kidney medulla (0.8 g) and cortex (1.0 g) from Sephadex G-100 chromatography. The protein concentration is indicated by the broken line. The activity with DL-glyceraldehyde, D-glucuronate, D-xylose and n-butyrophenone is indicated by closed circle, open circle, open square and open triangle, respectively. When n-butyrophenone was used as substrate, half the volume of sample was used in order to fit the peak on the graph.

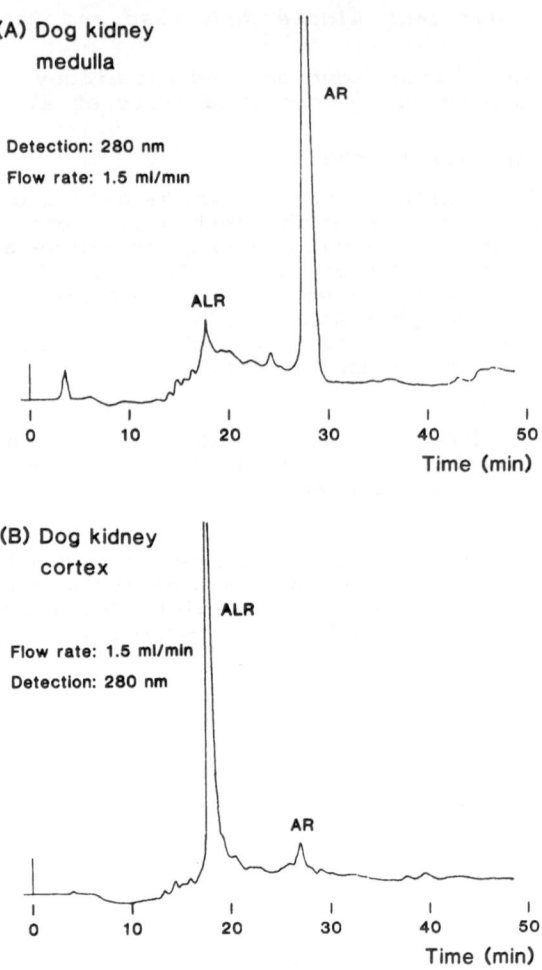

Fig. 2. Elution profiles of chromatofocusing on Mono P. NADPH eluted fraction from Matrex Gel Orange A of dog kidney medulla (A) and cortex (B) was applied to a Mono P column and the protein concentration was monitored at 280 nm.

RESULTS

In the kidney aldose reductase is predominantly present in the medulla while the cortex contains aldehyde reductase (Terubayashi et al., 1989). Using this fact, aldose reductase and aldehyde reductase were purified from dog kidney medulla or cortex, respectively, in steps summarized in Table 1 and 2. In the first chromatographic step of gel filtration on Sephadex G-100, two peaks of NADPH-dependent reductase activity were observed with DL-glyceraldehyde as substrate (Fig. 1). The first peak, corresponding to an estimated molecular weight of $ca.$ 70K displayed labile activity while the second peak, corresponding to an approximate molecular weight of 40K displayed stable activity. This second peak, containing both aldose reductase and aldehyde reductase was further co-purified by affinity chromatography on Amicon Matrex Gel Orange A. A clear separation of aldose reductase and aldehyde reductase was obtained in the final step through chromatofocusing on Pharmacia Mono P. Figure 2 illustrates the separation of these two reductases along with the relative amounts of these enzymes present in the cortex and medulla.

Table 1. Summary of purification of aldose reductase from dog kidney medulla.

Steps	Total protein (mg)	Total activity (mU)	Specific activity (mU/mg)	Purification (fold)	Yield (%)
Homogenate	475.1	5318	11.2	1	100
G-100	112.0	4220	37.7	3.4	79.4
Orange A	2.1	2246	1097	97.9	42.2
Mono P	1.1	1449	1294	115.5	27.2

This table was based on 5 g of the medulla (ca. one dog kidney).

Table 2. Summary of purification of aldehyde reductase from dog kidney cortex.

Steps	Total protein (mg)	Total activity (mU)	Specific activity (mU/mg)	Purification (fold)	Yield (%)
Homogenate	635.7	7082	11.1	1	100
G-100	225.9	9063	40.1	3.6	128.0
Orange A	1.7	5429	3270	293.6	76.7
Mono P	0.7	1797	2567 ·	230.4	25.4

Based on 5 g of the cortex of dog kidney as original materials.

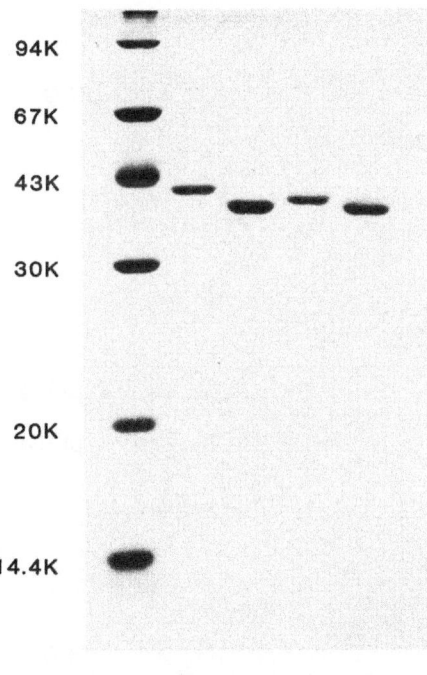

Fig. 3. SDS-polyacrylamide gel electrophoresis of purified rat kidney aldehyde reductase (lane 2), rat lens aldose reductase (lane 3), dog kidney aldehyde reductase (lane 4) and dog kidney aldose reductase (lane 5). Lane 1, molecular weight standards.

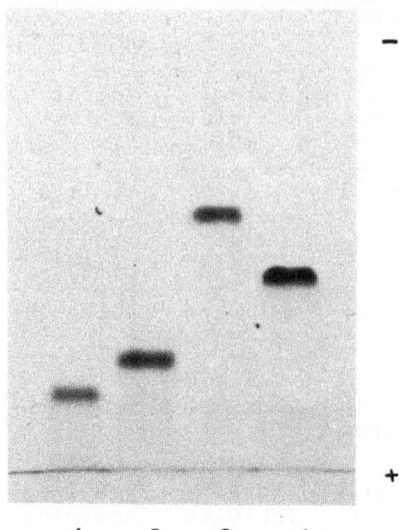

Fig. 4. Polyacrylamide gel (10-15% gradient gel) electrophoresis of purified rat lens aldose reductase (lane 1), dog kidney aldose reductase (lane 2), dog kidney aldehyde reductase (lane 3), and rat kidney aldehyde reductase (lane 4).

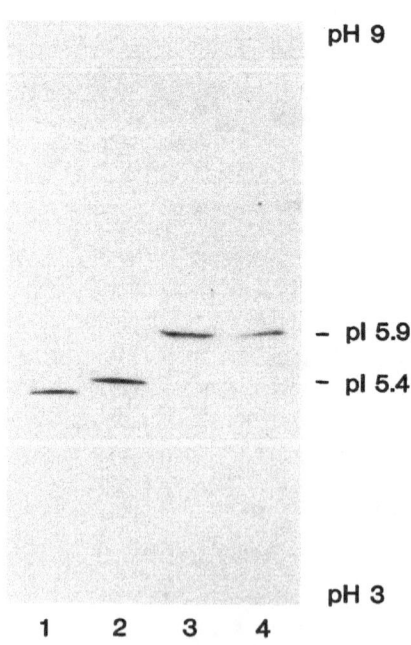

Fig. 5. Isoelectric focusing of purified rat lens aldose reductase (lane 1), dog kidney aldose reductase (lane 2), dog kidney aldehyde reductase (lane 3), and rat kidney aldehyde reductase (lane 4).

Both purified aldose reductase and aldehyde reductase appeared as single protein bands on either SDS-PAGE (Fig. 3) or 10-15% gradient PAGE (Fig. 4). The migration of aldose reductase was clearly distinct from aldehyde reductase on SDS-PAGE, with a migration corresponding to 38K versus 39K for dog kidney aldehyde reductase. The migration of dog kidney aldose reductase appeared to closely mirror rat lens aldose reductase; however, dog kidney aldehyde reductase has a slightly lower molecular weight than rat kidney aldehyde reductase (40K). All four enzymes displayed distinct migrations on 10-15% gradient PAGE. Isoelectric focusing (Fig. 5) of the purified enzymes on acrylamide gel also resulted in the appearance of a single major band corresponding to 5.4 and 5.9 for aldose reductase and aldehyde reductase, respectively. The isoelectric focusing patterns for the dog kidney reductases were essentially identical to those of rat lens aldose reductase and rat kidney aldehyde reductase (Sato et al., 1988; Shiono et al. 1987).

As summarized in Tables 3 and 4, dog kidney aldose reductase displayed substrate specificities similar to those of rat lens aldose reductase while dog kidney aldehyde reductase displayed substrate specificities essentially identical to rat kidney aldehyde reductase. Aldehyde reductase of either dog or rat kidney displayed the ability to oxidize L-gulonate to D-glucuronate while no activity was observed with aldose reductase. Both aldose reductase and aldehyde reductase displayed the ability to reduce aldoses; however, the Km's of aldehyde reductase for D-xylose, D-glucose and D-galactose were estimated to be at least 10-fold greater than those of aldose reductase.

Table 3. Kinetic properties of aldose reductase from dog kidney and rat lens.

Substrate	Aldose reductase (dog kidney)		Aldose reductase (rat lens)	
	Km (mM)	%Vmax	Km (mM)	%Vmax
DL-Glyceraldehyde	0.08	100	0.08	100
D-Glucuronate	11	25	19	25
L-Gulonate		trace		trace
D-Xylose	11	30	14	26
D-Glucose	187	3.0	204	2.1
D-Galactose	82	5.7	83	4.6

%Vmax was expressed as the activity at the substrate concentration of 10 mM.

Table 4. Kinetic properties of aldehyde reductase from dog kidney and rat kidney.

Substrate	Aldehyde Reductase (dog kidney)		Aldehyde Reductase (rat kidney)	
	Km (mM)	%Vmax	Km (mM)	%Vmax
DL-Glyceraldehyde	3.4	100	5.0	100
D-Glucuronate	1.6	111	6.2	85
L-Gulonate	2.1	4.5	7.6	3.3
D-Xylose	604	3.5	1027	0.7
D-Glucose	2000*	0.2	2000*	0.1*
D-Galactose	864	0.4	2000*	0.1*

%Vmax was expressed as the activity at the substrate concentration of 10 mM. * estimated number.

Both dog kidney aldose and aldehyde reductases were inhibited by micromolar concentrations of the aldose reductase inhibitors Al 1576, Statil, tolrestat and sorbinil (Table 5). Among these inhibitors Al 1576 was the most potent but least selective inhibitor of kidney aldehyde reductase. Inhibition of aldehyde reductase by this compound was essentially equal to that of aldose reductase (IC_{50} 0.030 *vs.* 0.039 μM) while all other compounds more strongly inhibit aldose reductase than aldehyde reductase. Statil was the most selective with 56-fold greater inhibition for aldose reductase. These results for the dog kidney enzymes are similar to those of rat lens aldose reductase and rat kidney aldehyde reductase (Sato and Kador, 1990).

Table 5. Inhibition of aldose and aldehyde reductases by aldose reductase inhibitors

| Inhibitors | IC50 (μM) | | | | |
| | Aldose reductase | | Aldehyde reductase | | Carbonyl reductase |
	(dog)	(rat)	(dog)	(rat)	(dog)
Al 1576	0.030	0.024	0.039	0.046	>10
Tolrestat	0.059	0.011	0.6	0.24	>10
Statil	0.025	0.016	1.4	1.9	>10
Sorbinil	0.63	0.07	9.8	1.9	>10

Antibodies were raised in goats against purified dog kidney aldose reductase and aldehyde reductase and these indicated that these enzymes are immunologically distinct (Fig. 6). Moreover, no cross-reactivity between the same enzymes from rat and dog were observed indicating species differences. Antibodies against dog kidney aldose reductase gave a single line of identity against purified dog kidney aldose reductase and crude dog kidney homogenate. No cross-reaction with either dog kidney aldehyde reductase or crude dog kidney carbonyl reductase was observed. No reaction with rat lens aldose reductase and rat kidney aldehyde reductase was also observed. Similarly, antibodies raised against dog kidney aldehyde reductase reacted only with purified dog kidney aldehyde reductase and the crude homogenate of dog kidney.

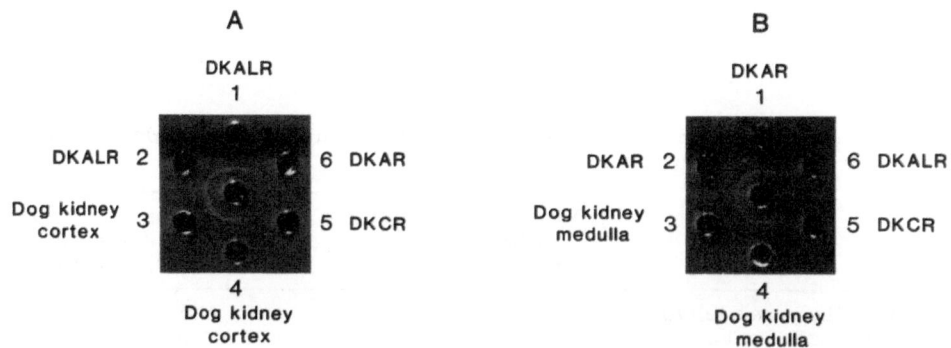

Fig. 6. Double immunodiffusion on Ouchterlony plate of purified dog kidney aldose reductase (DKAR), dog kidney aldehyde reductase (DKALR), dog kidney carbonyl reductase (DKCR). Center wells contain the antibodies raised against purified dog kidney aldehyde reductase (A) and aldose reductase (B), respectively.

While gel filtration on Sephadex G-100 of crude kidney homogenate under high salt conditions afforded NADPH-dependent reductase activity distinct from either aldose reductase or aldehyde reductase, its instability precluded further purification. Based on limited substrate kinetics, the crude enzyme was identified as carbonyl reductase. In the presence of NADPH, the crude enzyme preferentially reduced the aromatic ketones 4-benzoylpyridine > n-butyrophenone to a greater extent than the aldehyde DL-glyceraldehyde (Table 6). Only trace activity was observed with either D-glucuronate or the aldose sugars D-glucose, D-galactose, or D-xylose. The enzyme was not inhibited by micromolar concentrations of aldose reductase inhibitors (Table 5). Its estimated molecular weight of 70K is similar to that of the liver enzyme (Hara et al., 1986)

Table 6. Kinetic properties of carbonyl reductase of the dog kidney

Substrate	Carbonyl reductase (dog kidney)	
	Km (mM)	%Vmax (%)
DL-Glyceraldehyde	10.7	100
D-Glucuronate		trace
D-Xylose		trace
D-Glucose		trace
D-Galactose		trace
n-Butyrophenone	0.147	516*
4-Benzoylpyridine	0.403	760*

%Vmax was expressed as the relative activity at the concentration of 10 mM except that the values marked by asterisk were measured at 1 mM.

DISCUSSION

The dog has been extensively utilized as an animal model for investigating kidney function. With the growing importance of diabetic and galactose-fed dogs as animal models for diabetic complications, and the growing experimental link between sugar alcohol accumulation in kidney tissues and nephropathy, NADPH-dependent reductases in the dog kidney have been investigated. Two NADPH-dependent reductases, aldose reductase and aldehyde reductase have been purified from the dog kidney and the presence of a third NADPH-dependent reductase, identified as carbonyl reductase, has been observed. The predominant distribution of aldose reductase in the medulla and aldehyde reductase in the cortex is similar to that observed in either the rat or human kidney (Terubayashi et al., 1989). The presence of carbonyl reductase, which appears in both the cortex and medulla, has not been reported in either the rat or human kidney. However, an enzyme with similar properties has been observed in the dog lens (Sato and Kador, 1990).

Purified dog aldose reductase and aldehyde reductase displayed substrate kinetics (Table 3 and 4) and susceptibility to inhibition to aldose reductase inhibitors (Table 5) that were essentially identical to those observed for rat lens aldose reductase and rat kidney aldehyde reductase. Both enzymes are monomers with iso-electric points that are similar to those of the rat enzyme; however, these enzymes differ in molecular weight and in that they are immuno-logically distinct. The presence of the carbonyl reductase in kidney tissue prohibits the use of crude dog kidney medulla or cortex homogenates for the standard spectrophotometric screening of aldose reductase inhibitors with glyceraldehyde as substrate since carbonyl reductase utilizes glyceraldehyde but is not inhibited by aldose reductase inhibitors.

While the physiological roles of these NADPH-dependent reductases remain unknown, the predominant localization of aldose reductase in the medulla is consistent with the hypothesis that physiologically sorbitol serves as an osmolyte in the kidney medulla. In the rat, studies suggest that aldehyde reductase contributes to sugar alcohol production in the cortex (Sato, 1990). The similar distribution and substrate specificities of aldose reductase and aldehyde reductase in both the rat and dog kidney suggest that both kidneys should undergo similar biochemical changes under hyperglycemic or galactosemic conditions. Moreover, aldose reductase inhibitors inhibit the rat and dog enzymes to a similar extent.

REFERENCES

Beyer-Mears, A., 1986, The polyol pathway, sorbinil, and renal dysfunction, *Metabolism.* **35**(suppl 1):46.

Beyer-Mears, A., Cruz, E., Edelist, T. and Varagiannis, E., 1986, Diminished proteinuria in Diabetes mellitus by sorbinil, an aldose reductase inhibitor, *Pharmacology*, **32**:52

Bradford, M.M., 1976, A rapid and sensitive method for the quantitation of microgram qualities of protein utilizing the principal of protein-dye binding, *Anal. Biochem.*, **72**:248.

Burg, M. and Kador, P.F., 1988, Sorbitol, osmoregulation, and complication of diabetes, *J. Clin. Invest.*, **81**:635.

Culp, H.W. and McMahon, R.E., 1968, Reductase for aromatic aldehydes and ketones. The partial purification and properties of a reduced triphosphopyridine nucleotide-dependent reductase from rabbit kidney cortex, *J. Biol. Chem.*, **243**:848.

Felsted, R.L. and Bachur, N.R., 1980, Mammalian carbonyl reductases, *Drug Metabolism Reviews*, **11**:1.

Hara, A., Nakayama, T., Deyashiki, Y., Kariya, K. and Sawada, H., 1986, Carbonyl reductase of dog liver: Purification, properties, and kinetic properties, *Arch. Biochem. Biophys.*, **244**:238.

Hers, H.G., 1960, L'Aldose-Reductase, *Biochem. Biophys. Acta*, **37**:120.

Kador, P.K., 1988, The role of aldose reductase in the development of diabetic complications, *Medical Research Review*, **8**:325.

Kador, P.F., Akagi, Y., Terubayashi, H., Wyman, M. and Kinoshita, J.H., 1988, Prevention of pericyte ghost formation in retinal capillaries of galactose-fed dogs by aldose reductase inhibitors, *Arch. Ophthalmol.*, **106**:1099.

Kador, P.F., Goosey, J.D., Sharpless, N.E., Kolish, J. and Miller, D.D., 1981, Stereospecific inhibition of aldose reductase, *Eur. J. Med. Chem.*, **16**:293.

Kinoshita, J.H, 1974, Mechanism initiating cataract formation, *Invest. Ophthalmol.*, **13**:713.

Laemmli, U.K., 1970, Cleavage of structural proteins during the assembly of the head of bacteriophage T4, *Nature (London)*, **15**:680.

Mano, Y., Suzuki, K., Yamada, K. and Shimazono, N., 1961, Enzymatic studies on TPN L-hexonate dehydrogenase from rat liver, *J. Biochem.*, **49**:618.

O'Brien, M.M., Schofield, P.J. and Edwards, M.R., 1982, Inhibition of human brain aldose reductase, *J. Neurochem.*, **39**:810.

Sato, S., 1990, Polyol formation in rat kidney, *in*: "U.S.-Japan Aldose Reductase Workshop," N. Hotta, ed., in press

Sato, S. and Kador, P.F., 1990, NADPH-dependent reductases in the dog lens, *Exp. Eye Res.*, **50**:629.

Sato, S. and Kador, P.F., 1990, Inhibition of aldehyde reductase by aldose reductase inhibitors, *Biochem. Pharmacol*, in press.

Sato, S., Kador, P.F. and Kinoshita, J.H., 1988, Rat kidney aldehyde reductase: Purification and comparison with rat lens aldose reductase, *in*: "Polyol Pathway and its Role in Diabetic Complications," N. Sakamoto, J.H. Kinoshita, P.F. Kador, and N. Hotta, eds, Excepta Medica, Elsevier Science Publishers B.V., Amsterdam.

Shiono, T., Sato, S., Reddy, V.N., Kador, P.F. and Kinoshita, J.H., 1987, Rapid purification of rat lens aldose reductase, *Prog. Clin. Biol. Res.*, **232**:317.

Srivastava, S.K., Petrash, J.M., Sadana, I.J., Ansari, N.H. and Partridge, C.A., 1982, Susceptibility of aldehyde and aldose reductases of human tissues to aldose reductase inhibitors, *Curr. Eye Res.*, **2**:407.

Terubayashi, H., Sato, S., Kador, P.F. and Kinoshita, J.H., 1988, Aldose and aldehyde reductase in the kidney, *in:* "Polyol Pathway and its Role in Diabetic Complications," N. Sakamoto, J.H. Kinoshita, P.F. Kador, and N. Hotta, eds, Excepta Medica, Elsevier Science Publishers B.V., Amsterdam.

Terubayashi, H., Sato, S., Nishimura, C., Kador, P. and Kinoshita, J.H., 1989, Localization of aldose and aldehyde reductase in the kidney, *Kidney International*, **36**:843.

Turner, A.J. and Flynn, T.G., 1982, Nomenclature of aldehyde reductase, *Prog. Clin. Biol. Res.*, **114**:401.

Wermuth, B, 1981, Purification and properties of an NADPH-dependent carbonyl reductase from human brain, *J. Biol. Chem.*, **256**:1206.

STUDY ON DIHYDRODIOL DEHYDROGENASE (I) MOLECULAR FORMS OF THE ENZYME AND THE PRESENCE OF A DIHYDRODIOL SPECIFIC ENZYME IN BOVINE LIVER CYTOSOL

Tohru Nishinaka, Tomoyuki Terada, Toshifumi Umemura,
Hirofumi Nanjo, Tadashi Mizoguchi, and Tsutomu Nishihara

Laboratory of Biochemistry, Faculty of Pharmaceutical
Sciences, Osaka University, Suita, Osaka 565, Japan

INTRODUCTION

It has been known that benzo(a)pyrene and benzo(a)anthracene, typical carcinogenic polycyclic aromatic hydrocarbons, are metabolized in microsome to the corresponding dihydrodiols via epoxides (Yang, et al., 1976; Thakker, et al., 1982) and then converted to the ultimate carcinogens (Buening, et al., 1978). Dihydrodiol dehydrogenase is an enzyme which catalyzes the dehydrogenation of the dihydrodiols of benzo(a)pyrene and benzo(a)anthracene in the presence of $NADP^+$ and forms o-quinone (Vogel, et al., 1980; Smithgall, et al., 1988). The addition of this enzyme to the Ames test significantly reduced the mutagenicity of benzo(a)pyrene, suggesting that this enzyme might detoxify the trans-dihydrodiols which were formed in situ by oxidizing them to the less reactive o-quinones (Glatt, et al., 1979). In addition, similar experiments showed that the purified enzyme reduced the mutagenicity of other polycyclic aromatic hydrocarbons (Smithgall, et al., 1986). On the basis of these facts, Penning and coworkers suggested that dihydrodiol/ 3α-hydroxysteroid dehydrogenase might play an important role in the detoxification of these carcinogenic polycyclic aromatic hydrocarbons in rat liver (Smithgall, et al., 1988). Recently, many dihydrodiol dehydrogenases were purified from various animals and tissues, and these enzymes were identified as 3α-hydroxysteroid dehydrogenase, 17β-hydroxysteroid dehydrogenase and aldehyde reductase from their substrate specificities and inhibitor sensitivities (Smithgall, et al., 1988; Sawada, et al., 1988; Terada, et al., 1990).

In the course of investigation about carbonyl metabolism in the bovine liver, we found three dihydrodiol dehydrogenases by using benzenedihydrodiol (trans-1,2-dihydrobenzene-1,2-diol) which is a typical substrate for dihydrodiol dehydrogenase. In this paper, we described the purification and characterization of multiple forms of dihydrodiol dehydrogenases from the bovine liver.

MATERIALS AND METHODS

Materials

NADPH and $NADP^+$ were purchased from Oriental Yeast, Co. Steroids and

D-glucuronate were purchased from Sigma Chemical Company. The following materials were obtained from the stated sources : Matrex gel Red A (Amicon), Blue Cellulofine (Seikagaku-Kogyo), Freund's complete adjuvant (Difco), Q-Sepharose (Pharmacia-LKB Biotechnology), Ultrogel AcA 44 and HA-Ultrogel (IBF) and peroxidase-labeled goat anti-rabbit IgG antibody (BioMakor). The other substrates (ketones, quinones, aldehydes) and inhibitors were obtained from Wako Pure Chemical Industries or Nacalai Tesque. POD-Kit Wako was kindly supplied from Wako Pure Chemical Industries. Benzenedihydrodiol (trans-1,2-dihydrobenzene-1,2-diol) was synthesized according to the method of Platt and Oesch (1977).

Subcellular Distribution

The experiment on the subcellular distribution of the enzyme was carried out according to the method of Hogeboom (1955). Lactate dehydrogenase, cytochrome c reductase, glutamate dehydrogenase, and DNA were determined as markers of cytosol, microsome, mitochondria, and nuclei by the methods of Bergmeyer (1974), Phillips (1962), Strecker (1955), and Burton (1956), respectively.

Purification of Dihydrodiol Dehydrogenases

All the procedures were carried out below 6 °C, and 10 mM 2-mercaptoethanol and 0.1 mM EDTA were added to all the buffers used in the procedures unless otherwise specified. Detailed purifications of dihydrodiol dehydrogenases are described in the RESULTS. The fresh bovine liver obtained from the local slaughter house was cut into small pieces and rinsed with 3 mM Tris/HCl buffer, pH 7.4, containing 155 mM KCl and 1 mM EDTA. The liver was homogenized (25%) in a Waring Blender at 10,000 rpm for 3 min. PMSF was added to the resulting homogenate at a final concentration of 500 μM. This homogenate was centrifuged first at 10,000 x g for 60 min and then at 100,800 x g for 60 min. The resulting supernatant (cytosol) was subjected to the further purification procedure. Protein was determined by the Lowry method (1951).

Enzyme Assay

The enzyme activity was measured by monitoring the increase in absorbance at 340 nm of the following reaction mixture: 500 μM NADP$^+$ and 1.8 mM benzenedihydrodiol in 100 mM glycine/NaOH buffer, pH 9.5, where the final methanol concentration was less than 2 %. The reaction was initiated by the addition of the enzyme. The amount of the enzyme was expressed in units: one unit is defined as the amount to catalyze the dehydrogenation of 1 mol substrate per min at 25°C. Reductase activity of the enzymes was also measured under the following conditions: 100 mM sodium phosphate buffer, pH 7.0, 67 μM NADPH and an appropriate concentration of substrate. Assay was initiated by the addition of the enzyme.

Preparation of Antiserum

Antiserum was raised in female Japanese white rabbits (weighing about 2 kg) by successive subcutaneous injections of 500 μl of the purified enzyme solution which contained 150 μg enzyme with the same volume of Freund's complete adjuvant every 2 weeks. After 7 days from the 4th injection, blood was collected from the ear vein, and the antiserum was obtained and stored.

Western-blot Analysis

Sample solutions (1 μg of protein) were subjected to 12.5% SDS-polyacrylamide gel electrophoresis. After electrophoresis, proteins were transferred to an Immobilon P membrane at 100 mA for 60 min. The membrane was reacted successively with a 100-fold dilution of the antiserum against the purified enzyme and a 2,000-fold dilution of peroxidase-labeled goat anti-rabbit IgG antibody, and then, the reactive protein was visualized by POD Kit Wako at 25 °C for 5 min.

RESULTS

Subcellular Distribution of Dihydrodiol Dehydrogenase

Fig. 1 shows the subcellular distribution of dihydrodiol dehydrogenase activity in bovine liver cells. The activity was recovered only from the cytosol fraction, suggesting that dihydrodiol dehydrogenase was located mainly in cytosol of bovine liver.

Purification of Three Dihydrodiol Dehydrogenases

In this purification procedure, the following buffers which, unless otherwise specified, contained both 0.1 mM EDTA and 10 mM 2-mercapto-ethanol were used: buffer A, 10 mM Tris/HCl buffer, pH 8.0; buffer B, 30 mM sodium phosphate buffer, pH 7.0; buffer C, 2.5 mM sodium phosphate buffer, pH 7.5; and buffer D, 2.5 mM sodium phosphate buffer, pH 7.5, without EDTA. All the procedures were performed below 6 °C. The enzyme solutions were concentrated by using TOYO Ultrafiltration apparatus with a PM-10 membrane. First, cytosol obtained from the bovine liver was subjected to 35-75% ammonium sulfate fractionation. The precipitate was dissolved in a minimum volume of buffer A and then the suspension was dialyzed against 3 L of buffer A for 3 h. The dialysate was applied to a column (2.6 x 129 cm) of Ultrogel AcA 44 equilibrated with buffer A, eluted with the same buffer at a flow rate of 30 ml/h, and collected into fraction tubes (10 ml/tube). The active fractions were pooled and concentrated. The concentrated enzyme solution was then applied to a column (2.6 x 40 cm) of Q-Sepharose anion exchanger. The column, after washing out the unadsorbed fractions with 300 ml of buffer A, was eluted with a linear gradient of NaCl (0-200 mM) in 300 ml of buffer A at a flow rate of 30 ml/h. The elution was collected into fraction tubes (5 ml/tube). The enzyme activity assay showed one peak in the unadsorbed fractions and two peaks in the adsorbed fractions (at about 50 mM and 75 mM, respectively). These peaks were tentatively named DD1, DD2 and DD3 (data not shown).

Each peak was concentrated and further purified. DD1 and DD3 solutions were separately applied to a column (1.2 x 5 cm) of Matrex gel Red A equilibrated with buffer A. After the unadsorbed fraction was washed out with 30 ml of buffer A, then the adsorbed fractions were eluted with a linear gradient of 0-3 M NaCl in 60 ml of buffer A at a flow rate of 5 ml/h and the active fractions were collected into fraction tubes (5 ml/tube). The enzymatically active fractions (DD1 and DD3 at about 750 mM and 2.5 M, respectively) were concentrated and dialyzed against 3 L of buffer D for 6 h. The enzyme solutions were further applied to an HA-Ultrogel column (1.5 x 10 cm). The column, after washed with 35 ml of buffer D, was eluted with a linear gradient of 2.5-200 mM phosphate buffer, pH 7.0, without EDTA. Eluates (DD1 and DD3 at about 130 and 100 mM, respectively) were collected and concentrated.

DD2 was purified, after dialysis against 3 L of buffer B, by Blue Cellulofine column chromatography (1.2 x 15 cm). The column was washed with 30 ml of buffer B at a flow rate of 10 ml/h and DD1 was eluted at about 100 mM KCl in a linear gradient of 0-0.8 M KCl. The active enzyme

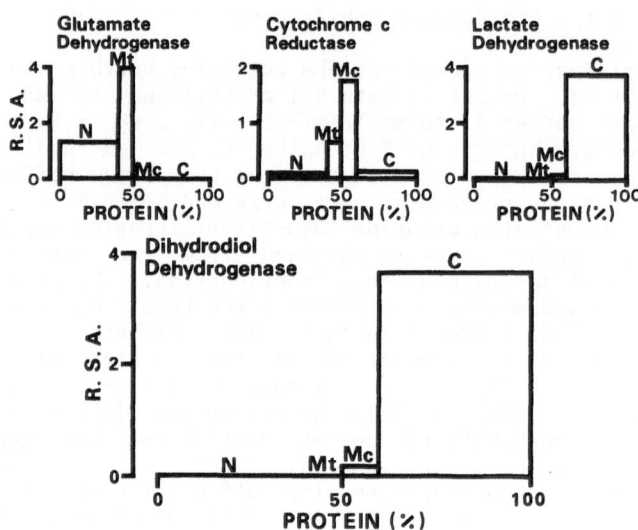

Fig. 1. Subcellular distribution of dihydrodiol dehydrogenase in bovine liver.
Relative specific activity (R.S.A.) is calculated by (percentage of enzyme
activity in a given fraction) / (percentage of protein in that fraction).

Table 1. Purification of dihydrodiol dehydrogenases (DD1, DD2 and DD3) from bovine liver cytosol. The results in the table represent typical values in the purification of three enzymes from 100 g bovine liver.

Step	Total protein (mg)	Total activity (units)	Specific activity (munits/mg)	Yield (%)	Purifi-cation (fold)
Cytosol	5,400	24	4	100	1.0
35-75% $(NH_4)_2 SO_4$ precipitate	2,500	16	6	67	1.5
Ultrogel AcA44	840	15	18	63	4.5
Q-Sepharose 4B					
DD1	30	0.4	13	1.6	3.3
DD2	430	3.2	7	13	1.8
DD3	430	8.6	20	36	5.0
Matrex gel Red A					
DD1	1.0	0.03	30	0.13	7.5
DD3	7.5	6.2	830	26	210
Blue Cellulofine					
DD2	16	0.92	58	3.8	15
HA-Ultrogel					
DD1	0.4	0.03	75	0.13	19
DD2	3.7	0.43	120	1.8	30
DD3	3.1	4.0	1,300	17	330

fractions (5 ml/tube) were collected, concentrated and then dialyzed against 3 L of buffer D for 6 h. Then, the dialysate was applied to a column (1.5 x 10 cm) of HA-Ultrogel at a flow rate of 10 ml/h and the enzyme activity was detected in the unadsorbed fractions (3 ml/tube).

A typical run of the purification procedure is summarized in Table 1. The final preparations of DD1, DD2 and DD3 were purified 19-, 30- and 330-fold, respectively, over the cytosolic fraction, and exhibited a single band on 12.5% SDS-polyacrylamide gel electrophoresis with silver staining.

Molecular Weight

The molecular weights of the three enzymes estimated by SDS-polyacrylamide gel electrophoresis were in good agreement with the values obtained by Ultrogel AcA 44 column chromatography, indicating that all of the three enzymes were monomeric enzymes with molecular weights of 35,000, 36,500 and 35,500, respectively.

Substrate Specificity

Substrate specificities of the three dihydrodiol dehydrogenases are shown in Table 2. DD1 was more active for androsterone than for benzene-dihydrodiol in the presence of $NADP^+$ and also catalyzed the reduction of 5β-androstan-17β-ol-3-one, menadione and 4-benzoylpyridine in the presence of NADPH. DD2 could strongly reduce D-glucuronate and DL-glyceraldehyde in the presence of NADPH, both of which are the typical substrates for aldehyde reductase. On the other hand, DD3 did not catalyze the dehydrogenation and the reduction of any other alcohols, ketones, aldehydes and quinones except benzenedihydrodiol.

DD3 had a low Km value of 0.18 mM for benzenedihydrodiol, which was the lowest of all the values so far reported and the highest kcat value of the three bovine enzymes, suggesting that DD3 might catalyze the

Table 2. Substrate specificity of dihydrodiol dehydrogenases. Dehydrogenase activity was measured using 0.5 mM NADP$^+$ at pH 9.5. Reductase activity was measured using 0.067 mM NADPH at pH 7.0.

Substrate	Concentration (mM)	Relative velocity (%)		
		DD1	DD2	DD3
[Dehydrogenase activity]				
Benzenedihydrodiol	1.8	100	100	100
Androsterone	0.1	210	−	−
Testosterone	0.1	−	−	−
[Reductase activity]				
D-Glucuronate	5.0	−	4,300	2
DL-Glyceraldehyde	5.0	−	5,000	5
4-Benzoylpyridine	1.0	130	−	7
5β-Androstan-3α-ol-17-one	0.1	13	−	−
5β-Androstan-17β-ol-3-one	0.1	170	−	−
Menadione	0.1	200	−	−

− , Not detectable.

dehydrogenation of benzenedihydrodiol most efficiently in the bovine liver.

In order to characterize DD3 and compare with other enzymes, the substrate specificity was further investigated. DD3 did not show any enzymatic activity for other compounds including epiandrosterone which is a typical substrate for 3β-hydroxysteroid dehydrogenase, 17β-estradiol which is a substrate for estradiol-17β dehydrogenase, 4-nitrobenzaldehyde, camphorquinones and 9,10-phenanthrenequinone which are substrates for carbonyl reductase. These results suggest strongly that DD3 does not correspond to any enzymes thus far reported. Considering its Km value, DD3 was expected to be a novel enzyme having the activity only for benzenedihydrodiol.

Inhibitor Sensitivity

The three dihydrodiol dehydrogenases were shown to have characteristic properties in their inhibitor sensitivity (Table 3). Ethacrynate was a potent inhibitor of these three enzymes. Barbital and valproate, which are high-Km aldehyde reductase inhibitors, inhibited only DD2 strongly. DD3 was not inhibited by any compounds except acetylsalicylate. DD1 and DD3 were inhibited by sulfhydryl residue modifying reagents such as DTNB,

Table 3. Inhibitor sensitivity of three dihydrodiol dehydrogenases

Compound	Concentration (mM)	Inhibition (%)		
		DD1	DD2	DD3
Barbital	1.0	0	100	0
Valproate	1.0	0	90	0
Warfarin	0.1	0	29	0
Acetylsalicylate	1.0	0	0	36
Pyrazole	10.0	0	0	0
Ethacrynate	1.0	100	86	100
DTNB	0.1	100	0	60

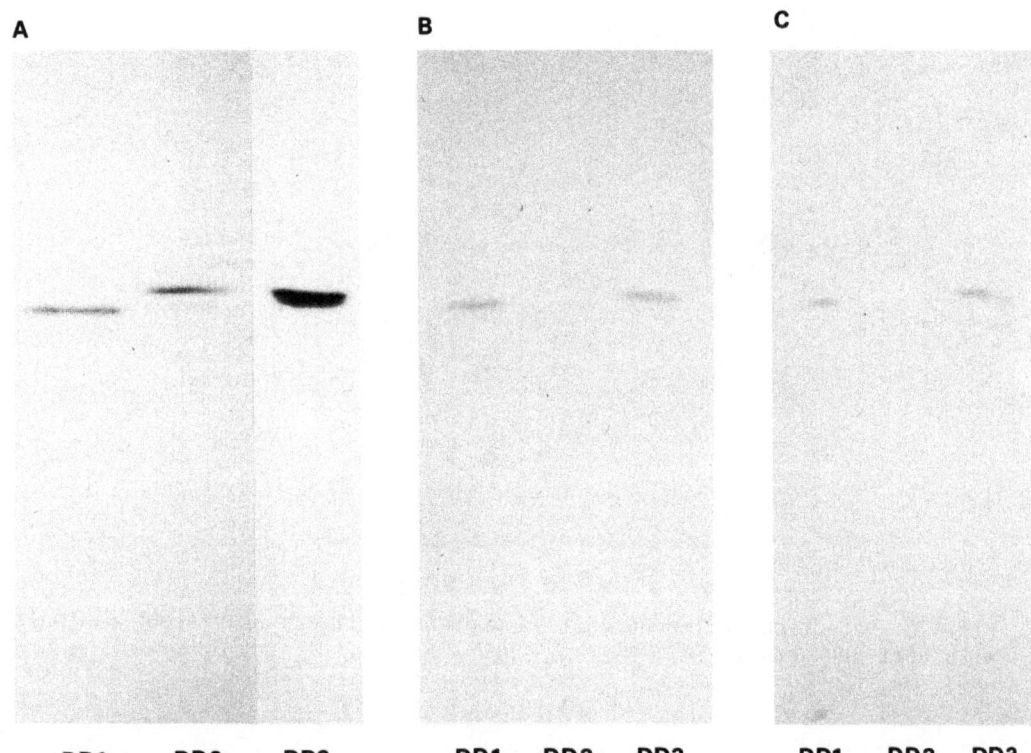

Fig. 2. Comparison of immunological properties between DD3 and the other enzymes, DD1 and DD2, by western-blot analysis.
A, Protein stained gel; B, Immunostaining with anti-DD3 serum;
C, Immunostaining with anti DD1 serum.

whereas DD2 was not. These results suggested that DD1 and DD3 might have cysteinyl residues at or near their catalytic sites.

Judging from the molecular weight, the substrate specificity and the inhibitor sensitivity, DD1 and DD2 were identified as 3α-hydroxysteroid dehydrogenase and high-Km aldehyde reductase, respectively. However, DD3 could not be identified as any enzymes so far found. Accordingly, DD3 may be a novel enzyme with a low Km value and high velocity, which specifically dehydrogenate benzenedihydrodiol.

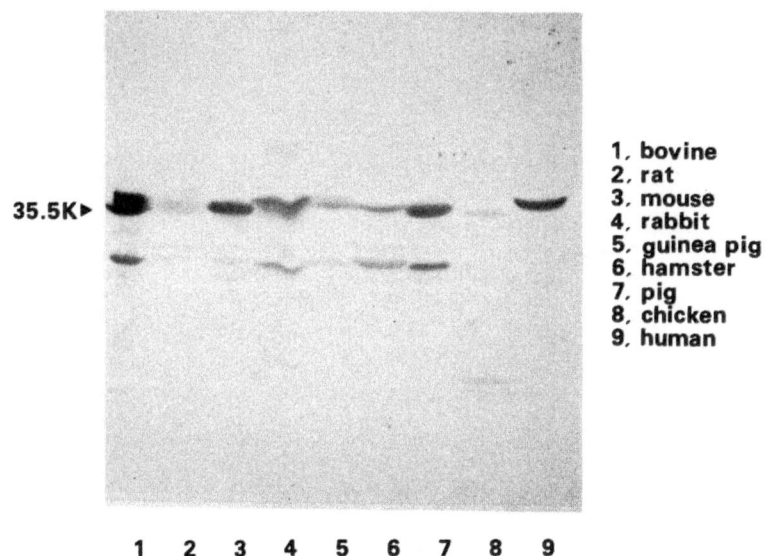

35.5K▶

1, bovine
2, rat
3, mouse
4, rabbit
5, guinea pig
6, hamster
7, pig
8, chicken
9, human

1 2 3 4 5 6 7 8 9

Fig. 3. Detection of immunoreactive protein in livers of various animals with anti-DD3 serum.

Immunocrossreactivity between Dihydrodiol Dehydrogenase 3 (DD3) and Other Enzymes

The immunocrossreactivity of DD3 with DD1 and DD2 was analyzed by the western-blotting technique using anti-DD1 and anti-DD3 sera. As shown in Fig. 2, DD3 and DD1 immunologically crossreacted with each other, whereas DD2 was not crossreactive with DD1 or DD3. These results suggested that DD3 was immunologically similar to DD1, that is 3α-hydroxysteroid dehydrogenase. Additionally, western-blot analysis using anti-DD3 serum was carried out for the detection of immunoreactive protein in livers of various animals. The presence of crossreactive proteins against this antiserum was demonstrated in all animals tested as shown in Fig. 3, suggesting the possibility of the existence of such a characteristic enzyme as DD3 in the liver cytosol of various animals.

DISCUSSION

Benzenedihydrodiol, a metabolite of benzene in the cell, is well known as a typical substrate for dihydrodiol dehydrogenase. Dihydrodiol dehydrogenase is supposed to participate in the detoxification of carcinogenic polycyclic aromatic hydrocarbons including benzo(a)pyrene and benzo(a)anthracene (Smithgall, et al., 1986). The addition of this enzyme reduced the mutagenicity of these polycyclic aromatic hydrocarbons in the Ames test which consisted of mouse liver microsome, NADPH and the test compound (Glatt, et al., 1979). Benzo(a)pyrene is metabolized in microsome to the ultimate carcinogen benzo(a)pyrene-7,8-dihydrodiol-9,10-epoxide, (Buening, et al., 1978). Dihydrodiol dehydrogenase has the ability to catalyze the dehydrogenation of benzo(a)pyrene-7,8-dihydrodiol which is a direct precursor of the ultimate carcinogen and interferes with the accumulation of this compounds (Thakker, et al., 1977; Smithgall, et al., 1988).

In the course of study on dihydrodiol dehydrogenase, we found the enzyme which catalyzes the dehydrogenation of benzenedihydrodiol. In bovine liver, dihydrodiol dehydrogenase was mainly located in the cytosolic fraction, from which we succeeded in the purification of three dihydrodiol dehydrogenases (DD1, DD2 and DD3) to be homogeneous by 19-, 30- and 330-fold. These three enzymes had similar molecular weights of around 36,000 (DD1, 35,000; DD2, 36,500; and DD3, 35,500). All of these enzymes could catalyze the dehydrogenation of benzene-dihydrodiol in the presence of $NADP^+$. DD1 also catalyzed both the $NADP^+$-specific dehydrogenation of androsterone and the NADPH-specific reduction of 5β-androstan-17β-ol-3-one and menadione. DD2 was able to specifically reduce both D-glucuronate and DL-glyceraldehyde, in the presence of NADPH, and it was strongly inhibited by barbital and valproate. These results suggested that DD1 and DD2 were 3α-hydroxysteroid dehydrogenase and high-Km reductase, respectively (Terada, et al., 1985; 1989). On the other hand, DD3 was a highly specific enzyme for benzenedihydrodiol (with a low Km value and a high Vmax value) and failed to catalyze the dehydrogenation and reduction of all alcohols and carbonyls, but benzenedihydrodiol, tested in this paper. Ethacrynate was a common inhibitor to these enzymes. DD1 and DD3 were immunologically crossreactive with each other. Anti-DD2 serum did not crossreact with DD1 or DD3 (data not shown). It was deduced from the specicific activity for benzenedihydrodiol and the relative amount in the liver that among these three enzymes, DD3 might be a major enzyme in dihydrodiol metabolism and the others might be minor forms. Although DD3 had similar properties to DD1 (3α-hydroxysteroid dehydrogenase) in molecular weight and immunological crossreaction, DD3 did not catalyze the dehydrogenation of androsterone. But it was competitively inhibited by androsterone with low Ki value (1μM) which was in the same order as the Km value of DD1 (4.6μM). Furthermore, both DD1 and DD3 were inhibited by sulfhydryl group modifying reagents, suggesting that their catalytic sites were similar and their activities might be modulated by steroid hormones in the liver. It is also supposed that DD3 and DD1 might be related with each other in the steroid hormone metabolism and the detoxification of polycyclic aromatic hydrocarbons.

In this study, we found a new enzyme in bovine liver, which was specific for dihydrodiol and was different from liver enzymes in other animals (Smithgall, et al., 1988; Sawada, et al., 1988). We supposed that this enzyme (DD3) might play an important role in the detoxification of carcinogenic polycyclic aromatic hydrocarbons in bovine liver. Further studies on the regulation of these enzyme activities, the induction and expression of enzyme protein in the living cell are now in progress.

ACKNOWLEDGMENT

This work was supported in part by a Grant-in Aid for Encouragement of Young Scientists (62771901) from the Ministry of Education, Science and Culture of Japan.

REFERENCES

Bergmeyer,H.U.,Gawehn,K.,Grassl,M., 1974, Enzymes as biochemical reagent. in: "Methods of Enzymatic Analysis," H.U.Bergmeyer ed., Verlag Chemie Weinheim Academic Press,Inc. New York and London

Buening,M.K., Wislocki,P.G., Levin,W., Yagi,H., Thakker,D.R., Akagi,H., Koreeda,M.,Jerina,D.M.and Conney,A.H., 1978, Tumorigenicity of the optical enantiomers of the diasteromeric benzo(a)pyrene 7,8-diol-9,10-epoxides in newborn mice: Experimental activity of (+)-7,8-dihydroxy-9,10-epoxy-7,8,9,10-tetrahydrobenzo(a)pyrene, Proc. Natl. Acad. Sci. USA, 75: 5358

Burton,K., 1956, A study of the conditions and mechanism of the diphenylamine reaction for the colorimetric estimation of deoxyribonucleic acid, Biochem.J., 62: 315

Glatt,H.R., Vogel,K., Bentley,P. and Oesch,F., 1979, Reduction of benzo-(a)pyrene mutagenicity by dihydrodiol dehydrogenase, Nature, 277: 319

Hogeboom,G.H., 1955, Fractionation of cell components of animal tissues, Methods Enzymol., 1: 16

Lowry, O.H., Rosebrough, N.J., Farr, A.L. and Randall, R.J., 1951, Protein Measurement with the Folin phenol reagent, J. Biol. Chem., 193: 265

Phillips,A.H. and Llandon,R.G.,1962, Hepatic triphosphopyridine nucleotide cytochrome c reductase : Isolation, characterization, and kinetic studies, J.Biol.Chem.,237: 2656

Platt,K.L.,Oesch,F.,1977,An improved synthesis of trans-5,6-dihydroxy-1,3-1,3-cyclohexidine(trans-1,2-dihydroxy-1,2-dihydrobenzene), Synthesis, 7: 449

Sawada,H.,Hara,A.,Nakayama,T.,Nakagawa,M.,Inoue,K.,Hasebe.,K. and Zhang, Y.-P.,1988,Mouse liver dihydrodiol dehydrogenases. Identity of the predominant and a minor form with 17 β-hydroxysteroid dehydrogenase and aldehyde reductase, Biochem. Pharmcol.,37: 453

Smithgall, T.E.,Harvey, R.G. and Penning, T.M., 1986, Regio- and stereo-specificity of homogeneous 3 α-hydroxysteroid-dihydrodiol dehydrogenase for trans-dihydrodiol metabolites of poly cyclic aromatic hydrocarbons, J.Biol.Chem., 261: 6184

Smithgall,T.E.,Harvey,R.G, and Penning,T.M., 1988, Stereoscopic identification of ortho-quinones as the products of polycyclic aromatic trans-dihydrodiol oxidation catalyzed by dihydrodiol dehydrogenase, J.Biol.Chem., 263: 1814

Smithgall,T.E. and Penning,T.M., 1988, Electrophoretic and immunological characterization of 3α-hydroxysteroid/dihydrodiol dehydrogenases of rat tissues, Biochem.J., 254: 715

Strecker,H.J.,1955, L-Glutamic dehydrogenase from liver, Methods Enzymol., 2: 220

Terada,T.,Kohno,T.,Samejima,T.,Hosomi,S.,Mizoguchi,T. and Uehara,K., 1985, Purification and properties of beef liver aldehyde reductase catalyzing the reduction of D-erythrose 4-phosphate, J.Biochem.,97:79

Terada,T., Niwase,N., Shinagawa,K., Koyama,I., Hosomi,S. and Mizoguchi,T., 1989, Bovine liver cytosolic aldehyde reductase and carbonyl reductase. Purification and characterization, Prog.Clin.Biol.Res., 290: 293

Terada,T., Shinagawa,K., Umemura,T., Nishinaka,T., Nanjo,H., Hosomi,S., Mizoguchi,T. and Nishihara,T., Study on dihydrodiol dehydrogenase (II) Modulation of the enzyme activity by biological disulfides, Enzymology and molecular biology of carbonyl metabolism, in press.

Thakker,D.R., Yagi,H., Akagi,H., Koreeda,M., Lu,A.Y.H.,Levin,W.,Wood,A.W., Conney,A.H. and Jerina,D.M., 1977, Metabolism of Benzo(a)pyrene VI. Stereoselective metabolism of benzo(a)pyrene and benzo(a)pyrene 7,8-dihydrodiol to diol epoxides, Chem.Biol. Interactions,16: 281

Thakker,D.R., Levin,W., Yagi,H., Tada,M.,Ryan,D.E.,Thomas,P.E.,Conney,A.H. and Jerina,D.M.,1982, Stereoselective metabolism of the (+)- and (-)-enantiomers of trans-3,4-dihydroxy-3,4-dihydrobenz(a)anthracene by rat liver microsomes and by a purified and reconstituted cytochrome P-450 system, J.Biol.Chem., 257: 5103

Vogel,K., Bentley,P., Platt,K-L. and Oesch,F.,1980, Rat liver cytoplasmic dihydrodiol dehydrogenase purification to apparent homogeneity and properties, J.Biol.Chem., 255: 9621

Yang, S.K., McCourt, D.W., Roller, P.P. and Gelboin,H.V., 1976, Enzymic conversion of benzo(a)pyrene leading predominantly to the diol-epoxide r-7,t-8-dihydroxy-t-9,10-oxy-7,8,9,10-tetrahydrobenzo(a)pyrene through a single enantiomer of r-7,t-8-dihydroxy-7,8-dihydrobenzo(a)-pyrene, Proc. Natl. Acad. Sci. USA,73: 2594

STUDY ON DIHYDRODIOL DEHYDROGENASE (II) MODULATION OF DIHYDRODIOL

DEHYDROGENASE ACTIVITY BY BIOLOGICAL DISULFIDES

Tomoyuki Terada, Kazuhiko Shinagawa, Toshifumi Umemura,
Tohru Nishinaka, Hirofumi Nanjo, Saburo Hosomi, Tadashi
Mizoguchi and Tsutomu Nishihara
Laboratory of Biochemistry, Faculty of Pharmaceutical
Sciences, Osaka University, Suita, Osaka 565, Japan

INTRODUCTION

Rat liver cytosolic 3α-hydroxysteroid dehydrogenase (EC 1.1.1.50) (dihydrodiol dehydrogenase) which can catalyze the conversion between androsterone and androstanedione in the presence of NADP(H) has also been shown to catalyze the oxidation of benzenedihydrodiol to catechol (Penning et al, 1984, 1985; Hara et al., 1988). From the facts that can oxidize the trans-dihydrodiol of polycyclic aromatic hydrocarbons such as benzo(a)pyrene and benzo(a)anthracene, it has been suggested that dihydrodiol dehydrogenase plays an important role in the detoxification of the polycyclic aromatic hydrocarbon through the effective suppression in the formation of the ultimate carcinogenic anti-diol epoxides (Penning et al., 1984, 1985).

On the other hand, it has been that thiol/disulfide exchange reaction in the cell is dependent mainly on the ratio of cellular thiols to disulfides (Cappel & Gilbert, 1986), e.g. reduced/oxidized glutathione ([GSH]/[GSSG]) and proteins containing accessible thiols or disulfides. The posttranslational modification can lead to the conversion of protein redox state (Ziegler, 1985). Accordingly, there can be a mechanism of metabolic control by thiol/disulfide exchange reaction because many metabolic enzymes vary their activities according to the redox state. Benzo(a)pyrene induces the depletion of cellular GSH in the metabolic process (Morrison et al., 1983), which is responsible for the induction of carcinoma together with a direct attack to DNA by its metabolite (Everson et al., 1986).

However, it is well known that dihydrodiol dehydrogenase is strongly inhibited by SH-reagents (Penning et al., 1984). Therefore, in the present work, we studied the effect of disulfides on rat liver cytosolic dihydrodiol dehydrogenase in relation with dihydrodiol metabolism and propose that the [GSH]/[GSSG] ratio modulates the enzyme activity.

MATERIALS AND METHODS

Materials

GSSG, L-cystamine, L-homocystine and cysteamine were purchased from Sigma Chemicals. L-Cystine, GSH and dithiobis (2-nitrobenzoic acid) (DTNB)

were obtained from Wako Pure Chemical Industries, Ltd. Monoiodoacetic acid, N-ethylmaleimide (NEM) and dithiothreitol (DTT) were the products of Nacalai Tesque. NAP-column was obtained from Pharmacia-LKB Biotechnology. S-Sulfocysteine was prepared from L-cystine by the oxidation with hydroxyperoxide. Benzenedihydrodiol was also synthesized according to the method of Platt and Oesch (1977). The other reagents used were of the highest grade commercially available.

Enzyme Activity

Unless otherwise specified, dihydrodiol dehydrogenase activity was measured under the conditions as described previously (Penning et al., 1984): 1.0 mM benzenedihydrodiol or 100 μM androsterone, 67 μM NADP$^+$ and 100 mM glycine/NaOH buffer, pH 9.0. All the reaction mixtures contained methanol less than 2%. The assay was initiated by addition of the purified rat liver enzyme preparation to the reaction mixture kept at 25 °C and then, the reaction was followed by monitoring the decrease in absorbance at 340 nm (standard assay system). In the case of assay for the reductase activity, measurements were carried out under the following conditions: 67 μM NADPH, 100 mM sodium phosphate buffer, pH 7.0, and an appropriate concentration of the substrate. Activities of aldehyde reductase (high-Km aldehyde reductase: EC 1.1.1.2) and DT-diaphorase (EC 1.6.99.2) were determined by the spectrophotometric method at 340 nm as described by Flynn (1982) and Wallin (1979) using D-glucuronate and menadione, respectively.

Enzyme Preparation

Dihydrodiol dehydrogenase was purified from the cytosolic fraction of male Wister rat liver, essentially according to the method of Penning et al (1984), that is, ammonium sulfate fractionation, chromatography on DE-53, Ultrogel AcA 44, Blue-cellulofine, 2',5'-ADP Sepharose and hydroxyapatite and, then, isoelectric focusing. The purity was estimated by 12.5% SDS-PAGE. By similar methods, aldehyde reductase and DT-diaphorase were purified to apparent homogeneity on 12.5% SDS-PAGE.

Treatment of Dihydrodiol Dehydrogenase with Disulfide and SH-Reagent

Dihydrodiol dehydrogenase was reduced by the pretreatment with 1 mM DTT for 30 min at 25 °C. After a solution of the reduced enzyme was ultrafiltered through an NAP-column for elimination of DTT, it was incubated at 25 °C with a disulfide (5 mM GSSG, 5 mM L-cystamine or 5 mM S-sulfocysteine) or an SH-reagent (5 mM monoiodoacetic acid, 0.5 mM NEM or 0.1 mM DTNB). And then, aliquots of the reaction mixture were taken out at appropriate times for measurement of the activity.

Treatment of Dihydrodiol Dehydrogenase with Redox Buffer System

Dihydrodiol dehydrogenase was added to the redox buffer having a given ratio of [GSH]/[GSSG] (from 0 to 50) to fix the total glutathione concentration at 10 mM. Then the mixture was incubated at 25 °C for 24 h under nitrogen gas. Throughout the period of incubation under these conditions, the [GSH]/[GSSG] ratio did not alter.

RESULTS

Characterization of Purified Rat Liver Cytosolic Dihydrodiol Dehydrogenase

Dihydrodiol dehydrogenase was purified 300-fold from the cytosolic fraction of rat liver and the homogeneity was determined on SDS-PAGE (data not shown). The purified enzyme was estimated to be a monomeric enzyme of

Table 1. Substrate specificity of dihydrodiol dehydrogenase. The enzyme activity was measured as described in the text, using various substrates. The initial velocity data were compared.

Substrate	Conc (mM)	Relative velocity (%)
Dehydrogenase activity		
Benzenedihydrodiol	1.0	100
Androsterone	0.1	256
Testosterone	0.1	0
Reductase activity		
Androstanedione	0.1	241
4-Benzoylpyridine	1.0	67
4-Nitroacetophenone	0.5	675
Phenanthrenequinone	0.1	988
Camphorquinone	0.1	1,470
1,4-Benzoquinone	0.1	7
4-Nitrobenzaldehyde	0.1	554
D-Glucuronate	3.3	7

32,000 dalton. Substrate specificities of the purified rat liver enzyme are summarized in Table 1.

The purified enzyme had the ability to catalyze the NADPH-dependent reduction of various carbonyl compounds, including androsterone, 4-nitroacetophenone, camphorquinone, phenanthrenequinone and 4-nitrobenzaldehyde. The enzyme did not reduce 1,4-benzoquinone or D-glucuronate. On the other hand, the enzyme had ability to catalyze the $NADP^+$-dependent oxidation of androstanedione and benzenedihydrodiol, but not testosterone. The results of substrate specificity and molecular mass determination suggested that this enzyme closely resembled dihydrodiol dehydrogenase reported by Smithgall et al (1986).

Inactivation of Dihydrodiol Dehydrogenase by Disulfides

Dihydrodiol dehydrogenase was inactivated by the incubation with disulfides or SH-reagents, as shown in Table 2. Disulfides such as GSSG, L-cystamine and DTNB, when androsterone was used as a substrate, were highly effective inactivators. Similar results were obtained when benzenedihydrodiol was used as a substrate (data not shown). The enzyme inactivated by 5 mM GSSG was reactivated by the addition of 10 mM GSH or DTT after removal of GSSG from the enzyme solution on an NAP-column. During the disulfide treatment of 15 min duration, the enzyme lost about 50% of its initial activity, and this inactivated enzyme restored its full activity by the addition of 10 mM GSH or DTT. These results suggested that dihydrodiol dehydrogenase contained cysteine residue(s) essential for the activity and also that oxidation of the cysteine residue(s) by disulfides caused inactivation of the enzyme. However, aldehyde reductase and DT-diaphorase purified from rat liver cytosol were not inhibited by SH-reagents. Table 2 suggests that these aldehyde reductase and DT-diaphorase might have no cysteine residue(s) essential for the activities.

In addition, as described previously (Nishinaka et al., 1990), two bovine liver dihydrodiol dehydrogenases (DD1 and DD3) were inhibited by SH-reagents (monoiodoacetic acid, DTNB, GSSG, cystamine and cystine). The other bovine liver dihydrodiol dehydrogenase (DD2), classified into aldehyde reductase, was insensitive to SH-reagents like the rat liver aldehyde reductase.

Table 2. Inactivation of dihydrodiol dehydrogenase, aldehyde reductase and DT-diaphorase by disulfides and SH-reagents. Purified dihydrodiol dehydrogenase (DD), aldehyde reductase (ALR) and DT-diaphorase were treated with a disulfide or an SH-reagent for 15 min at 25 °C, and then the remaining activity was measured.

Compound	Concentration (mM)	Remaining activity (%)		
		DD	ALR	DT-Diaphorase
None	—	100.0	100.0	100.0
GSSG	5.0	46.8	100.0	100.0
L-Cystamine	5.0	49.9	—	—
S-Sulfocysteine	5.0	56.9	—	—
Monoiodoacetic acid	5.0	41.7	100.0	90.0
N-Ethylmaleimide	0.5	26.6	—	—
DTNB	0.1	8.0	—	—

—, Not tested.

Effects of Coenzymes and Substrates on Inactivation

When dihydrodiol dehydrogenase was incubated with GSSG for inactivation, coenzyme or substrate was added. Coenzyme (NADP$^+$) completely protected the enzyme against the inactivation at the concentration of 0.1 mM. However, substrates including benzenedihydrodiol and androsterone, were not effective at all at the concentrations of 1.0 and 0.1 mM. From the specificity of the enzyme for coenzyme (Penning et al., 1984) and the specific protective effect of coenzyme on the inactivation by disulfides, we deduced that cysteine residue(s) of the enzyme might be located nearby the coenzyme binding site of the enzyme and at some distance from the substrate binding site.

Change of Dihydrodiol Dehydrogenase Activity in Redox Buffer System

The above results show that dihydrodiol dehydrogenase is inactivated by GSSG and reactivated by GSH. In order to explore the possibility of alteration of the enzyme activity under physiological condition, the enzyme was added to the redox buffer system carrying various [GSH]/[GSSG] ratios, and incubated for 24 h under anaerobic conditions. When the GSSG concentration was determined by the spectrophotometric method using glutathione reductase (EC 1.6.4.2) after the incubation, the [GSH]/[GSSG] ratio did not alter throughout the incubation time. The activity of dihydrodiol dehydrogenase was reduced according to decrease [GSH]/[GSSG] ratio in the range from 0 to 15, as shown in Table 3.

Table 3. The alteration of dihydrodiol dehydrogenase activity in redox buffer. The purified enzyme was incubated with a total amount of 10 mM glutathione (GSH + GSSG) under anaerobic conditions. After 24 h, the remaining activity was measured.

[GSH]/[GSSG] ratio	0	4	15	100
Remaining activity (%)	26	44	66	100

DISCUSSION

We purified the dihydrodiol dehydrogenase from rat liver cytosol. In contrast to the bovine liver enzyme (Nishinaka et al, 1990), only a single enzyme was detected. Although multiplicity of the rat liver enzyme had been reported by Vogel et al (1980) and Ikeda et al (1984), the data generated in any steps of the purification in this study, as well as the results of Smithgall et al (1986) provided no evidence for supporting it. We proved that rat liver enzyme had higher activity for the reduction of quinones (camphorquinone and phenanthrenequinone) and aromatic aldehyde (p-nitrobenzaldehyde) than the dehydrogenation for androsterone and benzenedihydrodiol. This result suggested that this dihydrodiol dehydrogenase may play an important role on the metabolism of carbonyl compounds as reported by Wermuth (1986) and the enzyme may belong to aldo-keto reductase.

Thiol/disulfide exchange reaction has been found to be one of remarkable regulation systems for various metabolic enzymes, e.g. phosphofructokinase (Oshida et al., 1989), pyruvate kinase (Sugiyama, 1988; Oshida et al., 1989), glutathione S-transferase (Oshida et al., 1989), as listed in Table 4.

This paper has demonstrated that the rat liver cytosolic dihydrodiol dehydrogenase which may play an important role on the carbonyl metabolism can be also modulated by the thiol/disulfide exchange reaction.

It is well known that in the living cell there are many thiols including low-molecular-weight compounds and proteins. For example, the intracellular concentrations of GSH, cysteine and cysteamine were reported to be 7-8, 0.25 and 0.1-0.2 mM, respectively (Taniguchi, 1986). In contrast, the concentrations of disulfides were much lower; the concentrations of GSSG and proteins carrying mixed disulfides were about 0.16 and 1.7 mM, respectively (Sugiyama, 1988). Furthermore, both L-cystine and L-cystamine are insignificant in the participation in the exchange reaction because of the trace amounts of these disulfides in the cell. In the formation of protein mixed disulfides, the thiol/disulfide exchange reaction with GSH or GSSG must be the most important process. Cappel & Gilbert reported (1978) a "third messenger" theory from the view

Table 4. Enzyme activity that may be regulated by thiol/disulfide exchange reaction.

Enzyme	Disulfide	Regulation	References
Phosphofructokinase	GSSG, CoASSG, CoASSCoA	Inactivation	Oshida et al., 1989
Fructose 1,6-bisphosphatase	CoASSG, cystamine, cystine	Activation	Oshida et al., 1989
Pyruvate kinase	GSSG	Inactivation	Oshida et al., 1989
Fatty acid synthetase	GSSG, CoASSG, CoASSCoA	Inactivation	Walters & Gilbert, 1986
HMG-CoA reductase	GSSG, CoASSCoA	Inactivation	Ness et al., 1985
γ-Glutamyltransferase	Cystamine	Inactivation	Sugiyama, 1988
Glutathione S-transferase π	Cystine, cystamine	Inactivation	Oshida et al., 1989

point that the alteration of [GSH]/{GSSG} ratio would be induced through the alteration of the c-AMP level. In addition to c-AMP, various factors causing the imbalance of [GSH]/[GSSG] ratio, namely oxidative stress, have been reported: radical formation, lipid peroxidation and GSH conjugation (Sugiyama, 1988). Carbonyl compounds are also an inducer of [GSH]/[GSSG] imbalance (Watabe & Hiratsuka, 1988), because radicals produced by the autooxidation of carbonyl compounds consume GSH in the living cell, resulting in the depletion of GSH. In this oxidative stress, thiol-transferase has been suggested to play an important role in the restoration of protein thiols in the presence of low concentrations of GSH (Hatakeyama et al., 1984, 1985; Mizoguchi et al., 1988; Oshida et al., 1989). This restoration of protein thiols may lead to maintaining activities of various enzymes in vivo.

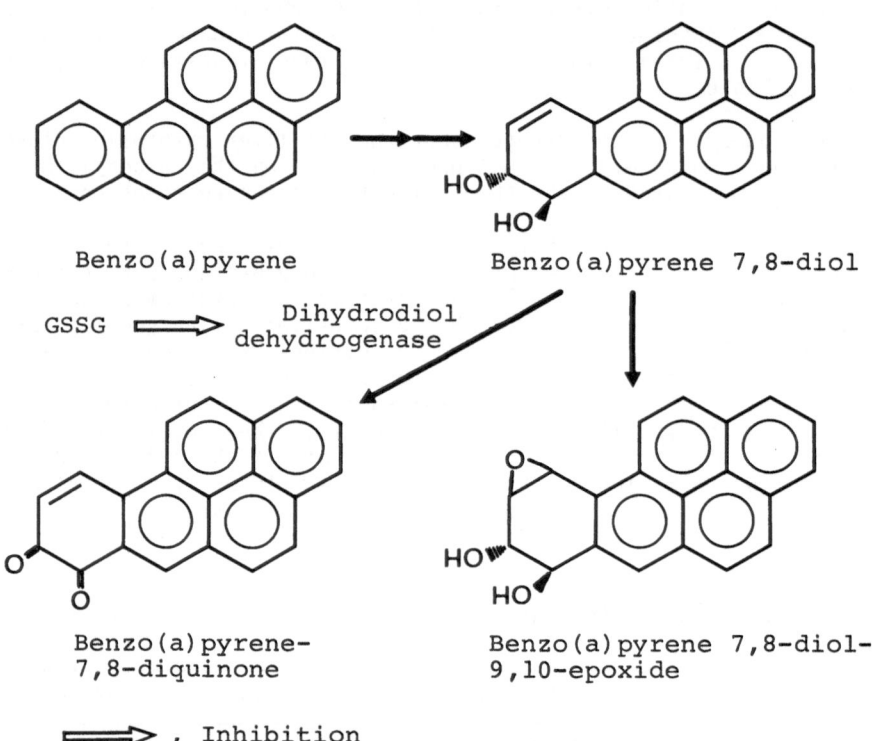

Scheme 1. Possible pathway for benzo(a)pyrene detoxification and participation of glutathione in the pathway. This pathway was originally proposed by Penning et al (1985).

Aldo-keto reductase is a general name for the enzymes which catalyze the NADPH-dependent reduction of various carbonyl compounds (Turner & Flynn, 1982) and it is widely distributed in various animals and tissues. Penning et al (1985) suggested the significant function of dihydrodiol dehydrogenase in the formation of 7,8-dicarbonyl benzo[a]pyrene from the 7,8-dihydrodiol form and they proposed the role of dihydrodiol dehydrogenase in benzo(a)pyrene detoxification as shown in Scheme 1.

We found in the bovine liver cytosol three dihydrodiol dehydrogenases (DD1, DD2 and DD3) which may belong to aldo-keto reductase and purified them (Nishinaka et al., 1990). These dihydrodiol dehydrogenases were assigned as 3α-hydroxysteroid dehydrogenase, high-Km aldehyde reductase and the unclassified enzyme from the results of substrate specificity, inhibitor sensitivity and relative molecular mass analysis. DD1 and DD3 activities were also highly sensitive to SH-reagents like the rat liver dihydrodiol dehydrogenase. Additionally, the purified chicken liver cytosol dihydrodiol dehydrogenases were characterized in that these enzymes also had cysteine residue(s) essential for the activity (unpublished data). The results of the present study and the reports by Keller and Jefacoate (1984) allowed to speculate that under the oxidative stress conditions, inhibition of dihydrodiol dehydrogenase might facilitate the accumulation of carcinogenic 7,8-dihydrodiol-9,10-epoxide form, an ultimate carcinogen. In fact, it has been reported that the addition of rat liver dihydrodiol dehydrogenase to the preincubation mixture reduces the mutagenic activity of benzo[a]pyrene in the Ames test (Glatt et al., 1979).

On the other hand, many researchers reported that [GSH]/[GSSG] ratio ranged between 20 and 300 under normal physiological conditions (Sugiyama, 1988), where no alteration in the enzyme activity might occur. However, abnormal physiological conditions such as oxidative stress would make the ratio smaller even in the living cell (Vina et al., 1978), where the repression of the dehydrogenase activity could occur to make the carcinogen more toxic.

In addition, this rat liver dihydrodiol dehydrogenase was inhibited competitively by synthetic steroid hormones such as megesterol acetate (Ki=0.1 nM), ethynylestradiol (Ki= 21 nM) and 17α-methyltestosterone (Ki= 800 nM) (data not shown). The fact of this inhibition suggests that the importance of the rat liver enzyme in the steroid metabolism is as much as that in the metabolism of carcinogenic dihydrodiols.

In this paper, it is postulated that the accumulation of harmful carbonyl compounds due to the inhibition of dihydrodiol dehydrogenase activity can occur by various factors including oxidative stress and the administration of some drugs, leading to the higher toxicity to the living cells.

ACKNOWLEDGMENT

This work was supported in part by a Grant-in-Aid for Encouragement of Young Scientists (01771977) from the Ministry of Education, Science and Culture of Japan.

REFERENCES

Cappel,R.E. and Gilbert,H.F.,1986, Cooperative behavior in the thiol oxida-
    tion of rabbit muscle glycogen phosphorylase in cysteamine-cystamine
    redox buffers, J.Biol.Chem., 261:15378
Everson,R.B.,Randerath,E.,Santella,R.M.,Cefalo,R.C.,Avitts,T.A. and
    Randerath,K.,1986, Detection of smoking-related covalent DNA adducts
    in human placenta, Science, 231:54
Flynn,T.G.,1982, Aldehyde reductase : monomeric NADPH-dependent oxidoreduc-
    tases with multifunctional potential, Biochem.Pharmacol., 31:2705
Turner,A.J. and Flynn,T.G.,1982, The nomenclature of aldehyde reductases,
    in : "Enzymology of Carbonyl Metabolism: Aldehyde dehydrogenase, aldo-
    keto reductase, alcohol dehydrogenase,", H.Weiner and B.Wermuth, eds.,
    Alan R. Liss, New York
Glatt,H.R.,Vogel,K.,Bentley,P. and Oesch,F.,1979,Reduction of benzo(a)-

pyrene mutagenicity by dihydrodiol dehydrogenase, Nature, 277:319

Hara,A.,Inoue,Y.,Nakagawa,M.,Naganeo,F. and Sawada,H.,1988, Purification and characterization of NADP-dependent 3α-hydroxysteroid dehydrogenase from mouse liver cytosol, J.Biochem., 103:1027

Hatakeyama,M.,Tanimoto,Y. and Mizoguchi,T., 1984, Purification and some properties of bovine liver cytosol thioltransferase, J.Biochem., 95:1811

Hatakeyama,M.,Lee,C.,Chon,C.,Hayashi,M.and Mizoguchi,T.,1985, Purification and some properties of rabbit liver cytosol thioltransferase, J.Biochem., 97:893

Ikeda,M.,Hattori,H. and Ohmori,S.,1984, Properties of NADPH-dependent carbonyl reductases in rat liver cytosol, Biochem.Pharmacol., 33:3957

Keller,G.M. and Jefcoate,C.R.,1984, Kinetic determinants of benzo(a)pyrene metabolism to dihydrodiol epoxides by 3-methylcholanthrene induced rat liver microsomes, Mol.Pharmacol., 22:451

Kosower,N.S.,1978,The glutathione status of cells, Int.Rev.Cytol.,54:109

Mizoguchi,T.,Uchida,G.,Oshida,T.,Terada,T. and Hosomi,S.,1988, Changes in mice liver thioltransferase activity and powerful inhibitors, in : "Glutathione Centennial" 114

Morrison,H.,Hammorskiold,V. and Jernstrom,B.,1983, Status of reduced glutathione in primary cultures of rat hepatocytes and the effect on conjugation of benzo(a)pyrene 7,8-dihydrodiol 9,10-epoxide, Chem-Biol.Interactions, 45:235

Ness,G.C.,McCreery,M.J.,Sample,C.E.,Smith,M. and Pendleton,L.C.,1985, Sulfhydryl-disulfide forms of rat liver 3-hydroxy 3-methylglutaryl coenzyme A reductase, J.Biol.Chem., 260:16395

Nishinaka,T.,Terada,T.,Umemura,T.,Nanjo,H.,Mizoguchi,T.and Nishihara,T., Study on dihydrodiol dehydrogenase (I) Molecular forms of the enzyme and the presence of a dihydrodiol specific enzyme in bovine liver cytosol, in: "Enzymology of Carbonyl Metabolism : Aldehyde dehydrogenase, aldo-keto reductase, alcohol dehydrogenase",in press

Oshida,T.,Maeda,H.,Hara,T.,Terada,T.,Hosomi,S.,Mizoguchi,T. and Nishihara, T.,1989, Human thioltransferase : Its purification and physiological role, J.Pharmacobio-Dyn., 12:s-52

Penning,T.M.,Mulcharji,I.,Barrows,S. and Talaley,P.,1984, Purification and properties of a 3α-hydroxysteroid dehydrogenase EC 1.1.1.50 of rat liver cytosol and its inhibition by anti-inflammatory drugs, Biochem.J., 222:601

Penning,T.M.,Sharp,R.B. and Krieger,N.R.,1985, Purification and properties of 3α-hydroxysteroid dehydrogenase from rat brain cytosol inhibition by nonsteroidal anti-inflammatory drugs and progestins, J.Biol.Chem., 260:15266

Platt,K.L. and Oesch,F.,1977, An improved synthesis of trans-5,6-dihydroxy-1,3-cyclohexadiene, Synthesis, 7:449

Smithgall,T.E.,Harvey,R.G. and Penning,T.M.,1986, Regio- and stereospecificity of homogeneous 3α-hydroxysteroid-dihydrodiol dehydrogenase for trans-dihydrodiol metabolites of polycyclic aromatic hydrocarbons, J.Biol.Chem., 261:6184

Sugiyama,T.,1988, Modification of enzyme activity by glutathione, Protein Nucleic Acid and Enzyme (in Japanese), 33:1423

Taniguchi,N., 1986, Modification of enzyme protein by glutathione, Metabolism and Disease (in Japanese), 23:1057

Vogel.K.,Bentley,P.,Platt,K.L. and Oesch,F.,1980, Rat liver cytoplasmic dihydrodiol dehydrogenase. Purification and apparent homogeneity and properties, J.Biol.Chem., 255:9621

Vina,J.,Hems,R. and Krebs,H.A.,1978, Maintenance of glutathione content in isolated hepatocytes, Biochem.J., 170:627

Wallin,R.,1979, Some molecular properties of NAD(P)H dehydrogenase from rat liver, Biochem.J., 181:127

Walters,D.W. and Gilbert,H.F.,1986, Thiol-disulfide redox equilibrium and kinetic behavior of chicken liver fatty-acid synthase, J.Biol.Chem. 261:13135

Watabe,T. and Hiratsuka,A.,1988, Glutathione cojugation reactions:
    Metabolic inactivation and activation of carcinogens and other
    toxicants by glutathione s-transferase, Protein Nucleic Acid and
    Enzyme (in Japanese), 33:1405
Wermuth,B.,Platt,K.L.,Seidel,A and Oesch,F.,1986, Carbonyl reductase
    provides the enzymatic basis of quinone detoxication in man,
    Biochem.Pharmcol., 35:1277
Ziegler,D.M.,1985, Role of reversible oxidation-reduction of enzyme
    thiol-disulfides in metabolic regulation, Ann.Rev.Biochem., 54:305

Wardak, T. and Niedzwiecki, A. (1983): Distribution behaviour, transitions,
metabolic characterization and desorption of catecholamines and other
catecholamine stimulating activity in skin. (see in Stein, H. et al. and
Thomas, H.W. Wiesbaden), 33, 1984.

Weinshilboum, R. and Raymond, F. and Desch, L., 1979. Catecholamine-synthesizing
enzymes in aromatic tests of purine metabolism in man.
(see in Acta Pharmacol), 5, 1983.

Zimmer, J.A., 1981. Role of physiological behaviour reactions of gross
stress determinants in adrenergic mechanisms. Am. Rev. Biochem., 34:101

DISTRIBUTION OF DIMERIC DIHYDRODIOL DEHYDROGENASE

IN PIG TISSUES AND ITS ROLE IN CARBONYL METABOLISM

Toshihiro Nakayama*, Hideo Sawada*, Yoshihiro Deyashiki*,
Takushi Kanazu*, Akira Hara*, Michio Shinoda**,
Kazuya Matsuura†, Yasuo Bunai† and Isao Ohya†

Department of Biochemistry, Gifu Pharmaceutical University*
Gifu 502, Japan; Gihoku General Hospital**, Gifu 501-21;
and Department of Legal Medicine, Gifu University School of
Medicine†, Gifu 500, Japan

INTRODUCTION

A cytosolic $NADP^+$-dependent dihydrodiol dehydrogenase (EC 1.3.1.20),
that oxidizes dihydrodiol derivatives of benzene and naphthalene to the
corresponding catechols, has been thought to play an important role in
metabolic detoxification of carcinogenic polycyclic aromatic hydrocarbons
(Oesch, et al., 1984) and in bioactivation of naphthalene in rabbit eye
(van Heyningen, 1976). Dihydrodiol dehydrogenase was first isolated from
rat liver (Vogel, et al., 1980) and has been subsequently identified as
3α-hydroxysteroid dehydrogenase (Penning, et al., 1984). The enzyme is a
monomer of Mr35,000 and shows dehydrogenase activity for xenobiotic alicyclic alcohols and carbonyl reductase activity, which indicate that it also
functions in carbonyl metabolism. Similar monomeric dihydrodiol dehydrogenases with broad substrate specificity for xenobiotics have been purified
from other mammalian livers, and have been reported to be identical with
17β-hydroxysteroid dehydrogenase in the guinea pig (Hara, et al., 1986a),
mouse (Sawada, et al., 1988) and rabbit (Hara, et al., 1986b), 3α(17β)-
hydroxysteroid dehydrogenases in the hamster (Ohmura, et al., 1990), 3(20)α-
hydroxysteroid dehydrogenase in the monkey (Hara, et al., 1989a), and
aldehyde reductase.

In addition to the striking species differences in nature of hepatic
dihydrodiol dehydrogenases, the enzymes, which have different substrate
specificity from the hepatic enzymes, have been described in guinea pig
testis (Matsuura, et al., 1987), mouse kidney (Nakagawa, et al., 1989a) and
monkey kidney (Hara, et al., 1987). The monkey kidney enzyme, in particular, differs from the monomeric enzymes obtained from rodent tissues and
monkey liver in its dimeric structure, strict specificity for trans-dihydrodiols, and high reductase activity towards α-diketones and some aldehydes.
Similar dimeric dihydrodiol dehydrogenase has been recently investigated in
pig lens (Hara, et al., 1989b). Since in monkey tissues the dimeric dihydrodiol dehydrogenase has been shown to distribute only in kidney (Nakagawa,
et al., 1989b), we examined the occurrence of dimeric dihydrodiol dehydrogenase in pig tissues and its substrate specificity for carbonyl compounds.
We also studied immunochemical properties using the antibodies against

dimeric dihydrodiol dehydrogenase, aldehyde reductase and aldose reductase in order to elucidate its physiological function and relationship to monkey kidney dihydrodiol dehydrogenase and monomeric aldo/keto reductases.

EXPERIMENTAL

trans-1,2-Dihydrobenzene-1,2-diol (benzene dihydrodiol) and 1,2-dihydro-1,2-dihydroxynaphthalene (naphthalene dihydrodiol) were synthesized as described by Platt and Oesch (1977; 1983). Lactaldehydes were prepared from D- or L-threonine and ninhydrin by the method of Zagalak et al. (1966).

Pig tissues were homogenized in 3 volumes of 20 mM potassium phosphate, pH 7.5, containing 5 mM EDTA, 5 mM 2-mercaptoethanol and 0.15 M KCl (Buffer A), and the homogenate was centrifuged at 105,000 xg for 1 h. Proteins in the cytosol were fractionated by the addition of ammonium sulfate. The 35-70% precipitate was dialyzed against Buffer A and then subjected to Sephadex G-100 gel filtration in Buffer A. The dihydrodiol dehydrogenase fractions were further purified by using a procedure for the purification of the enzyme from pig lens (Hara, et al., 1989b) except that Q-Sepharose was employed instead of DEAE-Sephacel. Pig muscle aldose reductase, pig kidney aldehyde reductase, and monkey kidney dihydrodiol dehydrogenase were purified by the methods of Cromlish and Flynn (1983), Flynn et al. (1975) and Hara et al. (1987), respectively.

Antibodies against the purified dihydrodiol dehydrogenases, aldehyde reductase and aldose reductase were raised in female rabbits. IgG fraction was prepared by ammonium sulfate fractionation from the antiserum. Immunodiffusion, immunoprecipitation and inhibition by the antibodies against the enzymes were performed as described by Sawada et al.(1984).

Dehydrogenase and reductase activities were assayed spectrophotometrically or fluorometrically by measuring the production and oxidation of NADPH (Matsuura, et al., 1987). The standard assay mixture for dihydrodiol dehydrogenase contained 100 mM glycine-NaOH, pH 10.0, 0.25 mM NADP$^+$, 1.8 mM naphthalene dihydrodiol and enzyme in a total volume of 2.0 ml. One unit of activity was defined as the oxidation or production of 1 µmol of NADPH/min at 25°C.

Protein concentration was determined by the method of Bradford (1976) with bovine serum albumin as a standard. The concentration of methylglyoxal was determined with 2,4-dinitrophenylhydrazine reagent (Cooper, 1975). The relative molecular weight (Mr) of the enzyme was estimated as previously described (Hara, et al., 1989b).

RESULTS

Tissue Distribution of Dimeric Dihydrodiol Dehydrogenase

The tissue content of cytosolic dihydrodiol dehydrogenase activity for dihydrodiols of naphthalene and benzene is shown in Table 1. The enzyme activity was highest in liver followed by lens and kidney. Lung had the lowest amounts of the activity, but was more active towards benzene dihydrodiol than towards naphthalene dihydrodiol in contrast to the other tissues. Enzyme activity could not be detected in blood.

Gel chromatography profiles of dihydrodiol dehydrogenase activity in the ammonium sulfate fraction showed that two enzymes (Mr65,000 and Mr35,000) exist in the pig tissues with the exception that lung showed an

Table 1.   Tissue distribution of dihydrodiol dehydrogenase

| Tissue | Activity (units/g of tissue ± SE) for | |
	Naphthalene dihydrodiol	Benzene dihydrodiol
Liver	1.54 ± 0.33	0.71 ± 0.19
Kidney	0.38 ± 0.06	0.12 ± 0.02
Lens	0.35 ± 0.07	0.16 ± 0.04
Spleen	0.17 ± 0.02	0.092 ± 0.005
Brain	0.12 ± 0.02	0.059 ± 0.022
Heart	0.079 ± 0.006	0.043 ± 0.007
Muscle	0.073 ± 0.007	0.045 ± 0.005
Lung	0.007 ± 0.001	0.023 ± 0.004

activity peak of Mr80,000 (Fig. 1). Although the high molecular weight
enzyme activity was predominant in the tissues, high amounts of the low
molecular weight enzyme activity were found in the liver, kidney and spleen.

As evidenced by further purification, the high molecular weight enzymes
from liver, kidney, spleen, muscle, brain and heart showed almost identical
elution profiles in the subsequent steps (Table 2). Since the enzyme activity in lung was very low, further purification was difficult. The specific
activities of the purified enzymes obtained from the tissues, except heart,
are almost equal to that of the lens dimeric enzyme (Hara, et al., 1989b).
These enzymes showed single protein bands with the same molecular weight of
39,000 as that of the lens dimeric enzyme on SDS-polyacrylamide gel electrophoresis and the preparation of heart enzyme contained several proteins
in addition to a main protein band of Mr39,000. The results indicate that
dimeric dihydrodiol dehydrogenase distributes in all the tissues except
lung.

The low molecular weight enzyme activity in kidney was co-purified to
homogeneity with D-glucuronate reductase activity, which indicates that this
enzyme is aldehyde reductase. However, the low molecular weight activity of
liver was separated into three peaks on Q-Sepharose chromatography, one of
which was co-eluted with the aldehyde reductase activity. The other low
molecular weight enzymes of liver were not further studied.

Immunological Properties

We further examined the immunological identity of the dimeric dihydrodiol dehydrogenases from the tissues including lens using the antibody
against the liver dimeric enzyme. In the immunodiffusion test, the antibody
yielded a fused immunoprecipitin line against the enzymes from liver,
kidney, spleen, muscle, brain, heart and lens, but did not react with the
high molecular weight enzyme of lung. The antibody inhibited and
precipitated the dihydrodiol dehydrogenase activity of all the dimeric
enzymes from the different tissues (Table 3). In contrast, the activity of
the lung enzyme was not affected even in the immunoprecipitation test.

Since lens dimeric dihydrodiol dehydrogenase has been shown to resemble
the monkey kidney enzyme with respect to subunit structure and catalytic
properties (Hara, et al, 1989b), the immunological relationship between the
dimeric enzymes from pig liver and monkey kidney was examined. The pig

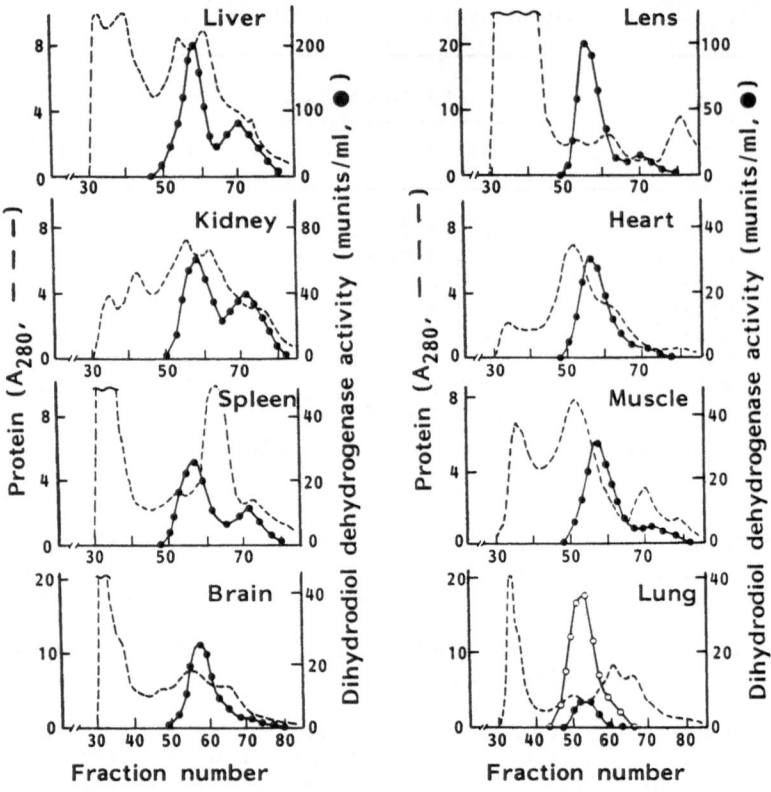

Fig. 1. Elution profiles of dihydrodiol dehydrogenase activity
of pig tissues on a Sephadex G-100 column. The activity
was assayed with 1.8 mM naphthalene dihydrodiol (●—●)
or benzene dihydrodiol (O—O) as a substrate.

Table 2. Purification of dihydrodiol dehydrogenases from pig tissues

Step	Activity, units/(specific activity, units/mg)					
	Liver	Kidney	Spleen	Brain	Muscle	Heart
$(NH_4)_2SO_4$ fraction	322	36.0	13.4	3.8	16.5	5.8
	(0.04)	(0.01)	(0.005)	(0.01)	(0.004)	(0.005)
Sephadex G-100	179	16.0	7.4	3.1	12.6	3.7
	(0.09)	(0.02)	(0.008)	(0.04)	(0.006)	(0.008)
Q-Sepharose	139	8.8	6.0	1.1	11.8	3.4
	(2.2)	(1.6)	(0.94)	(0.92)	(0.09)	(0.17)
Matrex Red A	84	5.8	3.8	0.95	8.1	1.8
	(10.4)	(9.3)	(7.4)	(7.3)	(2.5)	(2.6)
HA-Ultrogel	82	3.8	3.0		7.9	
	(10.8)	(10.2)	(9.1)		(6.7)	

enzyme cross-reacted with the monkey enzyme in the immunodiffusion test using the antibodies against the respective enzymes, but the precipitin lines gave spur formation (Fig. 2). Moreover, both of the antibodies against the pig and monkey enzymes inhibited the dehydrogenase activity of the two enzymes (Fig. 3). Thus the dimeric enzymes of pig tissues and monkey kidney have some common antigenic sites.

Table 3. Inhibition and immunoprecipitation of dihydrodiol dehydrogenase activity of the purified enzymes from pig tissues by the antibody against liver dihydrodiol dehydrogenase

Enzyme	Inhibition (%) by the IgG		Activity (%) in the supernatant after incubation with 150 μg IgG
	8 μg	150 μg	
Liver DD	61	82	0
Kidney DD	61	85	0
Lens DD	65	87	0
Spleen DD	71	87	0
Brain DD	65	85	0
Heart DD	72	97	0
Muscle DD	66	86	0
Lung DD	0	0	100
Aldehyde reductase	0	0	100
Aldose reductase	0	0	100

The dihydrodiol dehydrogenase (DD) or reductase was incubated at 25°C for 2 h with the IgG. The activity in the mixture or supernatant after centrifugation was determined, and is given as percentage relative to those obtained with control IgG. The activity of the lung enzyme was assayed with benzene dihydrodiol as a substrate.

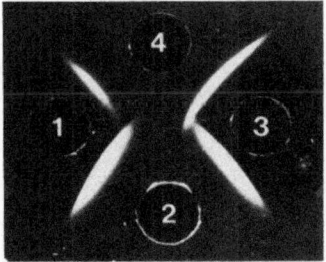

Fig. 2. Immunodiffusion of dimeric dehydrogenases from pig liver and monkey kidney using the antibodies against the respective enzymes. Well: (1) anti-pig enzyme IgG; (2) pig enzyme; (3) anti-monkey enzyme; and (4) monkey enzyme.

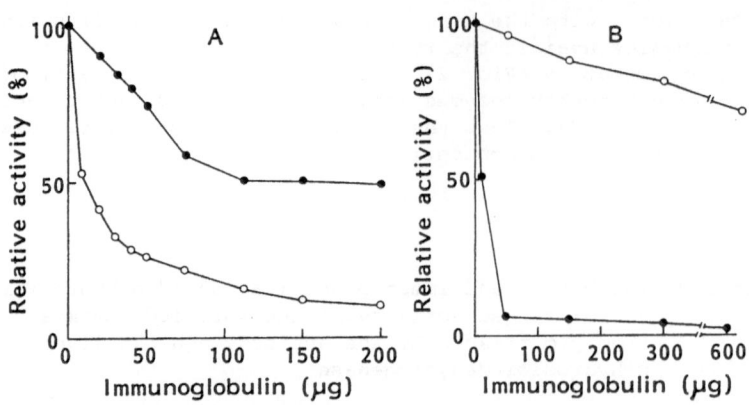

Fig. 3. Inhibition of dihydrodiol dehydrogenase activity by the antibodies against the dimeric enzymes from pig liver (A) and monkey kidney (B). The enzymes from pig liver (o) and monkey kidney (●) were incubated at 25°C for 2 h with one of the antibodies, and then the residual activity, which represents the percentage of those incubated with control IgG, was assayed.

Pig dimeric dihydrodiol dehydrogenase reduced some aldehydes including glyceraldehyde (Table 4), and homogeneous aldehyde reductase and aldose reductase from pig kidney and muscle exhibited dihydrodiol dehydrogenase activities of 1.28 and 0.28 units/mg, respectively. However, no precipitin line was observed between the dimeric enzyme and the antibodies against aldehyde and aldose reductases in the immunodifusion test, and dihydrodiol dehydrogenase activity of aldehyde and aldose reductases was neither inhibited nor immunoprecipitated by the antibody against the dimeric enzyme (Table 3). These results revealed that dimeric dihydrodiol dehydrogenase is a distinct protein from aldehyde and aldose reductases.

Catalytic Properties

The maximal rates of naphthalene dihydrodiol oxidation by the purified dimeric enzymes were observed at pH 10.0, but that by lung enzyme was at pH 10.5. The pH optima of dihydrodiol dehydrogenase activities of aldehyde and aldose reductases were broad at 8.0-9.5 and 7.5-8.0, respectively.

The dimeric enzymes from all the tissues except lung specifically oxidized trans-dihydrodiol derivatives of naphthalene and benzene, exhibiting similar apparent Km values of 0.1-0.2 mM for naphthalene dihydrodiol and 0.3-0.6 mM for benzene dihydrodiol at pH 10.0. The enzymes were inactive towards cis-benzene dihydrodiol, prostaglandin $F_{2\alpha}$, alicyclic alcohols such as 1-acenaphthenol and 1-indanol, and steroids with 3α-, 3β-, 17α-, 17β-, 20α-, or 21-hydroxy group.

In the reverse reaction using NADPH as a cofactor, the dimeric enzymes from pig tissues reduced α-diketones, aromatic aldehydes, α-ketoaldehydes, dihydroxyacetones and glyceraldehyde, and showed a broad pH optimum of 6.0-8.0. The enzyme was inactive towards glyceraldehyde-3-phosphate, indole-3-acetaldehyde, succinic semialdehyde, D-glucose, acetol, acetone and acetaldehyde. The representative kinetic constants of the liver enzyme for carbonyl compounds are shown in Table 4. Although camphorquinone was the best model substrate among artificial substrates, the enzyme exhibited the highest Vmax/Km value for methylglyoxal of the biogenic carbonyl substrates, and the Km value was much lower than that of aldehyde reductase (6.0 mM).

To identify the reaction products of the methylglyoxal reduction by the dimeric enzyme, the reaction mixture was treated for 2 h with 0.1% 2,4-dinitrophenylhydrazine and 0.5 M HCl and then the hydrazones of the reaction products were analyzed on thin-layer chromatography (Silica gel) with methylene chloride as a developing solvent. The Rf value (0.09) of the reaction product was the same as that of authentic lactaldehyde but different from that (0.13) of acetol.

The dimeric dihydrodiol dehydrogenases from the pig tissues were inhibited by SH-reagents but not by barbital, sorbinil and pyrazole, whereas the high molecular weight enzyme from lung was specifically inhibited by 1 mM pyrazole.

DISCUSSION

Tissue distribution of dihydrodiol dehydrogenase has been studied in several animals in order to understand the significance of trans-dihydrodiol oxidation in the metabolic activation and/or detoxification of xenobiotic aromatic hydrocarbons. In rat (Vogel, et al., 1980) and guinea pig (Hara, et al., 1986a), liver has much higher activity than extrahepatic tissues, whereas, in mouse (Nakagawa, et al., 1989a) and monkey (Hara, et al., 1987), kidney possesses high activity which is comparable with or superior to the hepatic activity. The species difference in tissue distribution has been thought to be due to the existence of tissue-specific dihydrodiol dehydroge-nases as reported in monkey kidney (Nakagawa, et al., 1989b), mouse kidney (Nakagawa, et al., 1989a) and guinea pig testis (Matsuura, et al., 1987).

Table 4. Substrate specificity for carbonyl compounds

Substrate	Vmax (units/mg)	Km (mM)	V/Km (units/mg/mM)
Camphorquinone	8.44	0.16	52.8
3-Nitrobenzaldehyde	6.42	0.38	16.9
Phenylglyoxal	5.53	0.59	9.4
2,3-Pentanedione	4.45	1.2	4.7
2,3-Butanedione	6.17	1.7	3.6
2,3-Hexanedione	3.21	1.5	2.1
Pyridine-4-aldehyde	3.49	1.9	1.8
Glyoxal	4.90	5.7	0.86
Methylglyoxal	2.17	0.15	18.0
D-Glyceraldehyde	2.47	0.74	3.3
Dihydroxyacetone	2.72	2.5	1.1
DL-Glyceraldehyde	2.48	4.8	0.52
D-Lactaldehyde	5.94	14	0.43
Dihydroxyacetone phosphate	1.80	14	0.13

The activity of the liver enzyme was assayed at pH 7.5.

In pig lung and liver we also confirmed the existence of different enzymes from either the dimeric dihydrodiol dehydrogenase or aldehyde reductase. However, the dimeric enzyme was predominant in all the pig tissues except lung, which is in contrast to the specific occurrence of the dimeric enzyme in monkey kidney (Nakagawa, et al., 1989b) despite the immunological similarity of the enzymes in the two species. Based on the activity distribution and gel filtration analysis, lens had the second highest amounts of the dimeric enzyme of pig tissues, which further supports the significance of the enzyme in the pathogenesis of naphthalene cataract (Hara, et al., 1989b). On the other hand, lung contained another high molecular weight enzyme which oxidized benzene dihydrodiol much higher than naphthalene dihydrodiol and was inhibited by pyrazole. It has been reported that a pyrazole sensitive carbonyl reductase of Mr86,000 in mouse and guinea pig lung oxidizes benzene dihydrodiol (Nakayama, et al., 1986).

The high reactivity and specificity for trans-dihydrodiols in the dehydrogenation is characteristic for the dimeric enzymes of pig tissues and monkey kidney in comparison with the monomeric enzymes with broad substrate specificity in other animal tissues. In any case, the dimeric enzyme belongs to a group of aldo/keto reductases, because it reduced several carbonyl compounds. Although Hara et al. (1989b) described the difference in specificity for carbonyl compounds and inhibitor sensitivity between the dimeric enzyme and four aldehyde reductases from pig brain (Cromlish, et al., 1985) or dimeric aldehyde reductase in human tissues (Srivastava, et al., 1984), the present immunochemical study clearly demonstrated that the dimeric enzyme is a distinct protein from aldehyde and aldose reductases. The pig dimeric enzyme also resembles diacetyl reductase from hamster liver (Hara, et al., 1985) in specificity for α-diketones, but diacetyl reductase does not oxidize benzene dihydrodiol.

Nakagawa et al. (1989b) have reported that the monkey kidney enzyme reduces dihydroxyacetone phosphate, but its Km value for this substrate is much higher than that of glycerol-3-phosphate dehydrogenase. We found that the dimeric enzyme exhibited the lowest Km value for methylglyoxal among the endogenous carbonyl substrates. Biosynthesis of methylglyoxal from dihydroxyacetone phosphate, glyceraldehyde 3-phosphate, aminoacetone and acetone has been demonstrated in mammals (see Ohmori, et al., 1989, for review), and this metabolite has been suggested to be cytotoxic and to act as a regulator of cell growth (Schauenstein, et al., 1977). On the other hand, methylglyoxal is metabolized to lactate and pyruvate by glyoxalase (Mannervik, 1980) and α-ketoaldehyde dehydrogenase (Ray and Ray, 1982), and is reduced by several enzymes such as NADH-dependent methylglyoxal reductase from goat liver (Ray and Ray, 1984), aldehyde reductase, aldose reductase and carbonyl reductase (Hara, et al., 1983; Nakayama, et al., 1985). Since the Km value of dimeric dihydrodiol dehydrogenase for methylglyoxal is lower than those of the other methylglyoxal reducing enzymes, the dimeric enzyme may function in control of the potential toxicity or activity of this compound as well as in drug metabolism.

REFERENCES

Bradford, M. M., 1976, A rapid and sensitive method for the quantitation of microgram quantities of protein utilizing the principle of protein-dye binding, Anal. Biochem., 72:248.
Cooper, R. A., 1975, Methylglyoxal synthase, Methods Enzymol., 41:502.
Cromlish, J. A. and Flynn, T. G., 1983, Pig muscle aldehyde reductases: identity of pig muscle aldehyde reductase with pig lens aldose reductase and with the low-Km aldehyde reductase of pig brain and pig kidney, J. Biol. Chem., 258:3583.

Cromlish, J. A., Yoshimoto, C. K., and Flynn, T. G., 1985, Purification and characterization of four NADPH-dependent aldehyde reductases from pig brain, J. Neurochem., 44:1477.

Flynn, T. G., Shires, J., and Walton, D. J., 1975, Properties of the nicotin amide adenine dinucleotide phosphate-dependent aldehyde reductase from pig kidney, J. Biol. Chem., 250:2933.

Hara, A., Deyashiki, Y., Nakayama, T., and Sawada, H., 1983, Isolation and characterization of multiple forms of aldehyde reductase in chicken kidney, Eur. J. Biochem., 133:207.

Hara, A., Seiriki, K., Nakayama, T., and Sawada, H., 1985, Discrimination of multiforms of diacetyl reductase in hamster liver, in "Enzymology of Carbonyl Metabolism 2: Aldehyde dehydrogenase, aldo-keto reductase, and alcohol dehydrogenase," T. G. Flynn and H. Weiner, eds., Alan R. Liss, New York, p. 291.

Hara, A., Hasebe, K., Hayashibara, M., Matsuura, K., Nakayama, T., and Sawada, H., 1986a, Dihydrodiol dehydrogenases in guinea pig liver, Biochem. Pharmacol., 35:4005.

Hara, A., Kariya, K., Nakamura, T., Nakayama, T., and Sawada, H., 1986b, Isolation of multiple forms of indanol dehydrogenase associated with 17β-hydroxysteroid dehydrogenase activity from male rabbit liver, Arch. Biochem. Biophys., 249:225.

Hara, A., Mouri, K., and Sawada, H., 1987, Purification and partial characterization of dimeric dihydrodiol dehydrogenase from monkey kidney, Biochem. Biophys. Res. Commun., 145:1260.

Hara, A., Nakagawa, M., Taniguchi, H., and Sawada, H., 1989a, 3(20)α-hydroxy steroid dehydrogenase activity of monkey liver indanol dehydrogenase, J. Biochem., 106:900.

Hara, A., Harada, T., Nakagawa, M., Matsuura, K., Nakayama, T., and Sawada, H., 1989b, Isolation from pig lens of two proteins with dihydrodiol dehydrogenase and aldehyde reductase activities, Biochem. J., 264:403.

Mannervik, B., 1980, Glyoxalase I, in "Enzymatic basis of detoxification," Vol II, W. B. Jakoby, ed., Academic Press, New York, p. 263.

Matsuura, K., Hara, A., Nakayama, T., Nakagawa, M., and Sawada, H., 1987, Purification and properties of two multiple forms of dihydrodiol dehydrogenase from guinea-pig testis, Biochim. Biophys. Acta, 912:270.

Nakagawa, M., Tsukada, F., Nakayama, T., Matsuura, K., Hara, A., and Sawada, H., 1989a, Identification of two dihydrodiol dehydrogenases associated with 3(17)α-hydroxysteroid dehydrogenase activity in mouse kidney, J. Biochem., 106:633.

Nakagawa, M., Matsuura, K., Hara, A., Sawada, H., Bunai, Y., and Ohya, I., 1989b, Dimeric dihydrodiol dehydrogenase in monkey kidney. Substrate specificity, stereospecificity of hydrogen transfer, and distribution, J. Biochem., 106:1104.

Nakayama, T., Hara, A., Yashiro, K., and Sawada, H., 1985, Reductases for carbonyl compounds in human liver, Biochem. Pharmacol., 34:107.

Nakayama, T., Yashiro, K., Inoue, Y., Matsuura, K., Ichikawa, H., Hara, A., and Sawada, H., 1986, Characterization of pulmonary carbonyl reductases of mouse and guinea pig, Biochim. Biophys. Acta, 882:220.

Oesch, F., Glatt, H. R., Vogel, K., Seidel, A., Petrovic, P., and Platt, K. L., 1984, Dihydrodiol dehydrogenase: A new level of control by both sequestration of proximate and inactivation of ultimate carcinogens, in "Biochemical Basis of Chemical Carcinogenesis," H. Grem, R. Jung, M. Kramer, H. Marquardt and F. Oesch, eds., Raven Press, New York, p. 23.

Ohmura, M., Hara, A., Nakagawa, M., and Sawada, H., 1990, Demonstration of 3α(17β)-hydroxysteroid dehydrogenase distinct from 3α-hydroxysteroid dehydrogenase in hamster liver, Biochem. J., 266:583.

Ohmori, S., Mori, M., Shiraha, K., and Michi, K., 1989, Biosynthesis and degradation of methylglyoxal in animals, in "Enzymology and Molecular Biology of Carbonyl Metabolism 2," H. Weiner and T. G. Flynn, eds., Alan R. Liss, New York, p.397.

Penning, T.M., Mukharji, I., Barrows, S., and Talalay, P., 1984, Purification and properties of a 3α-hydroxysteroid dehydrogenase of rat liver cytosol and its inhibition by anti-inflammatory drugs, Biochem. J., 222:601.

Platt, K. L. and Oesch, F., 1977, An improved synthesis of trans-5,6-dihydroxy-1,3-cyclohexadiene (trans-1,2-dihydroxy-1,2-dihydrobenzene), Synthesis, 7:449.

Platt, K. L. and Oesch, F., 1983, Efficient synthesis of non-K-region trans-dihydrodiols of polycyclic aromatic hydrocarbons from o-quinone and catechols, J. Org. Chem., 48:265.

Ray, M. and Ray, S., 1982, Purification and characterization of NAD- and NADP-linked α-ketoaldehyde dehydrogenases involved in catalyzing the oxidation of methylglyoxal to pyruvate, J. Biol. Chem., 257:10566.

Ray, M. and Ray, S., 1984, Purification and partial characterization of a methylglyoxal reductase from goat liver, Biochim. Biophys. Acta, 802:119.

Sawada, H., Hara, A., Hayashibara, M., Nakayama, T., and Usui, S., 1984, Immunological identification of soluble carbonyl reductase with testosterone 17β-dehydrogenase (NADP) in guinea pig liver and kidney, Biochim. Biophys. Acta, 799:322.

Sawada, H., Hara, A., Nakayama, T., Nakagawa, M., Inoue, Y., Hasebe, K., and Zhang, Y., 1988, Mouse liver dihydrodiol dehydrogenases. Identification of the predominant and a minor form with 17β-hydroxysteroid dehydrogenase and aldehyde reductase, Biochem. Pharmacol., 37:453.

Schauenstein, E., Esterbauer, H., and Zollner, H., 1977, "Aldehydes in Biological Systems", Pion Limited, London.

Srivastava, S., Ansari, N. H., Hair, G. A., and Das, B., 1984, Aldose and aldehyde reductases in human tissues, Biochim. Biophys. Acta, 800:220.

van Heyningen, R., 1976, Experimental studies on cataract, Invest. Ophthalmol., 15:685.

Vogel, K., Bentley, P., Platt, K. L., and Oesch, F., 1980, Rat liver cytoplasmic dihydrodiol dehydrogenase, J. Biol. Chem., 255:9621.

Zagalak, B., Frey, P. A., Karabatsos, G. L., and Abeles, R. H., 1966, The stereochemistry of the conversion of D- and L-1,2-propanediols to propionaldehyde, J. Biol. Chem., 241:3028.

INHIBITION OF ALDEHYDE REDUCTASE BY CARBOXYLIC ACIDS

Bendicht Wermuth

Chemisches Zentrallabor
Inselspital
CH 3010 Bern

INTRODUCTION

Aldehyde reductase (EC 1.1.1.2) and aldose reductase (EC 1.1.1.21) are structurally related, monomeric oxidoreductases that catalyze the NADPH-dependent reduction of a variety of aliphatic, aromatic and sugar aldehydes. For most substrates Km values of aldose reductase are one to two orders of magnitude lower than those of aldehyde reductase. Exceptions are anionic aldehydes, e.g. succinic semialdehyde, 4-carboxybenzaldehyde and glucu-ronate, which are metabolized much more efficiently by aldehyde than aldose reductase. The partiality of aldehyde reductase for negatively charged sub-strates has been attributed to the presence of an anion binding site, most probably an arginine residue, in the substrate binding domain of the active site (Davidson and Flynn, 1979; Branlant et al., 1981; Bohren et al., 1987).

Carboxylic acids of great structural diversity, e.g. valproic acid (2-propylpentanoic acid), dimethylsuccinic acid and Alrestatin (1,3-dioxo-1H-benzo[d,e]isoquinoline-2(3H)-acetic acid), potently inhibit the enzyme activity, whereby the carboxylate moiety is thought to interact with the active site-arginine residue (Whittle and Turner, 1981; Branlant, 1982; Daly and Mantle, 1982; de Jongh et al., 1987). Typically, inhibition by carbo-xylic acids relative to both NADPH and aldehyde is uncompetitive at low concentrations becoming non-competitive at higher concentrations. Following the rules of classical kinetics (Cleland, 1963) this behavior is best ex-plained by the formation of an E-NADP-I complex for uncompetitive inhibition and additional E-I and/or E-NADPH-I complexes for non-competitive inhibition (Worrall et al., 1986; De Jongh et al., 1987). Alternatively, the lack of competition between substrates and inhibitor has been taken as evidence for the existence of a specific inhibitor binding site apart from the active site (Ward et al., 1990) as previously described for aldose reductase (Kador and Sharpless, 1983).

In order to discriminate between the two possibilities, the effect of carboxylic acids on the formation of the E-NADP-gulonate complex and the susceptibility of the active site-arginine residue to modification by phe-nylglyoxal was investigated. We show that carboxylic acids bind to the E-NADP complex and thereby protect the arginine residue from being modified.

Enzymology and Molecular Biology of Carbonyl Metabolism 3
Edited by H. Weiner *et al.*, Plenum Press, New York, 1990

EXPERIMENTAL PROCEDURE

Aldehyde reductase from human liver was purified as described previously (von Wartburg and Wermuth, 1982). Pyridine nucleotides, valproic acid and tetramethyleneglutaric acid were purchased from Sigma, St. Louis, Mo, USA. All other chemicals were from Fluka, Buchs, Switzerland or Merck, Darmstadt, Germany. L-gulonic acid was prepared by hydrolysis of L-gulono-lactone by one equivalent of NaOH. Spurious amounts of aldehydes which are present in commercial preparations of gulonolactone and inhibit the oxidation of gulonate were reduced by sodium borohydride.

Aldehyde reductase activity was assayed spectrophotometrically at 30° C by following the change in absorbance at 340 nm. The standard assay mixture consisted of 0.1 M Na phosphate (pH 7.0), 50 μM NADPH, 10 mM D-glucuronate and inhibitors as indicated in the text. For studies of pH-dependence the buffer consisted of 0.2 M phosphoric acid/0.2 M Tris adjusted to the desired pH by NaOH. Initial-rate data from inhibition studies were fitted by non-linear regression, and results are presented as double-reciprocal plots for illustrative purposes.

Inactivation of aldehyde reductase by phenylglyoxal was performed in 0.1 M Na phosphate buffer (pH 7.0) at 30° C according to Bohren et al. (1987). Modification of lysine residues by dimethylsuberimidate was carried out in 0.2 M triethanolamine-HCl buffer (pH 8.5) at room temperature as described previously (Wermuth et al., 1979).

RESULTS AND DISCUSSION

Inhibition Studies

A variety of organic acids, including the metabolites of the Krebs cycle, decreased the enzymatic activity of aldehyde reductase. The inhibitory potency was strongly dependent on the presence of hydrophobic substituents. Thus, dimethylsuccinate and tetramethyleneglutarate (TMG) inhibited the enzyme activity by 50 per cent (I-50) at 100 to 1000-fold lower concentrations than the parent unsubstituted acids and their oxo- and hydroxy-derivatives (Table 1). The requirement of hydrophobic substituents for aldehyde reductase inhibition is not restricted to carboxylic acid inhibitors, but also applies to barbiturates, hydantoins and structurally related compounds, which have long been known to inhibit aldehyde reductase (Tabakoff and Erwin, 1970; Erwin and Deitrich, 1973; Ris et al., 1975). This analogy suggests that, despite structural dissimilarities, inhibitors of both the carboxylic acid and barbiturate/hydantoin type bind to the enzyme at the same or closely spaced sites.

Detailed inhibition studies of both the reduction of glucuronate and oxidation of gulonate at varied concentrations of all four substrates were carried out with TMG and valproate. Fig. 1 depicts the results obtained with TMG, though essentially the same picture was observed with valproate. In agreement with results from studies with aldehyde reductase from other sources (Whittle and Turner, 1981; Branlant et al., 1981; Daly and Mantle, 1982; De Jongh et al., 1987) inhibition of human liver aldehyde reductase yielded uncompetitive to slightly non-competitive patterns relative to both NADPH and glucuronate. In the reverse reaction, which has not previously been studied, uncompetitive inhibition with respect to NADP[+] and competitive inhibition relative to gulonate were observed. To compare inhibitors of the carboxylic acid type with those of the barbiturate/hydantoin type, inhibition of the reverse reaction by diphenylhydantoin was also examined. Again, uncompetitive inhibition relative to NADP[+] and competition between inhibitor and gulonate were observed.

Table 1   Inhibition of aldehyde reductase by carboxylic acids

Acid	I-50[a]	Acid	I-50[a]
	(mM)		(mM)
Valproic	0.07	2-Oxoglutaric	1
Succinic	17	Tetramethyleneglutaric	0.04
Malic	10	Citric	1.1
Oxaloacetic	10	Isocitric	1.1
Dimethylsuccinic	0.015		

[a]Concentration of inhibitor causing 50% inhibition in 0.1 M
 sodium phosphate buffer (pH 6.5), containing 50 μM NADPH
 and 10 mM glucuronate.

   The results indicate that carboxylic acids and barbiturate/hydantoins
inhibit aldehyde reductase by basically similar mechanisms, and are consistent with the formation of an abortive E-NADP-I complex. To obtain more
detailed information on the interaction of the two types of inhibitor with
the E-NADP complex, double inhibition studies of gulonate oxidation were

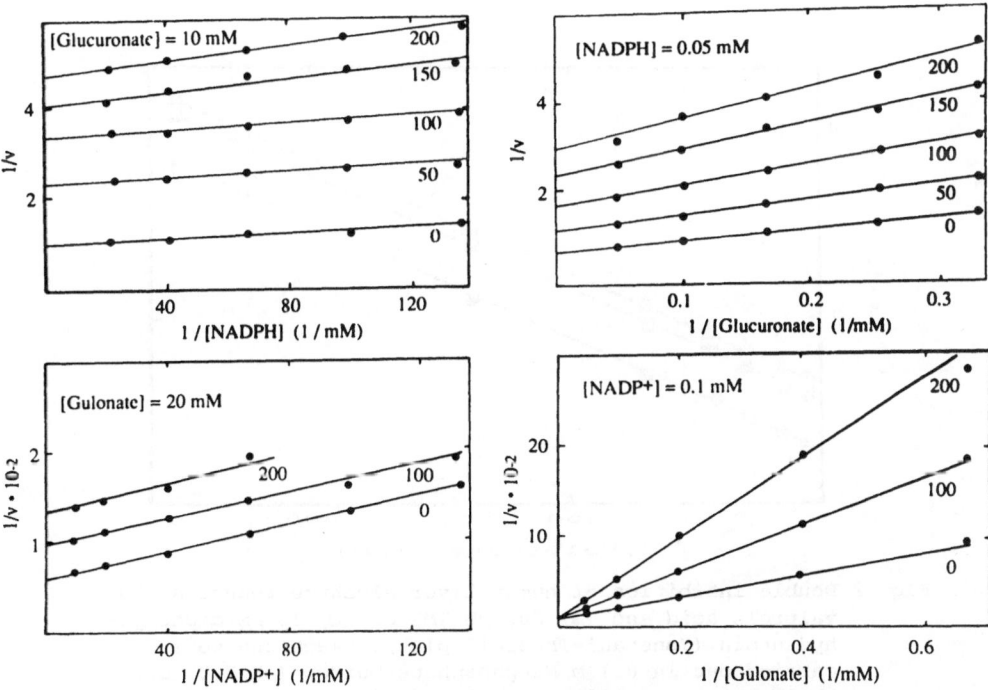

Fig. 1   Inhibition of human liver aldehyde reductase by TMG. Reactions were
          carried out in 0.1 M Na phosphate buffer (pH 7.0) in the presence of
          substrates and inhibitor (μM) as shown in the figure. V is indicated
          relative to the activity under standard assay conditions.

carried out. Representation of the data according to Yonetani and Theorell (Fig. 2) yielded parallel lines for all possible combinations of valproate TMG and diphenylhydantoin, indicating mutually exclusive binding of these compounds to the E-NADP complex.

Inactivation Studies

Phenylglyoxal inactivates aldehyde reductase in a time- and concentration-dependent manner. The loss of enzyme activity is prevented in the presence of coenzyme and substrates, which has led to the proposal of an arginine residue in the active site of the enzyme (Davidson and Flynn, 1979; Branlant et al., 1981; Bohren et al., 1987). In order to assess the role of this residue in the binding of carboxylic acids, aldehyde reductase was incubated with the arginine-specific reagent phenylglyoxal in the presence and absence of inhibitors and/or NADP$^+$. Assuming a reaction mechanism in which the inhibitor preferentially binds to the enzyme-NADP complex, protection against phenylglyoxal is expected to be afforded by a mixture of coenzyme and inhibitor rather than by the inhibitor alone. As shown in Fig. 3 the rate of inactivation was decreased more than tenfold in the presence of NADP$^+$ and saturating concentrations of TMG (50 x I-50), but was only halved by the same amount of TMG in the absence of the coenzyme and not affected by non-saturating concentrations of TMG (I-50). Similar results were obtained with valproate, dimethylsuccinate, succinate and citrate. Diphenylhydantoin, which is not ionized at pH 7, decreased the rate of inactivation 3 to 4 fold in the presence of NADP$^+$, but was ineffective even at saturating concentrations in the absence of the coenzyme.

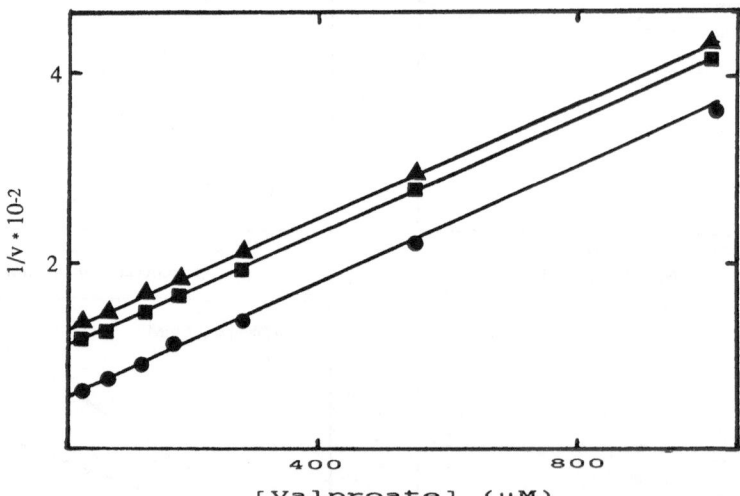

Fig. 2 Double inhibition of human liver aldehyde reductase by valproic acid and (▲) 200 μM TMG or (■) 15 μM diphenylhydantoin (Yonetani-Theorell plot). Reactions were carried out in 0.1 M Na phosphate buffer (pH 7.0) in the presence of 100 μM NADP$^+$ and 20 mM gulonate. V is indicated relative to the activity under standard assay conditions. (●) Inhibition without second inhibitor.

In addition to arginine, lysine residues which are thought to contribute to the binding of the coenzyme have been identified in the active site of aldehyde reductase (Wermuth et al., 1979). In order to test whether these residues interact with aldehyde reductase inhibitors, the effect of carboxylic acids and barbiturate/hydantoins on the inactivation of aldehyde reductase by dimethylsuberimidate was investigated. Saturating concentrations of TMG and diphenylhydantoin slightly but significantly decreased the rate of inactivation in the absence of coenzyme, and approximately doubled the protective effect of NADPH when added together with the coenzyme. This effect is significantly less pronounced than the one observed in the case of arginine modification.

The results of the inactivation studies corroborate the formation of an E-NADP-I complex as inferred from the kinetic analysis. In addition, the data show that for carboxylic acid inhibitors bonding to the enzyme occurs at least in part via the active site arginine residue. In the absence of the coenzyme, binding is negligible at low concentrations of inhibitor but becomes significant at higher concentrations. Barbiturates and hydantoins in their protonated form most likely do not directly interact with the arginine residue, but afford some protection against the action of phenylglyoxal probably by steric hindrance. Similarly, lysine residues seem not to be involved directly in the binding of either carboxylic acids or barbiturate/hydantoins.

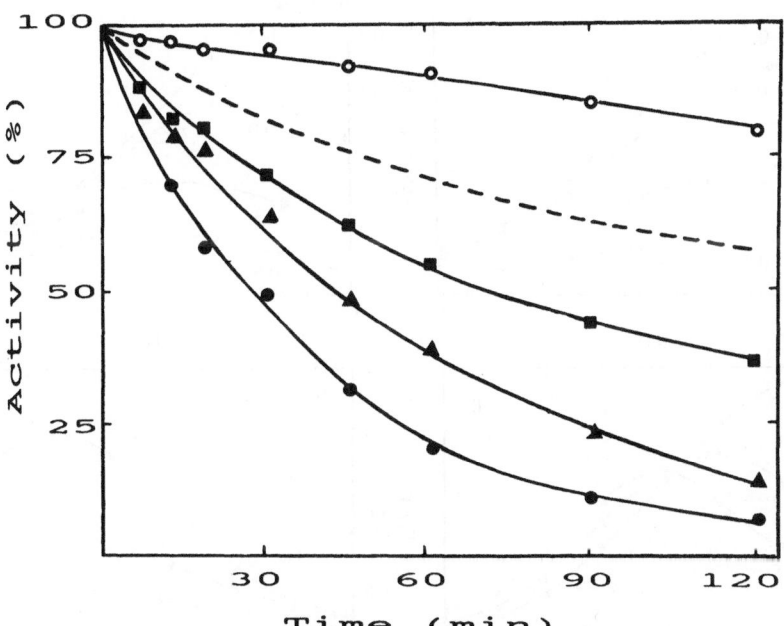

Fig. 3 Effect of NADP$^+$ and inhibitors on the inactivation of human liver aldehyde reductase by phenylglyoxal. The enzyme was incubated with 10 mM phenylglyoxal in the presence of (▲) 3 mM TMG, (■) 200 μM NADP$^+$, (○) 3 mM TMG + 200 μM NADP$^+$ and (---) 750 μM diphenylhydantoin + 200 μM NADP$^+$ in 0.1 M Na phosphate buffer (pH 7) at 30°C. (●) Control without additions.

The inhibition of aldehyde reductase by barbiturates and structurally related compounds strongly depends on the pH of the assay medium becoming more pronounced at higher pH (Erwin and Deitrich, 1973; de Jong et al., 1987). This pH dependence has been attributed to increased concentrations of the ionized form of the inhibitor at elevated pH, although this explanation has recently been questioned (Wermuth, 1987). Investigations of the effect of pH on the inhibition by carboxylic acids have so far been limited to valproic and other monocarboxylic acids, which inhibited aldehyde reductase from sheep liver to approximately the same degree between pH 5.5 and 8 (Schofield et al., 1987). On the other hand, when we determined the I-50 values for citrate and the other Krebs cycle metabolites listed in table 1, significantly higher values were obtained at pH 7 than 6.5. In order to more clearly define this effect, the inhibition by carboxylic acids was investigated systematically between pH 6 and 9.

As shown in Fig. 4D, and in contrast to the results obtained with barbiturates and hydantoins, inhibition by carboxylic acids decreased with increasing pH. In keeping with the results obtained with aldehyde reductase from sheep liver, inhibition by valproic acid only slightly decreased at higher pH. Inhibition by dicarboxylic acids, on the other hand, was markedly

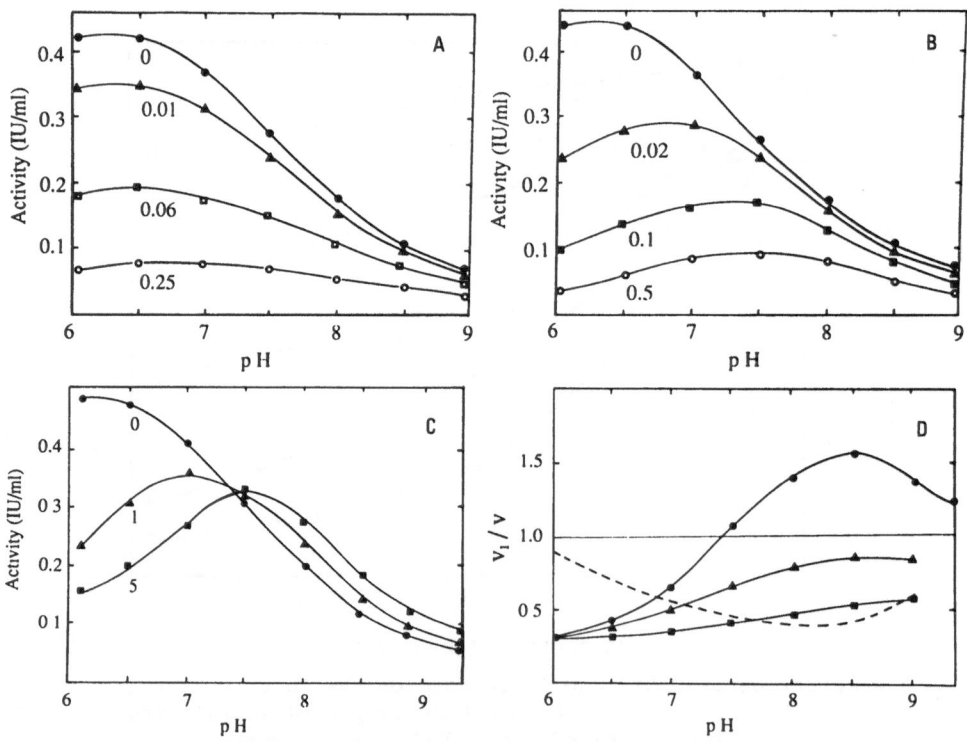

Fig. 4 Effect of pH on the inhibition of human liver aldehyde reductase by carboxylic acids. A-C: pH-optimum in the presence of (A) valproic acid, (B) TMG, (C) citric acid. Concentrations of inhibitor are indicated in mmol/l. D: pH-dependence of inhibition (v_i/v) by (●) 110 μM valproic acid, (▲) 70 μM TMG (●) 5 mM citric acid and (---) 10 μM diphenylhydantoin.

affected by pH. The potency of TMG and 2,3-dimethylsuccinate was cut approximately in half when the pH was increased from 6 to 9, and citrate, together with the other Krebs cycle metabolites, even activated aldehyde reductase at high pH. Replots of the data in the form of activity-pH profiles are depicted in Figs. 4A-C. They show that the pH-optimum was shifted towards higher values in the presence of the inhibitors. The shift was more pronounced for dicarboxylic acids and dependent on the concentration of inhibitor (Figs. 4A-C).

The change in pH-optimum caused by the addition of dicarboxylic acids readily explains the observed pH-dependence as well as the inhibition and activation of the enzyme by citrate and other Krebs cycle metabolites at low and high pH, respectively. Moreover, the shift of the pH-optimum causes non-linear Dixon and secondary intercept plots. This effect was particularly pronounced with the Krebs cycle metabolites which inhibit aldehyde reductase mainly by shifting the pH-optimum (Fig. 5), but may also explain some of the non-linear Dixon plots observed with strongly inhibiting dicarboxylic acids.

The molecular basis of the change in pH-optimum is not known. The acids used in this study have pKa values between 3.4 and 6.4 and exist essentially as anions above pH 7. Ionization of the inhibitor, therefore, cannot be responsible. For the same reason, the direct involvement of the active site-arginine residue can be excluded. Indirectly, however, the arginine residue might contribute to the observed effect by binding one of the carboxy groups of dicarboxylic acids while the other group would interact with a second, hitherto unidentified, positively charged residue at the catalytic site. In fact, Branlant and coworkers (1981) have predicted the presence of a second anion binding site in the active site of pig liver aldehyde reductase. More work, however, will be necessary to identify it.

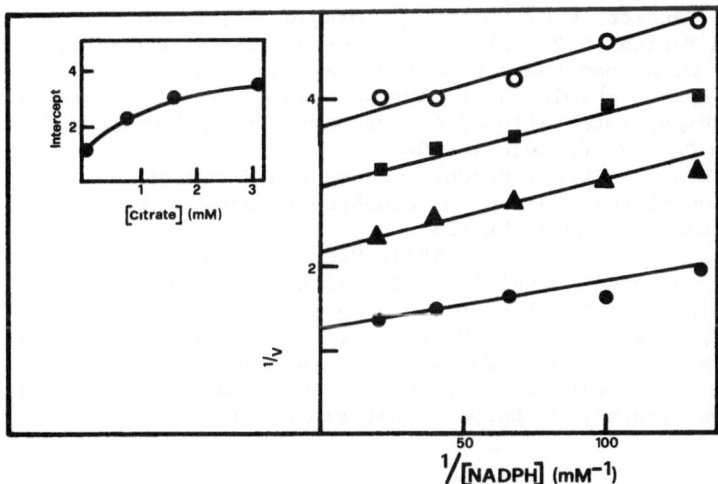

Fig. 5 Inhibition of human liver aldehyde reductase by citrate. Reactions were carried out in 0.1 M Na phosphate buffer (pH 6.5) in the presence of (●) 0, (▲) 0.8, (■) 1.6, and (○) 3.2 mM citrate. The concentration of glucuronate was 10 mM. The inset shows a replot of the ordinate intercepts versus citrate concentration.

REFERENCES

Bohren, K.M., von Wartburg, J-P., and Wermuth, B., 1981, Inactivation of carbonyl reductase from human brain by phenylglyoxal and 2,3-butane-dione: a comparison with aldehyde reductase and aldose reductase, *Biochim. Biophys. Acta*, **916**:185.

Branlant, G., 1982, The substrate binding site of aldehyde reductase from pig liver, *Eur. J. Biochem.*, **121**:407.

Branlant, G., Tritsch, D., and Biellmann, J-F., 1981, Evidence for the presence of anion-recognition sites in pig liver aldehyde reductase, *Eur. J. Biochem.*, **116**:505.

Cleland, W.W., 1963, The kinetics of enzyme-catalyzed reactions with two or more substrates or products. II, Inhibition: Nomenclature and theory, *Biochim. Biophys. Acta*, **284**:427.

Daly, A.K., and Mantle, T.J., 1982, The kinetic mechanism of the major form of ox kidney aldehyde reductase with D-glucuronic acid, *Biochem. J.*, **205**:381.

De Jongh, K.S., Schofield, P.J., and Edwards, M.R., 1987, Kinetic mechanism of sheep liver NADPH-dependent aldehyde reductase, *Biochem. J.*, **242**:143.

Davidson, W.S., and Flynn, T.G., 1979, A functional arginine residue in NADPH-dependent aldehyde reductase from pig kidney, *J. Biol. Chem.*, **254**:3724.

Erwin, V.G., and Deitrich, R.A., 1973, Inhibition of bovine brain aldehyde reductase by anticonvulsant compounds in vitro, *Biochem. Pharmacol.*, **22**:1067.

Kador, P.F., and Sharpless, N.E., 1983, Pharmacophor requirements of the aldose reductase inhibitor site, *Mol. Pharmacol.*, **24**:521.

Ris, M.M., Deitrich, R.A., and von Wartburg, J-P., 1975, Inhibition of aldehyde reductase isoenzymes in human and rat brain, *Biochem. Pharmacol.*, **24**:1865.

Schofield, P.J., de Jongh, K.S., Smith, M.M., and Edwards, M.R., 1987 Inhibition of aldehyde reductase, *Progr. Clin. Biol. Res.*, **232**:287

Tabakoff, B., and Erwin, V.G., 1970, Purification and characterization of an NADPH-linked aldehyde reductase from brain, *J. Biol. Chem.*, **245**:3263.

Von Wartburg, J-P., and Wermuth, B., 1982, Aldehyde reductase from human tissues, *Methods Enzymol.*, **89**:506.

Ward, W.H.J., Sennitt, C.M., Ross, H., Dingle, A., Timms, D., Mirrlees, D.J., and Tuffin, D.P., 1990, Ponalrestat: a potent and specific inhibitor of aldose reductase, *Biochem. Pharmacol.*, **39**:337.

Wermuth, B., 1987, Inhibition of aldehyde reductase: Structure-function relationships, *International Workshop on Aldose Reductase Inhibitors*, *Honolulu, Dec. 7-10*, Abstract C4.3.

Wermuth, B., Münch, J.D.B., Hajdu, J., and von Wartburg, J-P., 1979, Amidination of amino groups of aldehyde reductase from human liver, *Biochim. Biophys. Acta*, **566**:237.

Whittle, S.R., and Turner, A.J., 1981, Biogenic aldehyde metabolism in rat brain: Differential sensitivity of aldehyde reductase isoenzymes to sodium valproate, *Biochim. Biophys. Acta*, **657**:94.

Worrall, D.M., Daly, A.K., and Mantle, T.J., 1986, Kinetic studies on the major form of aldehyde reductase in ox kidney: A general kinetic mechanism to explain substrate-dependent mechanisms and the inhibition by anticonvulsants, *J. Enzyme Inhibition*, **1**:163.

REACTIVITY OF ENZYME MODIFICATION REAGENTS

WITH ALDOSE REDUCTASE AND ALDEHYDE REDUCTASE

Tadashi Mizoguchi, Hiroyuki Itabe, and Peter F. Kador

National Eye Institute
National Institutes of Health
Bethesda, MD

INTRODUCTION

Aldose reductase (EC 1.1.1.21) and aldehyde reductase (EC 1.1.1.2) are members of a broad family of NADPH-dependent reductases (Feldstead and Bachur, 1980). While the physiological role of either enzyme remains unknown, it is generally assumed that aldose reductase reduces aldose sugars to their respective sugar alcohols while aldehyde reductase reduces a broad range of aromatic and aliphatic aldehydes. Aldehyde reductase is also known as hexonate dehydrogenase EC 1.1.1.19) because of its ability to oxidize L-gulonic acid to glucuronic acid..

It has been experimentally demonstrated that the aldose reductase-catalyzed intracellular accumulation of polyols initiates complex biochemical changes that result in the formation of diabetic complications such as cataract, retinopathy, neuropathy and possibly nephropathy (Kador, 1988; Kador et al., 1990). Interest in aldehyde reductase has recently been sparked by studies that indicate that this enzyme can also be inhibited by aldose reductase inhibitors (ARIs, Sato and Kador, 1990a; Poulsom, 1986; Srivastava et al., 1982; O'Brien, Schofield, and Edwards, 1982). Since both the physiological and pathological role of aldehyde reductase remains unknown, the biological significance of its inhibition has not been established.

Studies indicate that, in general, both NADPH-dependent reductases are present in most tissues. Both enzymes are monomers with similar molecular weights, overlapping substrate specificities, and high protein sequence homology (Sato, 1990; Sato and Kador, 1989; 1990b; Tanimoto, Sato and Kador, 1989; Nakayama, Tanimoto and Kador, 1989; Flynn and Cromlish, 1985; Bohren et al., 1989). This latter point suggests that both enzymes are derived from the same gene (Carper et al., 1987). Despite these apparent similarities, evidence suggests that these enzymes differ in their secondary and tertiary structures. While both enzymes are inhibited by aldose reductase inhibitors, some inhibitors are more selective for aldose reductase (Sato and Kador, 1990a; Sato, 1990). For example, Al1576 (2,7-difluorospirofluorene-9,5'-imidazolidine-2',4'-diones) equally inhibits rat kidney aldehyde reductase and rat lens aldose reductase while Ponalrestat (Statil, 3-(4-bromo-2-fluorobenzyl-4-oxo-3- phthalazine-1-ylacetic acid) is 120-fold more selective and FK366 (FR74366, [3-(4'-bromo-2'-fluorobenzyl]-7-chloro-2,4-dioxo-1,2,3,4-tetrahydroquinazolin-1-yl]acetic acid) is 180-fold more selective for aldose reductase. Selectivity for aldose reductase is also observed with irreversible aldose reductase inhibitors containing isothiocyanate- or N-haloacetamido- functions which require nucleophilic attack for irreversibly binding to the enzyme. Inhibitors requiring nucleophilic attack to bind irreversibly inhibit aldose reductase to a greater extent than aldehyde reductase (Itabe, Sato and Kador, 1989). In addition to apparent differences in the site at which these inhibitors bind, differences in the secondary or tertiary structures are also suggested by antibody cross-reactivity studies. Antibodies against aldose reductase do not cross-react with aldehyde reductase and antibodies against aldehyde reductase do not cross-react with aldose reductase (Sato, 1990; Sato and Kador, 1989; Tanimoto, Sato and Kador, 1989; Nakayama, Tanimoto and Kador, 1989).

To gain further insight into the extent of similarities or differences between these two reductases, the reactivity of aldose reductase and aldehyde reductase toward protein modification reagents have been investigated. In addition, protection studies with substrate, nucleotide cofactor, and aldose reductase inhibitors have been conducted to determine the sites of reaction of these reagents.

EXPERIMENTAL

Chemicals Unless stated, all chemicals employed were of reagent grade. N-Ethylmaleimide (NEM), sulfosuccinimidobiotin (sulfo-NHS-biotin), sulfosuccinimidylacetate (sulfo-NHS-acetate), p-hydroxyphenylglyoxal, phenylmethanesulfonylfluoride (PMSF) were purchased from Pierce, Rockford, IL. Iodoacetamide, p-chloromercuriphenylsulfonate (p-CMPS), p-chloromercuribenzoate (p-CMB), dithiothreitol (DTT), β-nicotinamide adenine dinucleotide phosphate, oxidized form (NADP$^+$), β-nicotinamide mononucleotide, reduced form (NMNH), adenosine 5'-diphosphate (ADP), adenosine 5'-diphosphate ribose (ADP-ribose) were purchased from Sigma Chemical Co., St. Louis, MO. β-Nicotinamide adenine dinucleotide phosphate reduced form (NADPH) was purchased from Boehringer Mannheim Biochemicals, Indianapolis, IN.

Enzyme Assay Reductase activity was spectrophotometrically assayed on a Guilford Response spectrophotometer or a Shimadzu UV2100U spectrophotometer by following the decrease in the absorption of NADPH at 340 nm over a 4 min. period with DL-glyceraldehyde as substrate (Sato and Kador, 1990a). Each 1.0 ml cuvette contained equal units of enzyme, 0.10 M Na,K phosphate buffer, pH 6.2, 0.3 mM NADPH with/without 10 mM substrate and inhibitor. Appropriate controls were employed to negate potential changes in the absorption of nucleotide and/or protein modification reagents or aldose reductase inhibitors at 340 nm in the absence of substrate.

Kinetic studies were analyzed using the PROPHET computer system (Division of Research Resources, National Institutes of Health, Bethesda, MD and BBN Systems and Technologies Corporation, Cambridge MA) using the public procedure BINKIN2 by fitting the means of 3-6 determinations to the enzyme kinetic equation $v = V_{max} [S]/[S] + K_m$ were v represents the initial velocity of the enzyme reaction, V_{max} represents the maximum velocity, [S] represents the substrate concentrations and K_m represents the Michaelis constant.

Enzyme preparations Aldose reductase was prepared from frozen rat lenses as previously described (Shiono *et al.*, 1987; Sato and Kador, 1989). Aldehyde reductase was prepared from thawed cortices from dissected rat kidneys as previously described (Sato, Kador, and Kinoshita, 1988; Sato 1990). Both purification steps utilized a series of chromatographic steps which included gel filtration on Sephadex G-75, affinity chromatography on Amicon Matrex Gel Orange A, and chromatofocusing on Pharmacia Mono P. The highly purified enzymes appeared as single peaks on sodium dodecyl sulfate polyacrylamide gel electrophoresis (SDS-PAGE). The purified enzymes were either stored at -20°C in the presence of 10% glycerol and 10 mM 2-mercaptoethanol or at 4°C as a 50% saturated ammonium sulfate precipitate.

Irreversible inhibition Rat lens aldose reductase (RLAR) or rat kidney aldehyde reductase (RKALR) were passed through a NAP-5 desalting column (Pharmacia LKB, Piscataway, NJ) or an Excellulose GF-5 column (Pierce, Rockford, IL) to remove glycerol and 2-mercaptoethanol, or ammonium sulfate. The reaction was started by adding 10 µl of the solution of the modification reagents to the 490 µl of 100 mM phosphate buffer, pH 7.4 containing 10 µg of RKALR or RLAR. The reaction was allowed to proceed at the room temperature for 15 min and unreacted reagent was then removed from a 400µl aliquot of the reaction mixture by gel filtration through a desalting column with 0.1 M phosphate buffer, pH 7.0, containing 10 mM 2-mercaptoethanol. Immediately, 50 µl of 1 mg/ml BSA solution were added to the eluent to stabilize the reductase. The eluent was kept on ice until the residual reductase activity was measured.

Protection studies Studies were conducted by incubating 10 milliunits of purified enzyme with protein modification reagents in 100 mM phosphate buffer, pH 7.4. 20 mM glyceraldehyde substrate, 1.0 mM oxidized or reduced nucleotide cofactor, and 0.1 mM of the ARI, Al1576 was added to the purified enzymes 1 minute prior to the addition of protein modification reagents. Concentration and incubation times or the protein modification reagents utilized were: 0.1 mM p-chloromercuriphenylsulfonate, 15 min; p-chloromercuribenzoate, 15 min; 5 mM N-ethylmaleimide, 15 min; 1 mM iodoacetamide, 45 min; 0.1 mM sulfosuccinimidobiotin, 15 min; 0.1 mM sulfosuccinimidlyacetate, 30 min; 5 mM dithiothreitol, 45 min; 5 mM p-hydroxyphenylglyoxal 15 min. Enzyme-inhibitor mixtures containing Al1576 were passed through GF-5 desalting columns to remove reversibly bound aldose reductase inhibitor prior to assay.

RESULTS

A variety of protein modification reagents were utilized to probe the reactivities of rat lens aldose reductase and rat kidney aldehyde reductase. These reagents, summarized in Table 1, include the thiol reacting reagents p-chloromercuriphenylsulfonate, p-chloromercuribenzoate and N-ethylmaleimide, the histidine and thiol reacting reagent iodoacetamide, sulfosuccinimidobiotin and sulfosuccinimidylacetate, which react with primary amines, such as the epsilon amino group of lysine, dithiothreitol which reacts with disulfides, p-hydroxyphenylglyoxal which forms a Schiff's base with arginine, and phenylmethanesulfonylfluoride which reacts with serine.

The effects of these reagents on the activities of aldose reductase and aldehyde reductase are summarized in Table 2. In general, sulfosuccinimidobiotin, sulfosuccinimidylacetate, dithiothreitol, p-hydroxyphenylglyoxal, and phenylmethanesulfonylfluoride inhibited both reductases to a similar extent; however, aldehyde reductase was inhibited slightly more by sulfosuccinimidylacetate and dithiothreitol. p-Chloromercuriphenylsulfonate, a thiol reagent, inhibited both enzymes to the same extent; however, N-ethylmaleimide, a similar reagent inhibited aldose reductase to a greater extent. A difference between aldose and aldehyde reductase was also observed with iodoacetamide which in the time frame examined activated aldose reductase while inhibiting aldehyde

Table 1. Protein modification reagents.

Name	Structure	React with
p-Chloromercuriphenylsulfonate (p-CMPS)	Cl-Hg—⟨⟩—SO₃Na	thiol
p-Chloromercuribenzoate (p-CMB)	Cl-Hg—⟨⟩—CO₂Na	thiol
N-Ethylmaleimide (NEM)		thiol
Iodoacetamide (IA)		thiol histidine
Sulfosuccinimidobiotin (Sulfo-NHS-biotin)		1° amine
Sulfosuccinimidylacetate (Sulfo-NHS-acetate)		1° amine
Dithiothreitol (DTT)		disulfide
p-Hydroxyphenylglyoxal (HPG)		arginine
Phenylmethanesulfonylfluoride (PMSF)		serine

reductase. In addition to iodoacetamide, aldose reductase was also activated by sulfosuccinimidylacetate. Compared to the activation by iodoacetamide which increased over the 30 minute time period examined, this activation was more transient and rapidly followed by inhibition (Figure 1). Activation by both reagents was irreversible since increased aldose reductase activity was maintained after the enzyme was dialyzed by filtration through desalting columns. The transient activation observed with sulfosuccinimidylacetate appeared concentration dependent with higher concentrations of sulfosuccinimidylacetate resulting in increased inactivation of rat lens aldose reductase. With 5 mM sulfosuccinimidylacetate, a significant inactivation of aldose reductase activity occurred within 1 minute of incubation with enzyme. Excess amounts of bovine serum albumin (BSA) provided no protection against this inactivation, indicating that this inhibition is linked to a reaction between the protein modification reagent and a specific amino acid residue on the enzyme.

Table 2. Inactivation of rat lens aldose reductase and kidney aldehyde reductase protein modification reagents.

Reagent	Aldose Reductase			Aldehyde Reductase		
	Conc. (mM)	Remaining activity (%) 15	45 (min)	Conc. (mM)	Remaining activity (%) 15	45 (min)
Sulfosuccinimidobiotin	1	3	0	1	9	0
	0.1	40	12	0.1	49	25
Sulfosuccinimidylacetate	1	15	14	1	7	0
	0.1	104	39	0.1	71	8
Dithiothreitol	5	80	80	5	70	48
	1	95	100	1	95	85
p-Hydroxyphenylglyoxal	5	43	27	5	42	27
	1	58	60	1	72	63
Phenylmethanesulfonylfluoride	5	81	70	5	80	70
	1	94	89	1	92	91
p-Chloromercuriphenylsulfonate	1	0	0	1	0	0
	0.1	48	8	0.1	32	5
N-Ethylmaleimide	5	14	1	5	68	53
	1	42	19	1	90	85
Iodoacetamide	5	134	140	5	74	79
	1	116	143	1	94	86

Kinetic studies were conducted to gain insight into the nature of the activation of aldose reductase by iodoacetamide (Figure 2). Evaluation of aldose reductase irreversibly bound with iodoacetamide indicates that the modified enzyme is inhibited by increasing concentrations of glyceraldehyde substrate while Lineweaver-Burk analysis indicates a decreased affinity of aldose reductase for the glyceraldehyde substrate.

Inactivation of aldose reductase and aldehyde reductase by most protein modification reagents was linear with time. This is illustrated for p-chloromercuriphenylsulfonate and sulfosuccinimidobiotin which in Figure 3 appear to follow first order kinetics.

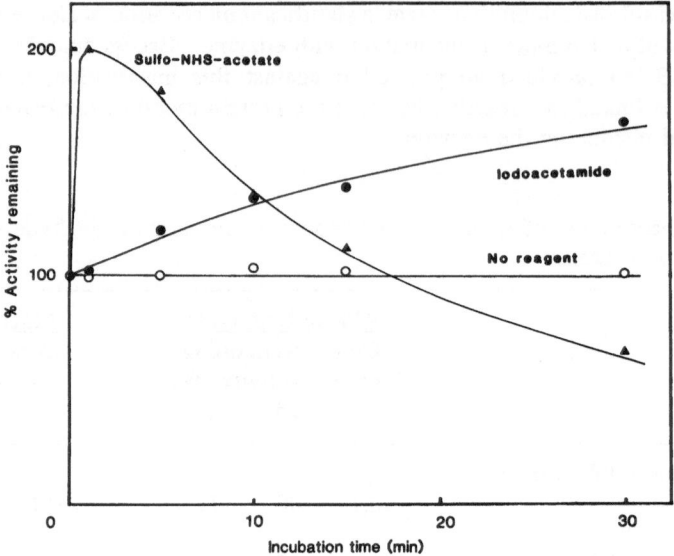

Figure 1. Effect of 1 mM iodoacetamide (▲), 1 mM sulfonylsuccinimidylacetate (●), or no reagents (O) on purified rat lens aldose reductase. Residual activity was determined after the removal of unreacted reagents by gel filtration through a desalting column.

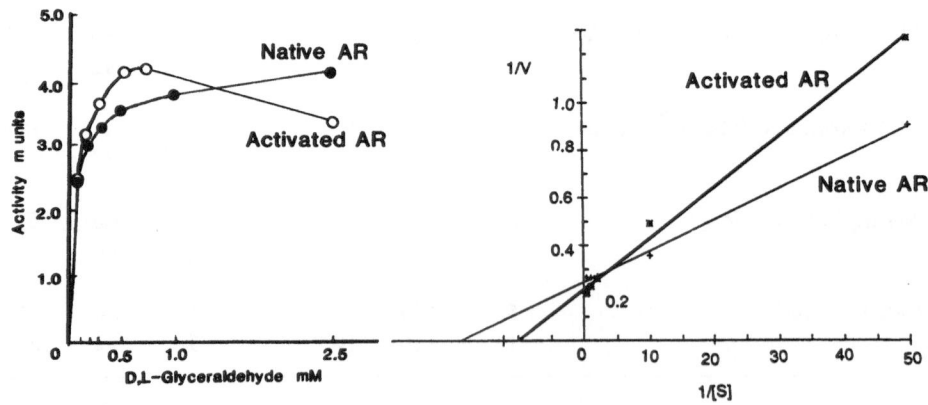

Figure 2. Kinetic analysis of "activated" versus native rat lens aldose reductase (AR). Aldose reductase was incubated with 1 mM iodoacetamide for 45 min and then filtered through a desalting column to remove unreacted reagents.

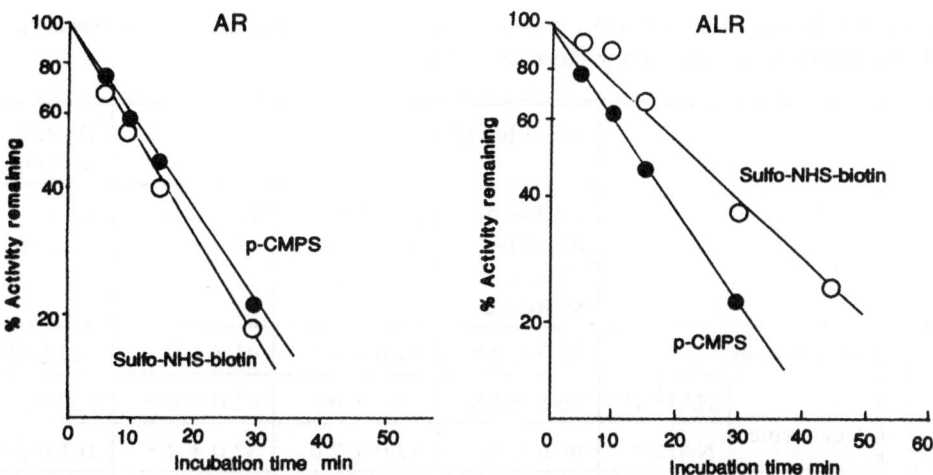

Figure 3. Time-dependent inactivation of rat lens aldose reductase (AR) and rat kidney aldehyde reductase (ALR) with 0.1 mM sulfosuccinimidobiotin and 0.1 mM chloromercuriphenylsulfonate.

To ascertain the potential site(s) on aldose reductase and aldehyde reductase reacting with these protein modification reagents, protection studies with glyceraldehyde substrate, oxidized and reduced nucleotide cofactor, and the aldose reductase inhibitor (ARI) Al1576 were investigated (Tables 3 and 4). With the thiol reagents, p-chloromercuriphenyl-sulfonate and benzoate, protection of aldose reductase was observed with either NADPH or NADP$^+$ while similar protection of aldehyde reductase was only afforded with NADPH. With the thiol reagent N-ethylmaleimide, no protection of aldose reductase was observed; however, the presence of NADPH and NADP$^+$ afforded protection for aldehyde reductase. With iodoacetamide, inhibition of aldose reductase was observed in the presence of glyceraldehyde or Al1576 while no adverse effects were observed with aldehyde reductase. With sulfosuccinimidobiotin, protection of aldose reductase was only observed in the presence of NADPH while with sulfosuccinimidlyacetate, slightly increased inhibition of aldose reductase was observed with either NADP$^+$ or glyceraldehyde. These reagents had no effect on the similar inhibition of aldehyde reductase. Increased inhibition of aldose reductase was also observed with a similar incubation of dithiothreitol in the presence of NADP$^+$ or glyceraldehyde. With the arginine reagent p-hydroxyphenylglyoxal, only slight protection of aldehyde reductase was observed in the presence of NADPH.

Further investigations of nucleotide protection against inactivation of these NADPH-dependent reductases by the protein modification reagents p-chloromercuriphenylsulfonate and sulfosuccinimidobiotin indicated that 1 mM NADPH displayed protective effects against inactivation of both enzymes while 1 mM NADP$^+$ afforded protection for aldose reductase against p-chloromercuriphenylsulfonate and aldehyde reductase against sulfosuccinimidobiotin (Table 5). No effect was observed with either 1 mM concentrations of ADP-ribose or NMNH, partial fragments of the NADPH molecule, or bovine serum albumin. The protection of aldose reductase and aldehyde reductase by NADPH against p-chloromercuriphenylsulfonate and sulfosuccinimidobiotin inactivation was concentration dependent with as little as 5 μM of NADPH affording protection (Figure 4). Similar effects were observed with NADP$^+$; however, this effect was more

Table 3. Percent activity (± S.D.) remaining after chemical modification studies in rat lens aldose reductase and rat kidney aldehyde reductase

			Thiol Reagents			Histidine Reagent
			p-Chloro-mercuri-phenyl-sulfonate	p-Chloro-mercuri-benzoate	N-Ethyl-maleimide	Iodo-acetamide
Aldose Reductase	No addition		36.3 ± 2.8	14.6 ± 1.7	41.5 ± 2.5	102.8 ±9.8
	Coenzyme	NADPH	96.8 ± 1.8	99.6 ± 0.8	42.0 ± 0.5	92.8 ± 7.4
		NADP$^+$	99.6 ± 2.5	97.0 ± 2.6	37.0 ± 2.5	101.0 ± 5.2
	Substrate	Glyc.	37.3 ± 1.5	13.0 ± 1.3	44.3 ± 4.3	58.1 ± 4.8
	ARI	Al1576	46.1 ± 2.6	18.6 ± 1.5	50.2 ± 2.3	45.7 ± 10.8
Aldehyde Reductase	No addition		54.3 ± 3.9	45.2 ± 4.4	58.1 ± 3.3	83.5 ± 9.5
	Coenzyme	NADPH	80.0 ± 3.4	64.5 ± 3.6	98.6 ± 2.4	94.0 ± 6.6
		NADP$^+$	33.2 ± 1.6	26.7 ± 3.1	80.4 ± 2.9	91.9 ± 7.3
	Substrate	Glyc.	43.2 ± 3.4	30.6 ± 2.4	49.8 ± 3.7	65.7 ± 6.6
	ARI	Al1576	55.2 ± 3.1	55.5 ± 5.8	63.6 ± 3.7	80.7 ± 8.6

Table 4. Percent activity (± S.D.) remaining after chemical modification studies in rat lens aldose reductase and rat kidney aldehyde reductase.

			Amino Reagents		Disulfide Reagent	Arginine Reagent
			Sulfo-succinimido-biotin	Sulfo-succinimidyl-acetate	Dithio-threitol	p-Hydroxy-phenyl-glyoxal
Aldose Reductase	No addition		36.9 ± 5.8	99.3 ± 6.2	79.5 ± 1.2	15.7 ± 5.6
	Coenzyme	NADPH	69.8 ± 7.4	87.4 ± 7.1	69.5 ± 4.6	8.0 ± 2.6
		NADP$^+$	26.1 ± 4.2	76.9 ± 7.6	55.5 ± 3.1	10.5 ± 6.6
	Substrate	Glyc.	41.5 ± 2.3	68.6 ± 4.9	51.9 ± 8.4	26.9 ± 8.7
	ARI	Al1576	37.0 ± 5.1	84.3 ± 4.6	77.6 ± 2.9	15.8 ± 3.6
Aldehyde Reductase	No addition		68.4 ± 1.6	84.1 ± 8.7	72.1 ± 3.5	57.5 ± 9.2
	Coenzyme	NADPH	70.6 ± 3.3	87.6 ± 3.3	69.2 ± 9.3	79.3 ± 3.2
		NADP$^+$	70.5 ± 2.8	93.0 ± 6.0	72.7 ± 6.5	58.4 ± 1.4
	Substrate	Glyc.	62.6 ± 1.3	71.8 ± 6.3	68.8 ± 5.9	54.3 ± 9.6
	ARI	Al1576	69.8 ± 1.2	80.0 ± 5.6	70.1 ± 1.5	50.6 ± 9.1

variable. Enzyme activities were not altered by the addition of nucleotides and NADPH was not oxidized in the assay system by any of the nucleotides or protein modification reagents examined.

Table 5. Effect of nucleotides and bovine serum albumin (BSA) on the inactivation of rat lens aldose reductase and rat kidney aldehyde reductase by sulfosuccinimidobiotin and p-chloromercuriphenylsulfonate. Enzyme (10 μg) was incubated for 15 min with 0.1 mM of sulfo-NHS-biotin and p-CMPS in the presence of 1 mM nucleotides or 50 μg of BSA.

| Compound | Percent Activity Remaining | | | |
| | Aldose Reductase | | Aldehyde Reductase | |
	Sulfo-NHS-biotin	p-CMPS	Sulfo-NHS-biotin	p-CMPS
None	37.9	43.9	34.7	23.5
NADPH	59.8	96.8	81.2	92.8
NADP$^+$	26.1	99.6	90.2	28.5
ADP-ribose	38.5	36.5	35.5	21.2
NMNH	37.5	34.7	30.3	21.4
BSA	29.6	54.7	49.9	28.4

DISCUSSION

A number of protein modification reagents have been utilized to investigate similarities and differences between purified rat lens aldose reductase and rat kidney aldehyde reductase. Greatest inhibition was observed with p-chloromercuriphenylsulfonate, sulfosuccinimidobiotin, and sulfosuccinimidylacetate while moderate inactivation was observed with dithiothreitol, phenylmethanesulfonylfluoride, p-hydroxyophenylglyoxal, and N-ethylmaleimide. These observations suggest that cysteine, lysine and arginine residues are important in maintaining conformations required for enzyme activity.

Inactivation of bovine lens aldose reductase (Flynn et al., 1989), pig kidney aldose and aldehyde reductase (Flynn et al., 1981) and human liver aldehyde reductase (Wermuth, Munch and von Wartburg, 1977) has been reported by the modification of lysine residue(s) with pyridoxal-5'-phosphate. Modification of cysteine residue(s) with p-chloromercuriphenylsulfonic acid has been reported to result in inhibition of bovine lens aldose reductase (Halder and Crabbe, 1985), rabbit muscle aldose reductase (Cromlish and Flynn, 1983) and human liver aldehyde reductase (Wermuth, Munch and von Wartburg, 1977). Modification of arginine residue(s) with phenylglyoxal, butadienone, or camphorquinone-10-sulfonic acid resulted in inactivation of bovine lens aldose reductase (Halder and Crabbe), bovine liver aldehyde reductase (Terada et al., 1989), pig kidney aldehyde reductase (Davidson and Flynn, 1979) and human placental aldose reductase (Kador and Sharpless, 1983). These present observations are consistent with these results.

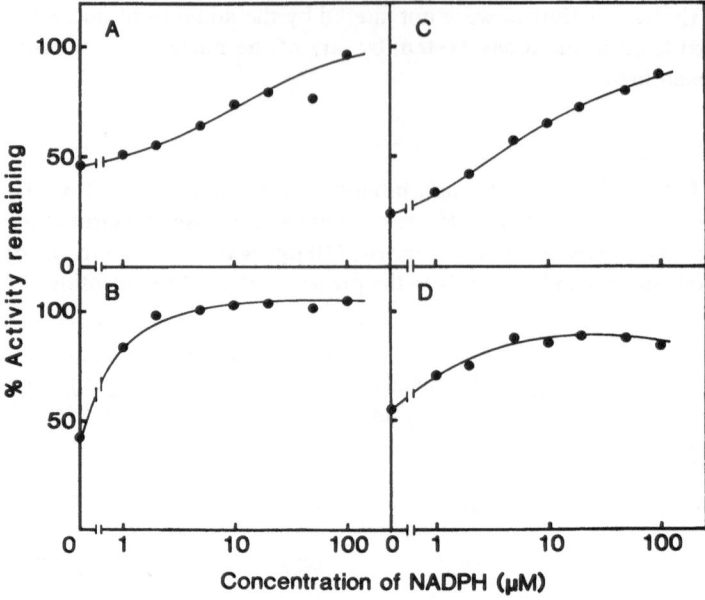

Figure 4. Effect of nucleotide on the p-chloromercuriphenylsulfonate and sulfosuccinimidobiotin mediated inhibition of aldose reductase (A and B, respectively) and aldehyde reductase (C and D, respectively).

Protection against inactivation of aldose reductase and aldehyde reductase by p-chloromercuriphenylsulfonate, p-chloromercuribenzoate and sulfosuccinimidobiotin was observed with the nucleotide cofactor, NADPH. Inactivation of both NADPH-dependent reductases by sulfosuccinimidobiotin was prevented by as little as 5 µM of NADPH, a concentration similar to the Km values of these enzymes for NADPH. However, protection against p-chloromercuriphenylsulfonate required 100 µM of NADPH. Preliminary spectroscopic and protein electrophoresis studies suggest that the conformation of aldose reductase and aldehyde reductase is altered by nucleotide binding. It is conceivable that these conformational changes could alter the availability of reactive amino acid residues. Alternatively, the enzyme-nucleotide cofactor complex may protect an essential lysine moiety from modification. This possibility is supported by the hypothesis that the amino acid sequence Pro-Ile-Lys-Ser is a common structure of the active sites of these enzymes (Bohren et al., 1989; Carper et al., 1987; Cromlish and Flynn, 1983). However, the observation that only partial restoration of original aldehyde reductase activity is achieved with NADPH suggests that additional lysine groups may also have been nonspecifically modified. This observation is not unexpected considering the nonspecific nature of protein modification reagents.

Apparent activation of rat lens aldose reductase was observed with iodoacetamide and, to a lesser extent, with sulfosuccinimidylacetate. Aldose reductase was protected from activation by the presence of the aldose reductase inhibitor Al1576, suggesting that acetamide exposure to the reacting enzyme moiety was either decreased by an altered conformation of the Al1576-enzyme complex or that the Al1576 inhibitor directly blocked access of the reactive moiety to iodoacetamide.   Although iodoacetamide has been observed to react with thiol amino acids, it is that doubtful that a sulfhydryl group is involved in this apparent activation since no activation occurred with other sulfhydryl reagents.   In addition to sulfhydryl residues, iodoacetamide has also been reported to selectively react with histidine in porcine malate dehydrogenase (Foster and Harrison, 1974).   Modification of lysine residues of human muscle aldose reductase by pyridoxal-5'-phosphate or pyridoxal-5'-diphospho-5'-adenosine have also been reported to result in up to 2-fold increases in aldose reductase activity (Flynn et al., 1989; Morjana, Lyones and Flynn, 1989).   The activation of enzyme activity linked to protein modification reagents appears to be a unique characteristic of aldose reductase which aldehyde reductase does not appear to possess.

REFERENCES

Bohren, K.M., Bullock, B., Wermuth, B. and Gabby, K.H., 1989, The aldo-keto reductase superfamily. CDNAS and deduced amino acid sequences of human aldehyde and aldose reductase, *J. Biol. Chem.* **264**:9547.

Carper, D., Nishimura, C., Shinohara, T., Dietzehold, B., Wistow, G., Craft, C., Kador, P. and Kinoshita J.H., 1987, Aldose reductase and p-crystallin belong to the same protein superfamily as aldehyde reductase, *FEBS Lett.* **220**:209.

Felsted, R.L. and Bachur, N.R., 1980, Mammalian carbonyl reductases, *Drug Metabolism Reviews*, **11**:1.

Cromlish, J.A. and Flynn, T.G., 1983, Purification and characterization of two aldose reductase isoenzymes from rabbit muscle, *J. Biol. Chem.*, **258**:3416.

Davidson, W.S. and Flynn, T.G., 1979, A functional arginine residue in NADPH-dependent aldehyde reductase from pig kidney, *J. Biol. Chem.* **254**:3724.

Flynn, T.G., Gallerneault, C., Ferguson, D., Cromlish, J.A. and Davidson, W.S., 1981, Studies on the active site of pig kidney aldehyde reductase, *Biochem. Soc. Trans.* **9**:273.

Flynn, T.G. and Cromlish, J.A., 1985, Pig brain aldehyde reductases, *Prog. Clin. Biol. Res.* **174**:265.

Flynn, T.G., Charington, B., Lyons, C., Chao, H., Hyndman, D. and Morjana, N., 1989, Chemical modification of aldehyde and aldose reductase by pyridoxal-5'-phosphate, *Prog. Clin. Biol. Res.*, **290**:251.

Foster, M. and Harrison, J.H., 1974, Characterization of porcine malate dehydrogenase II. Amino acid sequence of a peptide containing the active center histidine residue. *Biochim. Biophys. Acta.*, **351**:295.

Halder, A.B. and Crabbe, M.J.C., 1985, Chemical modification studies on purified bovine lens aldose reductase, *Ophthalmic Res.*, **17**:185.

Itabe, H., Sato, S. and Kador, P.F., 1989, Similarities and differences in the inhibition of rat lens aldose reductase and kidney aldehyde reductase by aldose reductase inhibitors. *Invest. Ophthalmol. Vis. Sci.* **30**:193.

Kador, P.F. and Sharpless, N.E., 1983, Pharmacophor requirements of the aldose reductase inhibitor site. *Mol. Pharmacol.*, **24**:521.

Kador, P.F., 1988, The role of aldose reductase in the development of diabetic complications, *Medical Research Review*, **8**:325.

Kador, P.F., Akagi, Y., Takahashi, Y., Ikebe, H., Wyman, M. and Kinoshita, J.H., 1990, Prevention of retinal vessel changes associated with diabetic retinopathy in galactose-fed dogs by aldose reductase inhibitors. *Arch. Ophthalmol.*, in press.

Morjana, N.A., Lyons, C. and Flynn, T.G., 1989, Aldose reductase from human psoas muscle: Affinity labeling of an active lysine by pyridoxal 5'-phosphate and pyridoxal 5'-diphospho-5'-adenosine, *J. Biol. Chem.* **264**:2912.

Nakayama, T., Tanimoto, T. and Kador, P.F., 1989, Human erythrocyte aldose reductase, *Prog. Clin. Biol. Res.*, **290**:265.

O'Brien, M.M., Schofield, P.J. and Edwards, M.R., 1982, Inhibition of human brain aldose reductase, *J. Neurochem.*, **39**:810.

Poulsom, R., 1986, Inhibition of hexonate dehydrogenase and aldose reductase from bovine retina by sorbinil, Statil and valproate. *Biochem. Pharmacol.*, **35**:2955.

Sato, S., Kador, P.F. and Kinoshita, J.H., 1988, Rat kidney aldehyde reductase: Purification and comparison with rat lens aldose reductase, *in*: "Polyol Pathway and its Role in Diabetic Complications," N. Sakamoto, J.H. Kinoshita, P.F. Kador, and N. Hotta, eds, Excepta Medica, Elsevier Science Publishers B.V., Amsterdam. pp.72-81.

Sato, S. and Kador, P.F., 1989, Rat lens aldehyde reductase, *Invest. Ophthalmol. Vis. Sci.*, **30**:1618.

Sato, S., 1990, Polyol formation in rat kidney, *in*: "U.S.-Japan Aldose Reductase Workshop," N. Hotta, ed., in press.

Sato, S. and Kador, P.F., 1990a, Inhibition of aldehyde reductase by aldose reductase inhibitors, *Biochem. Pharmacol*, in press.

Sato, S. and Kador, P.F., 1990b, NADPH-dependent reductases in the dog lens, *Exp. Eye Res.*, **50**:629.

Shiono, T., Sato, S., Reddy, V.N., Kador, P.F. and Kinoshita, J.H., 1987, Rapid purification of rat lens aldose reductase, *Prog. Clin. Biol. Res.*, **232**:317.

Srivastava, S.K., Petrash, J.M., Sdana, I.J., Ansari, N.H. and Partridge, C.A., 1982, Susceptibility of aldehyde and aldose reductases of human tissues to aldose reductase inhibitors, *Curr. Eye Res.*, **2**:407.

Tanimoto T., Kador P.F., 1989, Purification of aldose and aldehyde reductases from EHS tumor cells. *Biochem Pharmacol* **39**:445.

Terada, T., Niwase, N., Shinagawa, K., Koyama,I., Hosomi, S. and Mizoguchi, T., 1989, Bovine liver cytosolic aldehyde reductase and carbonyl reductase. Purification and characterization, *Prog. Clin. Biol. Res.* **290**:293.

Wermuth, B., Munch, J.D.B. and von Wartburg, J.-P., 1977, Purification and Properties of NADPH-dependent aldehyde reductase from human liver, *J. Biol. Chem.* **252**:3821.

A KINETIC PERSPECTIVE ON THE PECULIARITY OF ALDOSE REDUCTASE

Charles E. Grimshaw

Department of Molecular and Experimental Medicine
Research Institute of Scripps Clinic
La Jolla, CA 92037

INTRODUCTION

The reduction of D-glucose to D-sorbitol catalyzed by aldose reductase (ALR2; alditol:NADP$^+$ 1-oxidoreductase; EC 1.1.1.21) has been implicated in the pathogenesis of diabetic complications of the eye, kidney, and nervous system (Dvornik, 1987; Kador, 1988), and as a result, much research effort has been expended in order to understand the function of this enzyme both *in vitro* and *in vivo*. Understanding the mechanistic details of ALR2 catalysis has, however, proven to be rather difficult due to the peculiar character of this monomeric oxidoreductase. In particular, the kinetic properties of ALR2 have been a subject of controversy, with conflicting reports of either Michaelis-Menten type kinetics (Boghosian & McGuinness, 1979; Wermuth et al., 1982; Branlant et al., 1982; Cromlish & Flynn, 1983a,b; Morjana & Flynn, 1989) or nonlinear double-reciprocal plots displaying apparent negative cooperativity for the nucleotide and aldehyde substrate (Sheaff & Doughty, 1976; Hoffman et al., 1980; Daly & Mantle, 1982; Conrad & Doughty, 1982; Halder & Crabbe, 1984; Srivastava et al., 1985; Poulsom, 1986). In addition, changes in catalytic activity and in the susceptibility to inhibition by various aldose reductase inhibitors (ARI) have been reported that are dependent on the substrate assayed and on the purification state of the enzyme (Kador et al., 1983; Maragoudakis et al., 1984; Cromlish & Flynn, 1985; Poulsom, 1987). Recent studies of bovine kidney ALR2 provide a rationale for understanding the peculiar properties of this rather unique enzyme.

Nonlinear double-reciprocal plots

ALR2 can exist in two forms, an activated and an unactivated form, that differ markedly in their kinetic properties, susceptibility to inhibition by various ARI and secondary structure, but are indistinguishable by SDS-PAGE and narrow range isoelectric focusing (Grimshaw et al., 1989). We further showed that the individual contributions of these two forms to the overall reaction rate could quantitatively account for the biphasic double-reciprocal plots observed for a range of aldehyde substrates. Quantitation of the two forms was facilitated by the fact that two of the ARI tested, namely Statil (ICI-128,436; 3-[(4-bromo-2-fluorobenzyl)-4-oxo-3H-phthalazin-1-ylacetic acid]) and AL-1576 (spiro[2,7-difluorofluorene-9,4'-imidazolidine]-2',5'-dione), displayed a 200-fold lower affinity for binding to the activated relative to the unactivated enzyme form. Thus, analysis of the biphasic Dixon plot generated using either Statil of AL-1576 provided direct verification of the level of each of the two enzyme forms established by analysis of the biphasic double-reciprocal plot of $1/v_o$ versus $1/$[aldehyde] using the kinetic parameters determined for the individual forms. Figure 1 shows a simulated Dixon plot for titration with AL-1576 of a mixture containing 90% E_{unact} and 10% E_{act}. (Each form contributes equally to V_{max} since at [glycolaldehyde] = 5 mM, E_{unact} is saturated (K_{unact} = 0.068 mM; $V_{unact} \cdot E_{unact} \approx 0.90$) while E_{act} is only half-saturated (K_{act} = 5 mM; 0.5 $V_{act} \cdot E_{act} \approx 0.87$).

By comparing the effect of ALR2 activation on K_m and V_{max} for a range of aldehyde substrates, we were able to explain the anomalous substrate-dependent changes in activity and ARI potency observed by other investigators. For example, the level of *para*-nitrobenzaldehyde routinely used (0.25 mM) is saturating for both E_{unact} and E_{act}, while 100 mM *D*-glucose is saturating for E_{unact} but only 33% of K_m for E_{act} (Grimshaw et al., 1989). Thus, the average 17-fold increase in V_{max} seen upon ALR2 activation will only be 25% expressed when 100 mM *D*-glucose is used in the assay, compared to 100% expression with 0.25 mM *para*-nitrobenzaldehyde. Similar arguments can be used to explain the substrate dependence of ARI potency. Analysis of the kinetic parameters affected by ALR2 activation led us to propose a kinetic mechanism in which isomerization of the free enzyme form following release of NADP+ was rate-limiting for reaction in the direction of aldehyde reduction (Grimshaw et al., 1989).

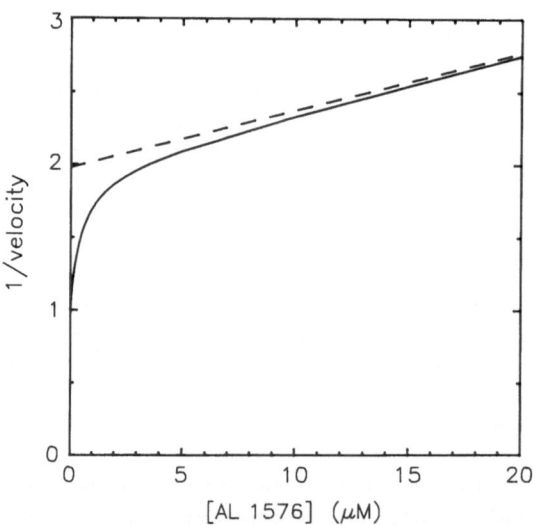

Figure 1. Dixon plot for inhibition by AL-1576. Computer simulation of $1/v_o$ versus [AL-1576] using K_i values determined previously for unactivated ($K_{i\ unact}$ = 0.25 μM) and activated ($K_{i\ act}$ = 50 μM) bovine kidney ALR2 (Grimshaw et al., 1989), and assuming a mixture of 90% E_{unact} and 10% E_{act}, with 5 mM glycolaldehyde, 0.16 mM NADPH, 50 mM Mops buffer (pH 7.0, 25°C). The contribution of the E_{act} component is also shown (- - -).

However, none of the results to this point explain why ALR2 activation has not been more commonly recognized, which is surprising in view of the apparent kinetic, immunologic and structural similarity of ALR2 isolated from several mammalian species (Conrad & Doughty, 1982; Mathur & Grimshaw, 1986; Morjana & Flynn, 1989; Bohren et al., 1989; Chung & LaMendola, 1989; Nishimura et al., 1989; Schade et al., 1990). To address this question we must consider the the mechanism of aldehyde substrate inhibition which leads to nonlinear progress curves for aldehyde reduction.

Substrate inhibition and nonlinear reaction progress curves

As shown in Figure 2, there is pronounced substrate inhibition of the *initial* steady-state rate (v_o) at high concentrations of the aldehyde substrate. Glycolaldehyde is given as

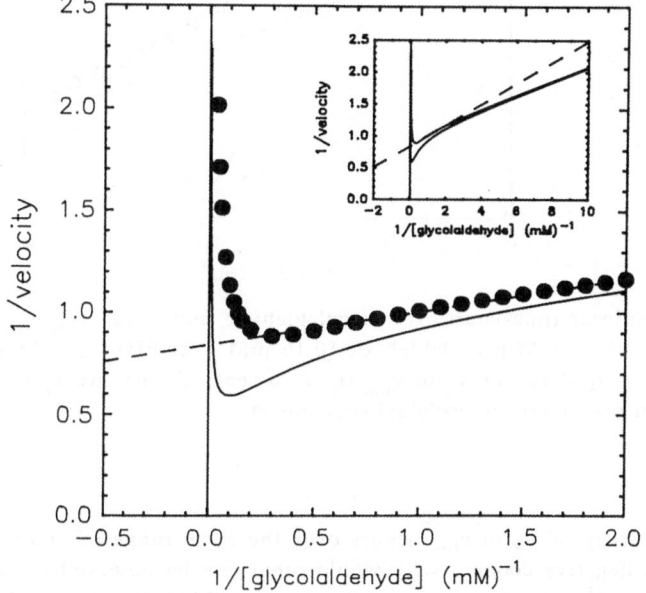

Figure 2. Double-reciprocal plot of $1/v_o$ and $1/v_{lim}$ for glycolaldehyde reduction. Computer simulation based on actual data obtained at pH 7.0, 25°C, for glycolaldehyde reduction at saturating NADPH (0.16 mM) (Grimshaw et al., 1989; 1990b). The curves for $1/v_o$ (——) and $1/v_{lim}$ (• • •) were calculated using an equation for the sum of two Michaelis-Menten terms (90% E_{unact} and 10% E_{act}; V_{unact} = 1.0, K_{unact} = 0.068 mM, V_{act} = 17.3, K_{act} = 5 mM) including substrate inhibition by glycolaldehyde. The additional inhibition seen for v_{lim} is accounted for by the lower K_I value for substrate inhibition (for v_o: K_I = 35 mM; for v_{lim}: K_I = 6.7 mM). The line (- - -) calculated from v_{lim} data for [glycolaldehyde] = 0.5-10 mM corresponds to an apparent K_m = 5.2 mM and V_{max} = 1.2 (actual V_{total} = 2.63).

the example; similar results are seen for essentially all aldehyde substrates tested (Grimshaw et al., 1989). However, for the 2- and 3-carbon aldose substrates, glycolaldehyde and glyceraldehyde, v_o rapidly decays to a lower, *limiting* steady-state rate (v_{lim}), with the $t_{\frac{1}{2}}$ for the first-order decay process decreasing with increasing aldehyde concentration (Grimshaw et al., 1990b). Figure 3 shows a representative progress curve for reduction of 50 mM glycolaldehyde at 15 °C (pH 7.0) and the insert shows a plot of $t_{\frac{1}{2}}$ *versus* [aldehyde]. For reaction of glyceraldehyde at 25 °C (pH 7.0), the decay is even more rapid with $t_{\frac{1}{2}} \approx$ 10 sec for transition from v_o to v_{lim} (McKercher et al., 1985; Grimshaw, 1987).

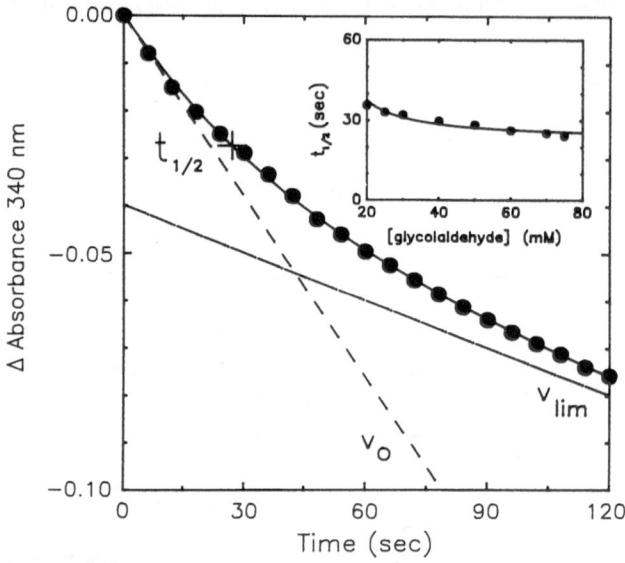

Figure 3. Nonlinear time-course for glycolaldehyde reduction. $A_{340\ nm}$ trace (•-•) for reduction of 50 mM glycolaldehyde (0.16 mM NADPH, 0.62 μM ALR2, pH 7.0, 15°C) showing decay of v_o to v_{lim} ($t_{\frac{1}{2}} \approx$ 28 sec). Shown are v_o (——), v_{lim} (- -), and a plot of $t_{\frac{1}{2}}$ *versus* [glycolaldehyde] (insert).

The rapid decay of v_o to v_{lim} occurs over the same range of aldehyde concentration where apparent negative cooperativity would otherwise be observed for a mixture of E_{act} and E_{unact}. Thus, if one is not aware of the decay problem the measured rate may actually be equal to v_{lim}, and this error can lead to incorrect values for K_m and V_{max} and a lack of recognition of the presence of two enzyme forms. Figure 2 shows a simulation, based on actual data, of the double-reciprocal plots for $1/v_o$ and $1/v_{lim}$ *versus* 1/[glycolaldehyde] for a mixture of 90% E_{unact} and 10% E_{act}. As is apparent from the two plots, assay data collected as v_{lim} over the limited aldehyde concentration range normally employed in such studies (e.g., 0.1 to 2 x K_m) can give an apparently linear double-reciprocal plot. The transient decay to v_{lim} is only seen for substrates containing an enolizable α-proton (see below). Thus, one must take care to determine the true initial rate, or use a nonenolizable substrate (e.g., *para*-nitrobenzaldehyde) or an enolizable substrate containing a low amount of the free aldehyde form (e.g., *D*-xylose, *D*-glucose).

Aldehyde substrate inhibition of v_o is uncompetitive *versus* NADPH (Grimshaw et al., 1990b), indicating that formation of the dead-end E•NADP•aldehyde complex prevents the release of $NADP^+$ (Cleland, 1963). The time-dependent onset of further inhibition occurs when the two substrates bound at the enzyme active site in this dead-end complex undergo a chemical reaction to generate a covalent NADP-aldehyde adduct. Substrate inhibition of ALR2 thus occurs via a mechanism similar to that described for pyruvate inhibition of lactate dehydrogenase (Burgner et al., 1978). More precisely, detailed kinetic and mechanistic studies of the ALR2-mediated and nonenzymic adduct formation reactions have shown that chemical reaction of the enol form of the aldehyde substrate with E•NADP to give E•adduct directly competes with the binding of the aldehyde (plus hydrate) form to generate the dead-end complex (Grimshaw et al., 1990a,b). At high enzyme concentrations the rate of enol formation in solution is rate-limiting for the overall adduct formation process. Adduct formation can be monitored in the absence of aldehyde turnover since ALR2, unlike alcohol dehydrogenase (Dalziel & Dickinson, 1965), does not catalyze NADP-dependent oxidation of aldehydes to their corresponding acids.

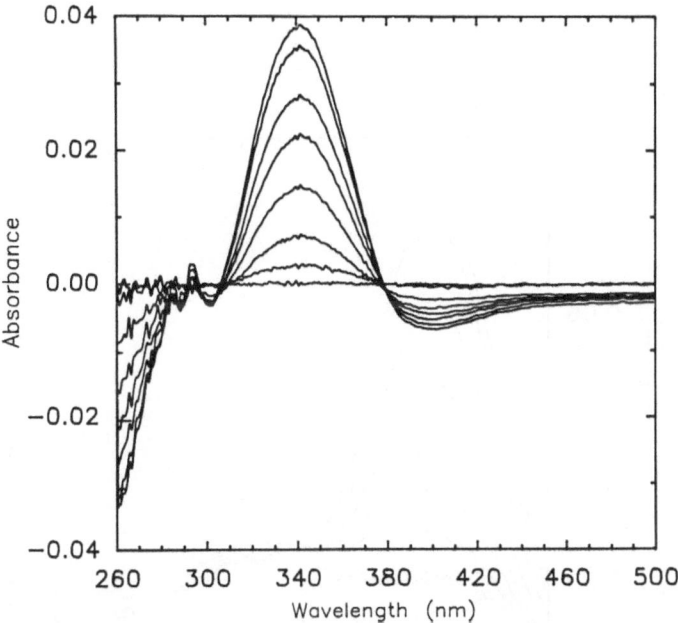

Figure 4. Spectrophotometric demonstration of ALR2-mediated NADP-glycolaldehyde adduct formation. The reaction mixture contained 20 μM $NADP^+$, 50 mM glycolaldehyde, and 7.5 μM ALR2 in Mops buffer (pH 7.0) at 15°C. Spectra shown are difference spectra corrected for the initial difference spectrum [(sample - reference)$_t$ - (sample - reference)$_{t=0}$] measured at t = 0, 1, 2, 4, 6, 8, 10, and 30 min after addition of glycolaldehyde. The rate ($\Delta A_{341\ nm}/\Delta t \approx 0.14$/min), corrected to 0.62 μM ALR2 (cf. Figure 3) yields a $t_\frac{1}{2} \approx 26.5$ sec.

Figure 4 shows a UV-visible reaction time-course for E•adduct formation (λ_{max} = 341 nm) from E•NADP and glycolaldehyde in the absence of aldehyde turnover. The $t_{\frac{1}{2}}$ obtained from analysis of these data corresponds to the $t_{\frac{1}{2}}$ for decay of v_o to v_{lim} determined in the presence of aldehyde turnover. Thus, the reaction process which accounts for the time-dependent onset of inhibition is clearly identified as adduct formation. The enhanced inhibition is a direct result of the extremely tight binding of the of what is essentially a bisubstrate analogue. The K_d for the NADP-glycolaldehyde adduct (Structure I) is estimated to be 2 x 10^{-12} M (Grimshaw et al., 1990b).

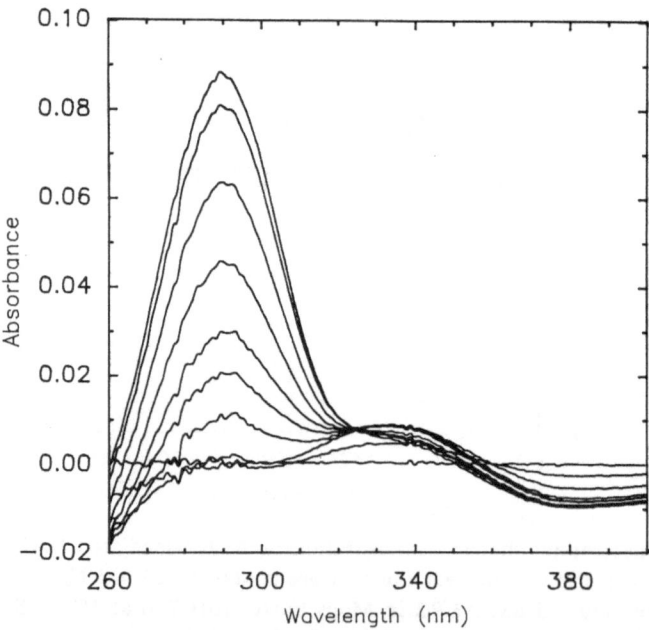

NADP-Glycolaldehyde Adduct
Structure I

Figure 5. Spectrophotometric demonstration of ALR2-mediated NADP-glyceraldehyde adduct formation. The reaction mixture contained 20 μM NADP$^+$, 50 mM glyceraldehyde, and 7.5 μM ALR2 in Na-phosphate buffer (pH 7.0) at 15°C. Spectra shown are difference spectra corrected for the initial difference spectrum [(sample - reference)$_t$ - (sample - reference)$_{t=0}$] measured at t = 0, 2, 8, 24, 32, 45, 60, 90, 145 and 180 min after addition of *DL*-glyceraldehyde.

For glyceraldehyde, NADP-aldehyde adduct formation is followed by a second reaction that generates a new chromophore absorbing maximally at 292 nm in the difference spectrum (Figure 5). This second product has not yet been identified, but its appearance is dependent on the configuration of the α-hydroxyl group since both *DL-* and *L*-glyceraldehyde display the 292 nm peak, while the preferred substrate, *D*-glyceraldehyde, does not (C. Grimshaw in preparation). In any case, the net effect is the same with v_o rapidly decaying to v_{lim} due to initial formation of the dead-end E•NADP•glyceraldehyde complex followed by E•adduct formation.

Nonenzymic side reactions

Several nonenzymic aldehyde-dependent side reactions have been analyzed to determine to what extent and under what conditions they contribute to the overall rate of NADPH consumption. Complex spectral changes result when glycolaldehyde and NADPH are allowed to react in the absence of enzyme (Figure 6). However, by examining the adenosine and nicotinamide fragments independently, we have been able to decipher the various contributing reactions.

The 272 nm peak is ascribed to reaction of the aldehyde with the 6-NH_2 moiety of adenine to generate a carbinolamine product (Structure II), since the observed spectral changes are similar to those reported for reaction of aldehydes (Fraenkel-Conrat & Singer,

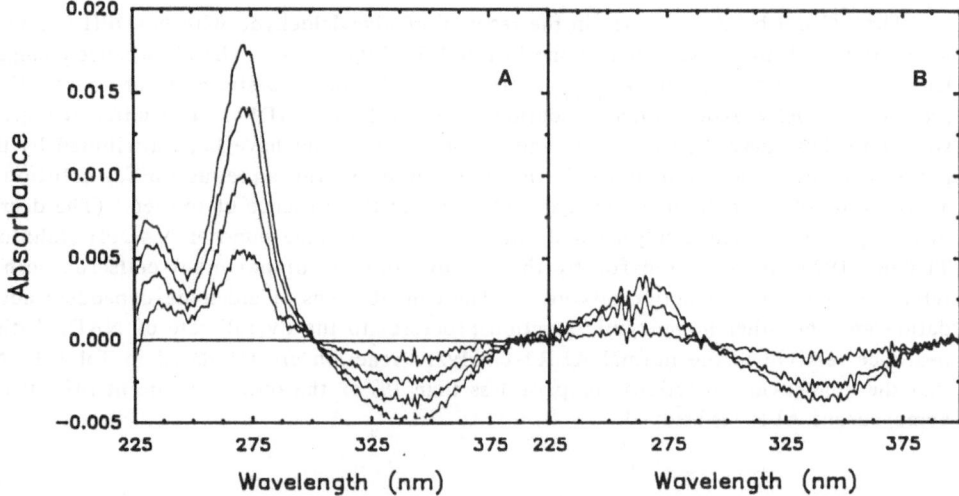

Figure 6. Spectrophotometric demonstration of the nonenzymic reaction of NADPH with glycolaldehyde. The reaction mixture contained 50 µM NADPH and 90 mM glycolaldehyde in Mops buffer (pH 7.0) at 25°C. Spectra shown are difference spectra corrected for the initial difference spectrum [(sample - reference)$_t$ - (sample - reference)$_{t=0}$]. Panel A: spectra measured at t = 0, 3, 6, 9, and 12 min after starting the reaction by addition of glycolaldehyde. The maxima occur at 272 nm (increase due to carbinolamine formation with adenine 6-NH_2) and 340 nm (decrease due to NADPH oxidation); Panel B: spectra measured at t = 0, 7.5 and 15 min after the difference spectrum was again zeroed at 45 min elapsed reaction time. At this point carbinolamine formation has reached equilibrium and only NADPH oxidation is observed.

1988) and epoxides (Windmueller & Kaplan, 1961) with the adenine ring. Figure 7A shows the reaction of glycolaldehyde with ADP-ribose; a similar spectral time-course was seen for reaction of any adenine containing nucleotide (Grimshaw et al., 1990b). However, replacement of the 6-NH_2 group by an -OH, as in the hypoxanthine containing analogue of $NADP^+$, eliminates the 272 nm product (Figure 7B). In that case, only nonenzymic adduct formation is observed ($\lambda_{max} \approx 345$ nm). ALR2 is quite sensitive to substitution at the 6-position of adenine (see below, Table II). As a result, incubation of glycolaldehyde with NADPH prior to starting the reaction by addition of enzyme leads to an apparent competitive component for substrate inhibition by glycolaldehyde versus NADPH simply due to depletion of the unmodified nucleotide substrate (Grimshaw et al., 1990b). Other aldehyde utilizing enzymes sensitive to adenine substitution may be similarly affected.

Adenine-glycolaldehyde Carbinolamine
Structure II

The 272 nm band seen early in the reaction of glycolaldehyde with NADPH is thus due to reaction with the 6-NH_2 of adenine in NADPH (Figure 6A). The absorbance changes at later times (Figure 6B), with $A_{340\ nm}$ decreasing above an isosbestic point of about 297 nm, are due to aldehyde-dependent oxidation of NADPH to $NADP^+$. The latter changes are similar to those described by Wolff and Crabbe (1985), and have been attributed by these authors to the generation of oxidizing equivalents in the aqueous buffer solution via autoxidation of the α-hydroxyaldehyde substrates in the presence of oxygen. (The decrease in $A_{340\ nm}$ due to acid-catalyzed hydration of the 5,6-double bond of NADPH (Johnson & Tuazon, 1977) is corrected for by the use of tandem compartment cells for both the reference and sample reaction mixtures.) The contributions of aldehyde-dependent autoxidation and the other nonenzymic reaction processes to the overall rate of NADPH disappearance relative to the normal ALR2-catalyzed reaction are tabulated in Table I. Note that the nonenzymic reactions comprise less than 1% of the observed rate at pH 7.0 when homogeneous ALR2 is assayed.

Table I. Relative contribution of nonenzymic side-reactions to NADPH consumption

reaction	percent of $\Delta A_{340\ nm}$[a]
ALR2 + NADPH + glycolaldehyde	99.1
NADPH (hydration)	0.7
NADPH + glycolaldehyde (autoxidation)	0.2

[a] Reaction conditions: 50 mM glycolaldehyde, 100 μM NADPH, 50 mM Mops buffer (pH 7.0) at 25°C.

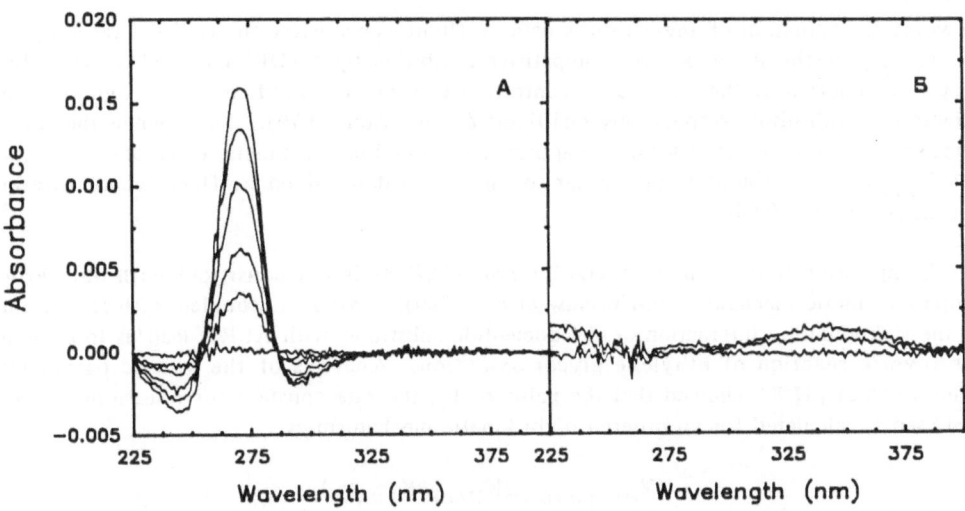

Figure 7. Spectrophotometric demonstration of the nonenzymic reaction of glycolaldehyde with adenine-containing nucleotides. Reaction mixtures contained 50 μM ADP-ribose (panel A) or hypoxanthine-NADP$^+$ (panel B) and 90 mM glycolaldehyde in Mops buffer (pH 7.0) at 25°C. Panel A: spectra measured at t = 0, 3, 6, 9, 12, and 15 min; the maximum occurs at 272 nm (increase due to carbinolamine formation with adenine 6-NH$_2$). Panel B: spectra measured at t = 0, 7.5 and 15 min; the maximum occurs at 345 nm (increase due to covalent adduct formation) with no change detected at 272 nm.

Tight-binding kinetics for NADP$^+$ and NADPH

The dissociation constants for NADP$^+$ and NADPH binding determined by fluorescence titration were both found to be less than 100 nM (Table II), making ALR2 unique among NAD(P)-dependent oxidoreductases. Furthermore, the K_d value for NADPH was much lower than the K_{NADPH} and $K_{i\ NADPH}$ values determined from initial velocity studies conducted using a conventional spectrophotometric assay method (Grimshaw et al., 1989).

Table II. Equilibrium dissociation constants for nucleotides

nucleotide	K_d (μM)[a]
NADP$^+$	0.080 ± 0.030
NADPH	0.046 ± 0.015
N(Hx)DP$^+$	290 ± 30
N(Hx)DPH	≥ 300[b]

[a] K_d values determined from quenching of protein fluorescence (Mops buffer, 15°C). N(Hx)DP(H) is the hypoxanthine-containing analogue of NADP(H).
[b] Lower limit only, due to large inner filter effect.

However, reevaluation of these results using a fluorescence assay showed that the K_{NADPH} and $K_{i\ NADP}$ (estimated as K_{is} for competitive inhibition by NADP$^+$ *versus* NADPH) values vary as a function of the enzyme concentration in a manner predicted for a tight-binding substrate or inhibitor, respectively (Williams & Morrison, 1979). Thus, when the initial velocity data were analyzed using the appropriate tight-binding kinetic equations, $K_{i\ NADPH}$ and $K_{i\ NADP}$ were found to be similar to the values determined by fluorescence titration (Grimshaw et al., 1990b).

Competitive inhibition by NADP$^+$ *versus* NADPH is not consistent with our earlier proposed kinetic mechanism (Grimshaw et al., 1989). This fact, coupled with recognition of the tight-binding interaction of the nucleotide substrates with ALR2, lead us to examine the reverse reaction of ethylene glycol oxidation. Analysis of the kinetic parameters determined at pH 8.0 showed that the value of k_7, the rate constant for release of NADP$^+$ (0.02 sec^{-1}) calculated for an ordered bi-bi kinetic mechanism as:

$$(V_{ethylene\ glycol}/K_{NADP} \cdot K_{i\ NADP})$$

was lower than $V_{glycolaldehyde}/E_t$ (0.18 sec^{-1}) measured for reaction in this direction (Grimshaw et al., 1990b). This clearly cannot be the case, since each first order rate constant must equal or exceed V_{max}/E_t for reaction in a given direction. However, Cleland (1963) has shown that if E•NADP isomerizes the above collection of kinetic constants is no longer equal to k_7, but instead $[k_7 k_9/(k_7 + k_{10})]$, which *can* have a value less than $V_{glycolaldehyde}/E_t$. The initial velocity results at pH 8.0 are thus consistent with the following kinetic mechanism:

NADPH	aldehyde		alcohol		NADP$^+$
$k_1 \downarrow k_2$	$k_3 \downarrow k_4$		$k_5 \uparrow k_6$		$k_7 \uparrow k_8$

$$E \qquad E\text{•}NADPH \qquad (E\text{•}NADPH\text{•}aldehyde \rightleftharpoons E\text{•}NADP\text{•}alcohol) \qquad E\text{•}NADP \underset{k_{10}}{\overset{k_9}{\rightleftharpoons}} {}^*E\text{•}NADP \qquad E$$

which includes isomerization of E•NADP as a kinetically significant step. Combined with the kinetic analysis of ALR2-mediated covalent NADP-aldehyde adduct formation in the presence and absence of aldehyde turnover (Grimshaw et al., 1990b), these data suggest that the net rate constant for isomerization of E•NADP and release of NADP$^+$ is at least 60% rate-limiting for glycolaldehyde reduction at pH 7.0. E•nucleotide isomerization is probably the slow step, however, since V_{max}/E_t determined for the hypoxanthine analogue of NADPH which binds much less tightly is equal to that for NADPH (Grimshaw et al., 1990b). Flynn and coworkers (this volume) have also obtained kinetic evidence using pig muscle ALR2 that support isomerization of E•NADP and possibly E•NADPH as kinetically important steps in the overall reaction mechanism.

CONCLUSION

ALR2 displays rather peculiar kinetic behavior with respect to both the aldehyde and nucleotide substrates. By combining a battery of techniques to decipher the kinetic and mechanistic basis for the various processes that contribute to the observed reaction velocity for the bovine kidney enzyme, we have been able to rationalize essentially all of the anomalous kinetic behavior reported to date. These results provide a framework which can be used in future studies to facilitate determination of the molecular basis for the unique properties of ALR2.

ACKNOWLEDGEMENTS

Supported by a grant from the National Institute for Diabetes, Digestive and Kidney Diseases (DK 32218). This is publication number 6499-MEM from the Research Institute of Scripps Clinic, Scripps Clinic and Research Foundation.

REFERENCES

Boghosian, R.A. & McGuinness, E.T. (1979) Affinity purification and properties of porcine brain aldose reductase. *Biochim. Biophys. Acta* **567**:278.

Bohren, K.M., Bullock, B., Wermuth, B. & Gabbay, K.H. (1989) The aldo-keto reductase superfamily. cDNAs and deduced amino acid sequences of human aldehyde and aldose reductases. *J. Biol. Chem.* **264**:9547.

Branlant, G. (1982) Properties of an aldose reductase from pig lens. *Eur. J. Biochem.* **129**:99.

Burgner, J.W., II, Ainslie, G.R., Jr., Cleland, W.W. & Ray, W.J., Jr. (1978) Bimodal substrate inhibition of lactate dehydrogenase. Factors affecting the enzyme *in vivo*. *Biochemistry* **17**:1646.

Chung, S. & LaMendola, J. (1989) Cloning and sequence determination of human placental aldose reductase. *J. Biol. Chem.* **264**:14775.

Conrad, S.M. & Doughty, C.C. (1982) Comparative studies on aldose reductase from bovine, rat and human lens. *Biochim. Biophys. Acta* **708**:348.

Cleland, W.W. (1963) The kinetics of multireactant enzyme-catalyzed reactions with two or more substrates. 1. Nomenclature and rate equations. *Biochim. Biophys. Acta* **67**:104.

Cromlish, J.A. & Flynn, T.G. (1983a) Purification and characterization of two aldose reductase isozymes from rabbit muscle. *J. Biol. Chem.* **258**:3416.

Cromlish, J.A. & Flynn, T.G. (1983b) Pig muscle aldehyde reductase. Identity of pig muscle aldehyde reductase with pig lens aldose reductase and with the low Km aldehyde reductase of pig brain and pig kidney. *J. Biol. Chem.* **258**:3583.

Cromlish, J.A. & Flynn, T.G. (1985) Identification of pig brain aldehyde reductases with the high-Km aldehyde reductase, the low-Km aldehyde reductase and aldose reductase, carbonyl reductase, and succinic semialdehyde reductase. *J. Neurochem.* **44**:1485.

Daly, A.K. & Mantle, T.J. (1982) Purification and characterization of the multiple forms of aldehyde reductase in ox kidney. *Biochem. J.* **205**:373.

Dalziel, K. & Dickinson, F.M. (1965) Aldehyde mutase. *Nature* **206**:255.

Dvornik, D. (1987) Hyperglycemia in the pathogenesis of diabetic complications, in *Aldose Reductase Inhibition* (Porte, D., Ed) pp 69-152, McGraw-Hill, New York.

Fraenkel-Conrat, H. & Singer, B. (1988) Nucleoside adducts are formed by cooperative reaction of acetaldehyde and alcohols. Possible mechanism for the role of ethanol in carcinogenesis. *Proc. Natl. Acad. Sci. (USA)* **85**:3758.

Grimshaw, C.E., Shahbaz, M., Jahangiri, G., Putney, C.G., McKercher, S.R. & Mathur, E.J. (1989) Kinetic and structural effects of activation of bovine kidney aldose reductase. *Biochemistry* **28**:5343.

Grimshaw, C.E., Shahbaz, M. & Putney, C.G. (1990a) Spectroscopic and kinetic characterization of nonenzymic and aldose reductase-mediated covalent NADP-glycolaldehyde adduct formation. *Biochemistry* in press.

Grimshaw, C.E., Shahbaz, M. & Putney, C.G. (1990b) Mechanistic basis for nonlinear kinetics of aldehyde reduction catalyzed by aldose reductase. *Biochemistry* in press.

Grimshaw, C.E. (1987) Analysis of enzyme-catalyzed nucleotide modification by aldose reductase. *Proc. Fed. Amer. Soc. Exp. Biol.* **46**:2226.

Halder, A.B. & Crabbe, M.J.C. (1984) Bovine lens aldehyde reductase (aldose reductase): purification, kinetics and mechanism. *Biochem. J.* **219**:33.

Hoffman, P.L., Wermuth, B. & von Wartburg, J.P. (1980) Human brain aldehyde reductases. Relationship to succinic semialdehyde reductase and aldose reductase. *J. Neurochem.* **35**:354.

Johnson, S.L. & Tuazon, P.T. (1977) Acid-catalyzed hydration of reduced nicotinamide adenine dinucleotide and its analogues. *Biochemistry* **16**:1175.

Kador, P.F., Shiono, T. & Kinoshita, J.H. (1983) Studies with purified aldose reductase. *Invest. Ophthamol. Vis. Sci.* **24 (Suppl)**:267.

Kador, P.F. (1988) The role of aldose reductase in the development of diabetic complications. *Medicinal Res. Rev.* **8**:325.

Maragoudakis, M.E., Wasvary, J., Hankin, H. & Garguilo, P. (1984) Human placenta aldose reductase. Forms sensitive and insensitive to inhibition by Alrestatin. *Molec. Pharmacol.* **25**:425.

Mathur, E.J. & Grimshaw, C.E. (1986) Phylogenetic conservation of epitopes in mammalian aldose reductase. *Arch. Biochem. Biophys.* **247**:321.

McKercher, S.R., Mathur, E.J. & Grimshaw, C.E. (1985) Bovine kidney aldose reductase: microheterogeneity and kinetic anomalies. *Proc. Fed. Amer. Soc. Exp. Biol.* **44**:472.

Morjana, N.A. & Flynn, T.G. (1989) Aldose reductase from human psoas muscle. Purification, substrate specificity, immunological characterization, and effect of drugs and inhibitors. *J. Biol. Chem.* **264**:2906.

Nishimura, C., Matsuura, Y., Kokai, Y., Akera, Y., Akera, T., Carper, D., Morjana, N.A., Lyons, C. & Flynn, T.G. (1990) Cloning and expression of human aldose reductase. *J. Biol. Chem.* **265**:9788.

Poulsom, R. (1986) Inhibition of hexonate dehydrogenase and aldose reductase from bovine retina by Sorbinil, Statil, M79175 and valproate. *Biochem. Pharmacol.* **36**:2955.

Poulsom, R. (1987) Comparison of aldose reductase inhibitors *in vitro*. Effects of enzyme purification and substrate type. *Biochem. Pharmacol.* **36**:1577.

Schade, S.Z., Early, S.R., Williams, R.T., Kézdy, F.J., Heinrikson, R.L., Grimshaw, C.E. & Doughty, C.C. (1990) Sequence analysis of bovine lens aldose reductase. *J. Biol. Chem.* **265**:3628.

Sheaff, C.M. & Doughty, C.C. (1976) Physical and kinetic properties of homogeneous bovine lens aldose reductase. *J. Biol. Chem.* **251**:2696.

Srivastava, S.K., Hair, G.H. & Das, B. (1985) Activated and unactivated forms of human erythrocyte aldose reductase. *Proc. Natl. Acad. Sci. (USA)* **82**:7222.

Wermuth, B., Bürgisser, H., Bohren, K. & von Wartburg, J.-P. (1982) Purification and characterization of human brain aldose reductase. *Eur. J. Biochem.* **125**:279.

Williams, J.W. & Morrison, J.F. (1979) The kinetics of reversible tight-binding inhibition. *Methods Enzymol.* **63**:437.

Windmueller, H.G. & Kaplan, N.O. (1961) The preparation and properties of N-hydroxyethyl derivatives of adenosine, adenosine triphosphate, and nicotinamide adenine dinucleotide. *J. Biol. Chem.* **236**:2716.

Wolff, S.P. & Crabbe, M.J.C. (1985) Low apparent aldose reductase activity produced by monosaccharide autoxidation. *Biochem. J.* **226**:625.

BEST-FIT ANALYSIS OF KINETIC SCHEME FOR THE STEPWISE REDUCTION OF THE "DIKETO" GROUP OF 6-PYRUVOYL TETRAHYDROPTERIN BY SEPIAPTERIN REDUCTASE

Terumi Sueoka, Harumi Hikita* and Setsuko Katoh

Department of Biochemistry, Meikai University
School of Dentistry, Sakado, Saitama 350-02, Japan
*Laboratory of Physics, Meikai University, Urayasu
Chiba, 279, Japan

INTRODUCTION

Sepiapterin reductase [EC 1.1.1.153](SPR) is an enzyme required in the biosynthesis of tetrahydrobiopterin (Katoh and Akino, 1986), an essential H-donor cofactor of aromatic amino acid hydroxylases such as tyrosine hydroxylase (Kaufman, 1986). Recently we found (Katoh and Sueoka, 1984; Sueoka and Katoh, 1985) that this enzyme belongs to the "Aldo-keto reductases" group (Tuner and Flynn, 1982; Wermuth, 1985). SPR can reduce various carbonyl compounds including "diketo-" compounds such as phenylpropanedione and diacetyl with NADPH (Katoh and Sueoka, 1984). The natural substrate of this enzyme, 6-pyruvoyl tetrahydropterin(6(R)-L-1',2'-dioxopropyl 5,6,7,8-tetrahydropterin)(PPH$_4$) is also a "diketo-" compound, and the vicinal "diketo" group (C1'-keto and C2'-keto) in the molecule was found to be reduced successively with NADPH by SPR to a "dihydrodiol" group to form tetrahydrobiopterin (BH$_4$) (Masada et al., 1985: Curtius et al., 1985; Milstien and Kaufman, 1985; Brown et al., 1985; Smith and Nichol, 1986). In the previous works, only one type of mono-keto derivative (C2'-keto type) was found as the intermediate of SPR during the reduction of PPH_4 at the usual assay pH (neutral pH). But recently we detected both types of the mono-keto derivatives (C1'-keto type and C2'-keto type) as the intermediate of the reaction if the reaction was performed at a pH lower than neutral (Katoh and Sueoka, 1990)(Fig. 1).

In the present study we investigated the kinetic scheme for this enzyme by using a computer program in order to elucidate the reaction mechanism that accounts for the appearence of these two types of the mono-keto intermediate during the reduction of PPH_4 to BH_4 (Fig. 1).

Assay of PPH_4 reduction by SPR

SPR was purified from rat erythrocytes (Sueoka and Katoh, 1982). PPH_4 was catalytically synthesized from dihydroneopterin triphosphate and PPH_4 synthase in the presence of $MgCl_2$ (Masada et al., 1985). Dihydroneopterin was synthesized from GTP by the function of GTP cyclohydrolase I prepared from E. coli (Yim and Brown, 1976), and was purified by column chromatography procedures (Yoshioka et al., 1983). Steady-state reaction of SPR against PPH_4 was carried out in the reaction mixture

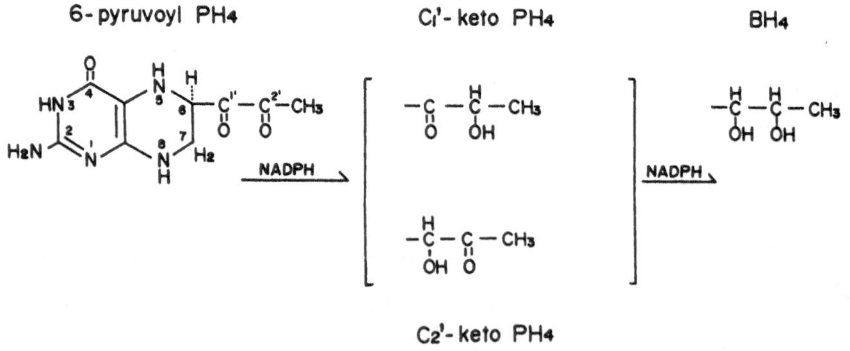

Fig. 1. Reduction of 6-puruvoyl tetrahydropterin (pyruvoyl PH_4) to tetrahydrobiopterin (BH_4) by sepiapterin reductase (SPR) and observed intermediates.

containing SPR, PPH$_4$, and an excess amount of NADPH in 50 mM Tris-HCl (pH 8.6) buffer or 0.3 M glycine-0.15 M NaCl-HCl (pH 3.8) buffer at 37 °C in the dark (Katoh and Sueoka, 1990, Sueoka et al.,in preparation). An aliquot of the mixture was separated after an appropriate time, and the reaction was stopped by the addition of 2 M trichloro-acetic acid equivalent to 0.14 vol. of the reaction mixture. The filtrate of the final solution was analyzed by HPLC. Tetrahydropterins of PPH$_4$, monoketo intermediate, and BH$_4$ were measured electrochemically (Milstien and Kaufman, 1985; Smith and Nichol, 1986).

Theory

SPR catalyzes the reduction of two carbonyl groups of PPH$_4$ with NADPH. This double reduction may occur stepwise through a mono-keto derivative that is produced from PPH$_4$ by reduction of one of the two carbonyl groups (Fig. 1). Thus two pathways, through C1'-keto or C2'-keto type of tetrahydropterin (PH$_4$), are possible for the sequential reduction of PPH$_4$ to the final product, BH$_4$. In addition, we recently found another new function of SPR, that is, catalysis of NADP-stimulated isomerization of C1'-keto PH$_4$ to C2'-keto PH$_4$ (Katoh and Sueoka, 1987, 1988; Sueoka and Katoh, 1989). Thus, another pathway, one proceeding from C1'-keto PH$_4$ to C2'-keto PH$_4$, has

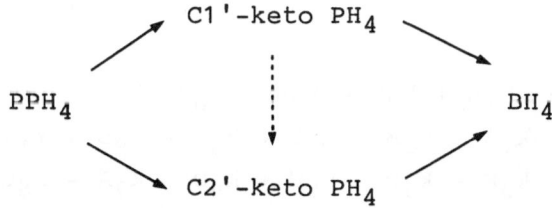

Fig. 2. Possible pathways in the conversion of 6-pyruvoyl tetrahydropterin (PPH$_4$) into tetrahydro-biopterin (BH$_4$)by sepiapterin reductase function.

(reduction, ⟶; isomerization, ┄►)

also been proposed besides the reducing pathways during the reduction of PPH_4 to BH_4, as shown in Fig. 2.

In the present discussion of the enzymatic reaction, only the interconversion of substrate, intermediate species, and final product of pteridine will be considered to aid in the understanding of the outline of this complicated reaction mechanism. The experimental data were analyzed by approximation of our system by a computer program to the compartmental model of the kinetic scheme:

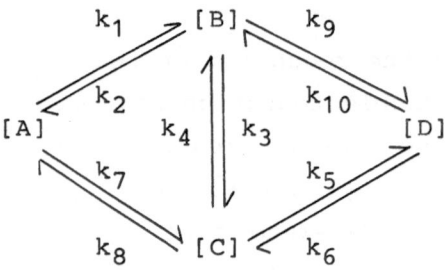

In this scheme [A], [B], [C] and [D] represent the concentration of each reactant in the reaction mixture, PPH_4, C1'-keto PH_4, C2'-keto PH_4, and BH_4, respectively; and k_1 - k_{10} are the first-order rate constants. Steady-state conditions give the following equations for the variations of PPH_4, C1'-keto PH_4, C2'-keto PH_4, and BH_4 concentrations with time:

$$dA/dt = -k_1A + k_2B - k_7A + k_8C \tag{1}$$
$$dB/dt = k_1A - k_2B - k_3B + k_4C - k_9B + k_{10}D \tag{2}$$
$$dC/dt = k_3B - k_4C - k_5C + k_6D + k_7A - k_8C \tag{3}$$
$$dD/dt = k_5C - k_6D + k_9B - k_{10}D \tag{4}$$

$$\text{at } t=0, \quad [A]=[A_0]=1, \ \& \ [B]=[C]=[D]=0 \tag{5}$$
$$[A] + [B] + [C] + [D] = 1 \tag{6}$$

Data analysis and simulation

Simulations were carried out in which the time course of the concentrations of reactants taken from the experimental data was modeled by use of the solution obtained by Laplace transformation of Eqns. (1)-(6). The resulting parameters were then fitted by non-linear least-squares regression (Marcardt algorithm and Damping Gauss Newton algorithm) by use of a numerical iterative procedure. Computations were performed with a program written in BASIC and run on an EPSON 286 VS computer. All the programs were developed in our laboratory. Then the time course curves were plotted and the best-fit parameters of rate constants were obtained.

Determination of kinetic scheme and rate constants

Data from steady-state experiments are shown as points in Figures 3 and 4, and the lines are the result of best-fit analyses. At neutral pH (Fig. 3), as reported by previous workers, only the C2'-keto PH_4 seemed to be the intermediate of the reaction (Katoh and Sueoka, 1990), for no significant amount of C1'-keto PH_4 was detected during the reaction at this pH. However, at acidic pH (Fig. 4) the C1'-keto PH_4 was clearly observed(Katoh and Sueoka, 1990, Sueoka et al.,in preparation). When the theoretical curves obtained by simulation of Eqns (1)-(4) and by use of kinetic parameters calculated by non-linear least-squares regression were superimposed on the data experimentally obtained, a good agreement between the predicted behaviour and the experimental results was obtained, as seen in both Figures 3 and 4.

From the best-fit calculated values of rate constants $(x10^{-3} \cdot sec^{-1})$, k_1 (=0.24), k_3 (=43) and k_5 (=0.59) were predominant over the other constants at neutral pH (Fig. 3); and at acidic pH k_1 (=3.2), k_3 (=0.74), k_5 (=0.55), and k_7(=6.4) were predominant (Fig. 4). These values indicate that the reaction at neutral pH proceeds in the course of A→B→C→D and that the process of B→C goes rapidly. The value of k_1/k_3

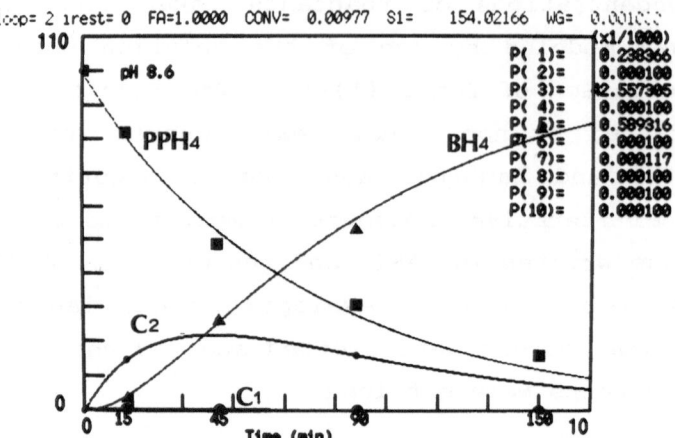

Fig. 3. Kinetics of SPR-reduction of PPH_4 at pH 8.6.
Points are experimental data showing the percent
of each reactant (vertical) as a function of time.
The lines were drawn by best-fit analysis. The
substrate used is shown by the lines. P(1)-P(10)
show the best-fit parameters of the rate constants
($x10^{-3}$ sec^{-1}). Points and the line for C1'-keto
compound are drawn on the abscissa.

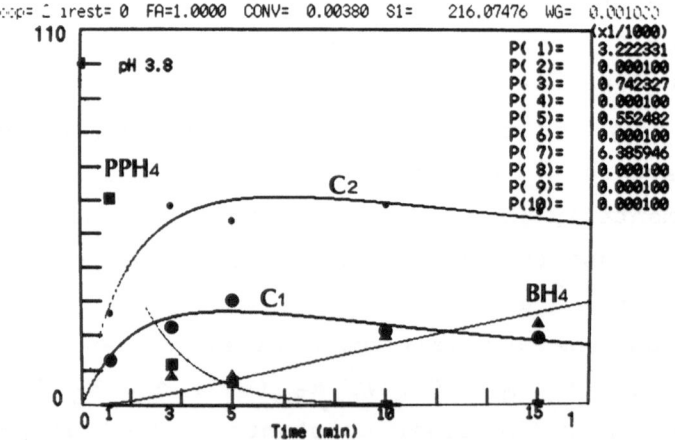

Fig. 4. Kinetics of SPR-reduction of PPH_4 at pH 3.8.
See caption to Figure 3 for details.

234

approximates the degree of the transient accumulation of B in the reaction (Table I). Since k_3 is quite larger than k_1, accumulation of B was hardly detectable at neutral pH (Fig. 3). At acidic pH, on the contrary, k_3 is smaller than k_1. The ratio at acidic pH was about 800-fold greater than that at neutral pH, as shown in Table I, which explains why accumulation of B occurs more easily at acidic pH than at neutral pH. Thus C1'-keto PH_4 could be clearly observed at acidic pH.

Table I. Value of k_1/k_3 in the reaction.

$$A \xrightarrow{\quad k_1 \quad} B \xrightarrow{\quad k_3 \quad} C$$

	pH 8.6	pH 3.8
k_1/k_3	0.0056	4.3

From the values of k constants obtained at acidic pH, besides the reaction of A→B→C→D with a quite slow B→C process, a direct course from A to C (A→C→D) is also available. The reason for the possibility of the two pathways may be the protonation of PPH_4 at the N atom of 5 position of the pteridine ring. This protonation may decrease the polarization of the keto group at C1' position and then the isomerization of protonated C1'-keto PH_4 to C2'-keto PH_4 (B→C reaction) may decline. The pKa value of PPH_4 is expected to be quite similar to that of C1'-keto PH_4 which was previously measured as 3.6 by Smith and Nichol (1986). In the reaction system at pH 3.8, both protonated and unprotonated PPH_4 can coexist, and then slow and rapid isomerization (B→C reaction) would occur simultaneously. The conversion from A to C including rapid B→C process would apparently be a direct conversion of A to C when it was compared with the other conversion of A to C including slow B→C process in the same system. Thus, even at acidic pH, the reaction may proceed basically in the course of A→B→C→D.

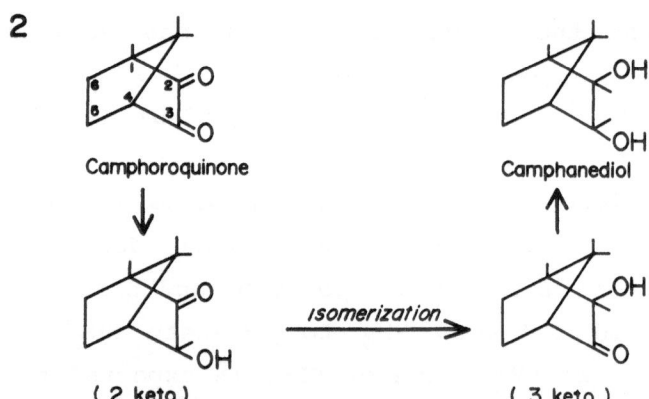

1

Pyruvoyl tetrahydropterin   Tetrahydrobiopterin

(1' keto)   *isomerization*   (2' keto)

2

Camphoroquinone   Camphanediol

(2 keto)   *isomerization*   (3 keto)

Fig. 5. Reaction mechanism of sepiapterin reductase
by best-fit analysis (1) and of 3-hydroxysteroid
dehydrogenase demonstrated by Boutin (1986)(2).

236

Therefore, it appears from the best-fit analysis of the experimental data that the reaction proceeds in the order of A→B→C→D. Thus, PPH_4 is first reduced to C1'-keto PH_4, followed by isomerization to C2'-keto PH_4, and then is reduced to BH_4 by SPR function (Fig. 5). This result coincides with that concluded in our previous work on SPR (Katoh and Sueoka, 1988 and 1990).

DISCUSSION

It had previously been believed that only the C2'-keto PH_4 is the intermediate of the SPR reaction. However, as mentioned above, best-fit analyses showed both C1'-keto PH_4 and C2'-keto PH_4 are the intermediate of the reaction. The reaction mechanism seems to be complicated, but it is not curious. There are some previous descriptions of enzymatic reactions like this by an "aldo-keto reductase" in which isomerization of a keto-hydroxy group occurs between the sequential reduction of a vicinal diketo group as in the reduction of campharoquinone by 3-hydroxysteroid dehydrogenase, (Boutin, 1986) shown in Fig. 5.

In the present study, only the interconversion of pteridine reactants was considered to know the outline of this complicated reaction mechanism of SPR. And pteridines measured in this experiment consisted of various forms such as free pteridine (protonated and unprotonated) and pteridine complexes with enzyme or coenzyme(NADPH or NADP+)-binding enzyme. Analysis based on a more precise kinetic scheme will be necessary in the future.

ACKNOWLEDGEMENT

We thank Dr. Naoto Sakamoto of Tsukuba University (Ibaragi, Japan) for invaluable advice on computer analysis of the enzyme reaction.

REFERENCES

Boutin, J.A., 1986, Camphoroquinone reductase: another reaction

catalyzed by rat liver cytosol 3-hydroxysteroid
dehydrogenase, Biochim. Biophys. Acta, 870:463-472.

Brown, G.M., Switchencho, A.C. and Primus, J.P., 1985, Enzymatic formation of H_4-biopterin in Drosophila melanogaster, In Biochemical and Clinical Aspects of Pteridines (Wachter, H., Curtius, H.-Ch., Pfleiderer, W. eds.), vol 4, pp119-131, Walter de Gruyter, Berlin, New York.

Curtius, H.-Ch., Heintel, D., Ghisla, S., Kuster, T., Leimbacher, W. and Niederwieser, A., 1985, Tetrahydrobiopterin biosynthesis. Studies with specifically labeled (^{2}H)NAD(P)H and ^{2}H$_2$O and of the enzymes involved, Eur. J. Biochem., 148:413-419.

Katoh, S. and Sueoka, T., 1984, Sepiapterin reductase exhibits a NADPH-dependent dicarbonyl reductase activity, Biochem. Biophys. Res. Commun., 118:859-866.

Katoh, S. and Akino, M., 1986, Biosynthesis of tetrahydrobiopterin in animals, Zool. Sci., 3:745-757.

Katoh, S. and Sueoka, T., 1987, Isomerization of 6-lactoyl tetrahydropterin by sepiapterin reductase, J. Biochem. 101: 275-278.

Katoh, S. and Sueoka, T., 1988, Coenzyme stimulation of isomerase activity of sepiapterin reductase in the biosynthesis of tetrahydrobiopterin, J. Biochem., 103:286-289.

Katoh, S. and Sueoka, T., 1990, Catalytic implication of the reductase and isomerase activity of sepiapterin reductase in the biosynthesis of tetrahydrobiopterin, In Chemistry and Biology of Pteridines (Curtius, H-Ch., Blau, N. eds.), Walter de Gruyter, Berlin, New York.(in press)

Kaufman, S., 1986, The metabolic role of tetrahydrobiopterin, In Chemistry and Biology of Pteridines (Cooper, B.A., Whitehead, V.M., eds.), pp185-200, Walter de Gruyter, Berlin, New York.

Masada, M., Akino, M., Sueoka, T. and Katoh, S., 1985, Dyspropterin, an intermediate formed from dihydroneopterin triphosphate in the biosynthetic pathway of tetrahydrobiopterin, Biochim. Biophys. Acta, 840:235-244.

Milstien, S. and Kaufman, S., 1985, Biosynthesis of tetrahydro-

biopterin: conversion of dihydroneopterin triphosphate to
tetrahydropterin intermediates, Biochem. Biophys. Res.
Commun., 128:1099-1107.

Smith, G.K., and Nichol, C.A., 1986, Synthesis, utilization,
and structure of the tetrahydropterin intermediates in the
bovine adrenal medullary de novo biosynthesis of tetrahydrobiopterin, J. Biol. Chem., 261:2725-2737.

Smith, G.K., 1987, On the role of sepiapterin reductase in the
biosynthesis of tetrahydrobiopterin, Arch. Biochem.
Biophys., 255:254-266.

Sueoka, T. and Katoh, S., 1982, Purification and characterization of sepiapterin reductase from rat erythrocytes, Biochim.
Biophys. Acta, 717:265-271.

Sueoka, T. and Katoh, S., 1985, Carbonyl reductase activity of
sepiapterin reductase from rat erythrocytes, Biochim.
Biophys. Acta, 843:193-198.

Sueoka, T. and Katoh, S., 1989,Enzymatic isomerization of
sepiapterin to 6-1'-hydroxy-2'-oxopropyl 7,8-dihydropterin,
Pteridines, 1(2):103-109.

Sueoka, T., Masada, M. and Katoh, S., (in preparation).

Tuner, A.J. and Flynn, T.G., 1982, The nomenclature of Aldehyde
reductases, In Enzymology of Carbonyl Metabolism: Aldehyde
Dehydrogenase and Aldo/keto Reductase (Weiner, H., and
Wermuth, B. eds.) pp401-402, Alan R Liss, New York.

Wermuth, B., 1985, Aldo-keto reductases, In Enzymology of Carbonyl Metabolism 2, Aldehyde dehydrogenase, Aldo-Keto Reductase, and Alcohol Dehydrogenase (Flynn, T.G., Weiner, H.
eds.), pp209-230, Alan R Liss, New York.

Yoshioka, S., Masada,M., Yoshida, T., Inoue, K., Mizokami, T.,
and Akino, M., 1983, Synthesis of biopterin from dihydroneopterin triphosphate by rat tissue, Biochim. Biophys. Acta,
756:279-285.

Yim., J.Y. and Brown, G.M., 1976, Characteristics of guanosine
triphosphate cyclohydrolase I purified from Escherichia
coli, J. Biol. Chem., 251:5087-5094.

CATALYSIS BY YEAST ALCOHOL DEHYDROGENASE

Bryce V. Plapp, Axel J. Ganzhorn, Robert M. Gould, David W. Green, Tobias Jacobi, Edda Warth, and Darla Ann Kratzer

Department of Biochemistry, The University of Iowa
Iowa City, IA 52242

INTRODUCTION

The structure and mechanism of alcohol dehydrogenases have been extensively studied (Brändén et al., 1975; Klinman, 1981; Pettersson, 1987). The three-dimensional structures of the horse liver enzyme in several ternary complexes have been solved at high resolution (Eklund et al., 1981, 1982). Amino acid sequences for more than 22 NAD^+-dependent alcohol dehydrogenases from 11 animal, plant and fungal species are known. Comparison of these sequences raises many questions about the structure-function relationships in these enzymes. How do the amino acid residues at the active site participate in catalysis? What is the basis of substrate specificity? What was selected for during the evolution of the different enzymes?

Since these enzymes are homologous, it is possible to build a model of yeast alcohol dehydrogenase I (ADH I: constitutive, cytoplasmic) based upon the three-dimensional structure of the horse liver enzyme. Figure 1 is a representation of the active site of the yeast enzyme. It shows that many amino acid residues contact the substrate and can participate in catalysis. We should like to know how each amino acid residue contributes. Some clues to the roles of these residues come from the inspection of the structure, from chemical modification studies, and from examining amino acid substitutions in this family of alcohol dehydrogenases.

Table 1 shows the substitutions that are known to occur in these alcohol dehydrogenases. Many of the residues are strictly conserved, whereas others only have conservative changes, and other positions have many substitutions. In particular, substitutions within the substrate binding pocket apparently give rise to the enzymes with greatly different specificities for substrates. For instance, the yeast enzyme is most active on ethanol, and the model in Figure 1 shows that the active site has two tryptophan residues, which could greatly restrict the access of larger substrates. In contrast, the horse liver enzyme is active on substrates such as cyclohexanol, apparently because it has the smaller leucine and phenylalanine residues. Other enzyme variants have residues with smaller side chains and are active on molecules as large as steroid alcohols.

Using site-directed mutagenesis and steady-state kinetic studies, we have explored the roles of some of these residues. Our results provide some information about the general principles concerning the function of the residues in catalysis. In addition, the work provides some surprising results that lead to new questions about enzyme structure and function.

Enzymology and Molecular Biology of Carbonyl Metabolism 3
Edited by H. Weiner *et al.*, Plenum Press, New York, 1990

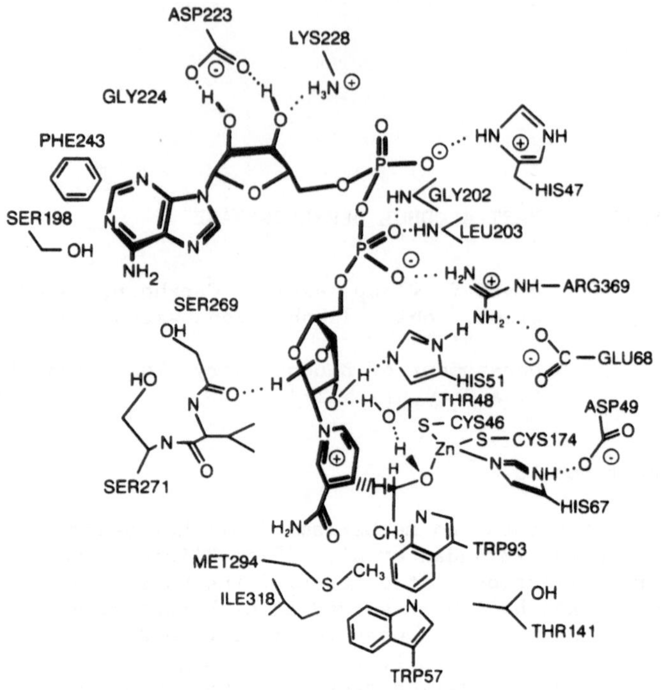

Figure 1. Model of Active Site of Yeast Alcohol Dehydrogenase

Table 1. Homologous Substitutions at Active Sites of 22 NAD$^+$-Dependent Ethanol Dehydrogenases.

Res. No.	Yeast I	Horse E	Other Structures
46	Cys		
47	His	Arg	Gly, Thr
48	Thr	Ser	
49	Asp		
51	His	His	Thr, Tyr
57	Trp	Leu	Phe, Met, Asp, Gly
68	Glu		
93	Trp	Phe	Tyr, Ala
141	Thr	Leu	Val, Met, Phe
174	Cys		
198	Ser	Phe	Val, Pro
202	Gly	Gly	Ala
203	Leu	Val	
223	Asp		
224	Gly	Ile	Ala, Ser, Thr, Phe
228	Lys	Lys	Arg
269	Ser	Ile	Ala, Thr, Leu
271	Ser	Arg	Gly, Ala, Asn
294	Met	Val	Leu
318	Ile	Ile	Ala, Val, Phe
369	Arg		

EXPERIMENTAL APPROACH

The gene for yeast alcohol dehydrogenase I has been cloned (Bennetzen and Hall, 1982), and mutations were made by mutagenesis of double stranded plasmid or by use of the bacteriophage M13 using synthetic oligonucleotides with the altered codon sequences. Sequencing of the DNA confirmed the mutations. In addition, sequence analysis of peptides derived from the expressed proteins showed that the amino acid substitutions were present and that other changes in the sequence were not apparent. The mutant enzymes were expressed in a host strain of yeast that did not produce the wild type enzyme, and tens of milligrams of enzymes could be purified in a few days (Ganzhorn et al., 1987). The concentration of enzyme was determined by titration with NAD^+ in the presence of pyrazole. A physiological buffer of 83 mM potassium phosphate and 40 mM KCl at pH 7.3 and 30 °C was used for most kinetics experiments (Cornell, 1983). Analysis of steady-state kinetic data has been described previously (Ganzhorn et al., 1987).

In addition to the steady-state kinetic studies, isotope effects with deuterated substrates were used to explore changes in rate-determining steps. Furthermore, pH dependency studies provided values of the kinetic constants as a function of pH, giving some information about the roles of ionizable groups in catalysis. In the results presented below, a selection of the most relevant kinetic constants are presented, so that it is possible to focus upon the most significant changes.

Steady-state kinetics were used to determine the product and dead-end inhibition patterns. Shown in Figure 2 are the results with wild type enzyme. Coenzymes inhibited competitively against one another, and ethanol inhibited non-competitively against varied acetaldehyde concentrations. Acetaldehyde inhibited non-competitively against varied ethanol concentrations, since there were small but significant intercept effects as determined by Student's t test. With some of the mutant enzymes, acetaldehyde appeared to be a competitive inhibitor against ethanol. NADH inhibited non-competitively against varied ethanol concentrations. Low concentrations of trifluoroethanol, a dead-end inhibitor that is competitive against varied concentrations of ethanol, inhibited uncompetitively against varied NAD^+ concentrations. These results are consistent with the Ordered Bi Bi mechanism, although the mechanism is probably partially random, especially at higher concentrations of alcohol, and thus is best described as a preferred ordered mechanism (Wratten and Cleland, 1963; Dickinson and Monger, 1973; Dickenson and Dickinson, 1978; Ganzhorn et al., 1987). The kinetic studies provide estimates of kinetic constants for the complete mechanism, and in particular allow one to evaluate turnover numbers (V_1 for the forward reaction with NAD^+, and V_2 for the reaction of NADH and acetaldehyde), catalytic specificity constants (V/K_m) for alcohols, and the dissociation constants for trifluoroethanol and the coenzymes NAD^+ and NADH.

SUBSTRATE SPECIFICITY

It is generally assumed that substrate specificity of an enzyme is determined by the size and shape of the substrate binding pocket. Thus, it was reasonable to suggest that the differences in K_m for ethanol between the yeast ADH isoenzyme I and the inducible isoenzyme II would be due to substitutions at position 294. ADH II has a K_m for ethanol that is more than 20-fold lower than is the K_m of ADH I. These isoenzymes differ by only 24 amino acids out of the 347 residues (Russell et al., 1983), and the only difference in the active site region is that ADH I has methionine and ADH II has leucine at residue 294. Site-directed mutagenesis of ADH I produced the Leu-294 mutant and kinetic studies showed that the K_m was essentially unchanged by this substitution (Table 2). Furthermore, the dissociation constant for trifluoroethanol is still at least 7 times higher with the ADH I enzyme, and the catalytic activity with ethanol (V/K_m) is 7 times lower with the ADH I enzymes as compared to ADH II. Nevertheless, catalytic efficiency with butanol was improved by a factor of 7 by the substitution with leucine. This indicates that residue 294 is in the active site and close enough to interact with butanol, even if it is not responsible for the different activities with ethanol.

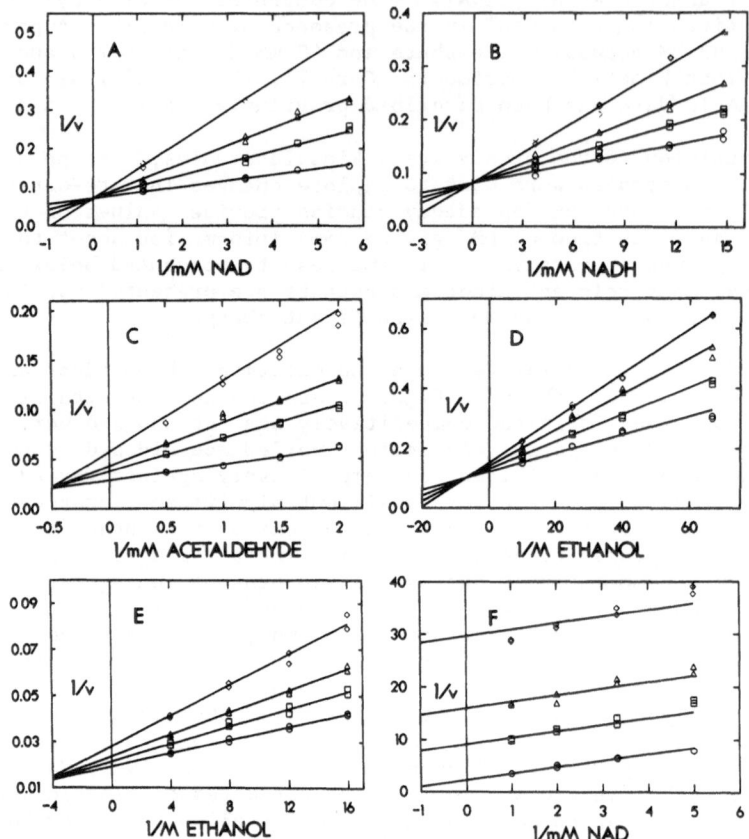

Figure 2. Inhibition studies of wild-type yeast ADH I. Velocities (v) have units of $100 \times \Delta A_{340}$/min. (A) Inhibition by NADH against varied concentrations of NAD^+ at 250 mM ethanol using 1.1 nN enzyme and NADH concentrations of 0 (O), 29 (□), 57 (Δ), and 120 μM (◊). (B) Inhibition by NAD^+ against varied concentrations of NADH at 4 mM acetaldehyde using 0.2 nN enzyme and NAD^+ concentrations of 0 (O), 0.52 (□), 1.0 (Δ), and 2.1 mM (◊). (C) Inhibition by ethanol against varied concentrations of acetaldehyde at 0.22 mM NADH using 0.8 nN enzyme. Ethanol concentrations were 0 (O), 63 (□), 130 (Δ), and 250 mM (◊). (D) Inhibition by acetaldehyde against varied concentrations of ethanol at 2.1 mM NAD^+ using 5.6 nN enzyme. Acetaldehyde concentrations were 0 (O), 0.25 (□), 0.5 (Δ), and 1 mM (◊). (E) Inhibition by NADH at 2 mM NAD^+ using 3.1 nN enzyme. NADH concentrations were 0 (O), 41 (□), 83 (Δ), and 170 μM (◊). (F) Inhibition by trifluoroethanol: 0 (O), 10 (□), 20 (Δ), and 40 mM (◊), at 10 mM ethanol and 8.4 nN enzyme. v is in units of ΔA_{340}/min.

244

Table 2. Kinetics of Met294Leu Mutant Enzyme[a]

Kinetic constant	ADH I Met294	ADH I-Leu Leu294	ADH II Leu294
K_m Ethanol (mM)	17	19	0.81
K_d CF$_3$CH$_2$OH (mM)	2.5	3.0	0.36
V/K_m Ethanol (mM^{-1}s^{-1})	20	26	160
V/K_m Butanol (mM^{-1}s^{-1})	0.93	7.2	34

[a] Ganzhorn et al., 1987

The important conclusion here is that a change in the active site of approximately the size of one methyl group can make a difference in activity of about 10-fold with the appropriate substrates. However, the results do not explain why ADH I-Leu and ADH II differ in V/K_m for ethanol. Apparently other substitutions outside of the active site must *indirectly* affect activity. Inspection of the three-dimensional model of the yeast enzyme leads us to suggest that other substitutions might be affecting the conformational changes involving movement of coenzyme and catalytic domains during the catalytic process. Which residues might be involved is of interest for elucidating allosteric mechanisms. How could a substitution outside of the active site affect activity?

One must qualify these conclusions, however, by noting that three-dimensional structure of the yeast alcohol dehydrogenase has not been determined. This enzyme is a tetramer, as compared to the dimeric horse liver enzyme. Perhaps some subunit interactions can bring other amino acid residues into the active site region. Nevertheless, it is important to note that the model we built for the yeast alcohol dehydrogenase does show residue 294 in the appropriate position to explain activity with ethanol or butanol, and thus the model is consistent with the kinetic results.

Another residue of interest in the active site is at position 48. The yeast enzyme and some of the mammalian enzymes have threonine at this position, but the horse liver enzyme has serine. Inspection of the model of the horse liver enzyme shows that residue 48 is directly involved in determining the high stereospecificity of these alcohol dehydrogenases for transferring the pro-*R* hydrogen of ethanol to NAD$^+$ (Eklund et al., 1982). This residue also should affect the activity on secondary alcohols such as the stereoisomers of 2-butanol. Changing Thr-48 to serine produced a yeast enzyme with 10-fold more activity on *S*-2-butanol as compared to *R*-2-butanol. The substitution also increased catalytic efficiency with 2-methylpropanol. These preferences for substrates are similar to those exhibited by the horse E isoenzyme. Somewhat different results have been obtained in a comparison of class I human isoenzymes, where the γ enzyme with Ser-48 has more activity on *R*-2-butanol than on *S*-2-butanol, whereas the β isoenzyme with Thr-48 has much more activity on *S*-2-butanol (Stone et al., 1989). These results suggest that there are many residues that participate in forming the productive complex with the substrate. The conclusion from such studies is that one methyl group can change activity by perhaps an order of magnitude. Thus, we can begin to imagine the results to be obtained from rational protein engineering.

COENZYME BINDING

The phosphate of the AMP moiety of coenzyme interacts with Arg-47 in the horse liver enzyme, but according to molecular modeling, this interac-

Table 3. Kinetics of His47Arg Enzymes

Kinetic Constant	Yeast ADH[a]		Human[b]	
	His	Arg	β_2-His	β_1-Arg
K_m NAD$^+$ (μM)	160	150	180	7.4
K_m Ethanol (mM)	21	66	0.94	0.049
K_d NAD$^+$ (μM)	950	260	340	90
K_d NADH (μM)	31	16	9.7	0.19
V_1 (s^{-1})	360	60	6.7	0.15
V_2 (s^{-1})	1800	460	65	4.0

[a] pH 7.3, 30°C (Gould and Plapp, 1990)
[b] pH 7.5, 25°C (Yin et al., 1984; Bosron et al., 1983)

tion would be replaced by one with His-47 in the yeast enzyme. As noted in Table 1, this position can also be occupied by glycine or threonine. It is often assumed that arginine is the most suitable residue for interacting with negatively charged ligands, and thus it is of interest to examine the effect of this substitution on coenzyme binding and upon the reaction mechanism. Site-directed mutagenesis of yeast ADH I produced a mutant arginine enzyme with the kinetic constants shown in Table 3 (Gould and Plapp, 1990). The dissociation constants for NAD$^+$ and NADH decreased by factors of 2- to 4-fold, at pH 7.3, and the turnover numbers decreased 4- to 6-fold. This result fits the expected behavior. Since the release of coenzyme is a slow step in catalytic turnover, tighter binding of coenzyme could decrease V_{max}. Interestingly, we also obtained evidence that the Arg-47 enzyme underwent a slow conformational change, which could partially limit turnover.

Nevertheless, it is important to emphasize that the *magnitudes* of the effects of a single amino acid substitution can be different in different isoenzymes. Binding of coenzyme is a cooperative process in that many amino acid residues participate in binding. Thus, the magnitudes will be affected by the other amino acid residues in the active site. This is illustrated by the data for the human β isoenzymes (Table 3), where it is noted that the dissociation constant for NAD$^+$ is decreased 50-fold and the turnover number in the forward direction (V_1) is decreased 44-fold by the substitution. Further data on the role of residue 47 in coenzyme binding is provided also by Hurley and coworkers elsewhere in this volume. Of particular interest is that a glycine residue at position 47 also produces an enzyme (like α isoenzyme) with affinity for coenzyme that is comparable to that with Arg-47. What features of the active site compensate for the lack of charge of the glycine residue?

SUBSTITUTIONS AT THE ZINC SITE

Figure 1 shows that the alcohol binds directly to the catalytic zinc. This interaction may facilitate release of the proton from the substrate, or during reduction of aldehyde may result in polarization of the carbonyl group, which facilitates hydride transfer. Interaction with the zinc can affect the pK value of the zinc-bound water or alcohol, and this can in turn affect the pH dependence of the enzyme and the ability of the enzyme to stabilize the transition state. Two residues that are conserved in all known zinc-dependent alcohol dehydrogenases are Asp-49 and Glu-68. These are not innersphere ligands to the zinc, but are in the second sphere. The carboxyl group of Asp-49 interacts with the imidazole group of His-67, which is in turn ligated directly to the zinc. The carboxyl group of Glu-68 has one oxygen in a hydrogen bond to the guanidino group of

Table 4. Amidation of Carboxylates near Catalytic Zinc[a]

Kinetic Constant	ADH I	Asp49Asn	Glu68Gln
K_m NAD$^+$ (mM)	0.17	9.0	0.41
K_m Ethanol (mM)	17	430	41
K_d NAD$^+$ (mM)	0.92	5.8	3.5
K_d NADH (mM)	0.031	0.24	0.029
V_1 (s^{-1})	340	7.5	9.9
V_2 (s^{-1})	1700	110	730
K_d CF$_3$CH$_2$OH (mM)	2.8	60	25
V/K_m ethanol	20	0.018	0.24

[a] Ganzhorn and Plapp, 1988

Arg-369, and the other oxygen is close to the zinc. The negative charges furnished by these residues may serve to modulate the polarizing effect of the metal in order to facilitate hydride transfer and to prevent substrates or products from binding too tightly. To evaluate the role of these conserved carboxyl groups, we replaced these residues with their respective amide analogs.

Although the Asn-49 enzyme no longer has the buried carboxyl group, the mutant enzyme still has good heat stability and essentially the same CD spectrum as the wild type enzyme (Ganzhorn and Plapp, 1988). The enzyme also has 2 zincs per subunit, just as with native enzyme. Nevertheless, these substitutions have significantly altered binding of substrates and catalytic activities. Both substitutions have decreased the affinity of the enzymes for NAD$^+$ and trifluoroethanol and have greatly decreased catalysis (Table 4). One possible explanation for these results is that the active site would be more positively charged, but then one would expect coenzyme and trifluoroethanol to bind more tightly. Another possible explanation is that conformational changes that require reorientation of groups around the zinc are hindered in the mutant enzymes so that productive enzyme-substrate complexes are not formed as readily. Based on substrate isotope effects, we have suggested that slow isomerizations of ternary complexes could account for the results and that carboxylate groups of Asp-49 and Glu-68 are necessary to create an electrostatic environment that is suitable for the structural changes. The nature of these structural changes is unknown at present, but it may resemble the conformational change observed with the horse liver alcohol dehydrogenase (Eklund et al., 1981).

PROTON RELAY SYSTEM

Inspection of the horse liver enzyme and the model of the yeast enzyme (Figure 1) shows that the hydroxyl group of the alcohol is linked to His-51 by hydrogen bonds through the hydroxyl group of Thr-48 and the 2'-hydroxyl group of the nicotinamide ribose. The structure suggests that His-51 acts as a base during alcohol oxidation, in that it can accept a proton through the proton relay system, facilitating formation of the zinc alkoxide. The role of this histidine residue was tested by replacing it with glutamine or glutamic acid. These residues were chosen since they have the appropriate size to form the hydrogen bond with the 2'-hydroxyl group of the nicotinamide ribose. Thus, binding of coenzyme in the mutant enzymes could resemble that found in the wild type enzyme. On the other hand, a glutamine residue could not participate in base catalysis, whereas a glutamate potentially could accept a proton.

Table 5. Kinetics of His-51 Mutant Enzymes

Kinetic Constant	His-51	Gln-51	Glu-51
K_d NAD$^+$ (μM)	950	460	77
K_d CF$_3$CH$_2$OH (mM)	2.5	33	130
V_1 Ethanol (s^{-1})	360	27	2
V/K_m Ethanol (mM^{-1}s^{-1})	17	1.5	0.26

Some of the kinetic constants for these mutant enzymes are displayed in Table 5. The dissociation constant for NAD$^+$ shows that the mutant enzymes bind NAD$^+$ as well or better than does the wild type enzyme. In contrast, the binding of trifluoroethanol, the turnover numbers and the catalytic efficiencies are decreased significantly. Note that each mutation changed these three parameters by about the same factor, 13-fold for His51Gln and 60-fold for His51Glu. These results are most simply interpreted by a mechanism in which the amino acid residue in the mutant enzyme hinders the deprotonation of the water or alcohol through the proton relay system. Thus, ethanol or trifluoroethanol are less able to deprotonate at the active site. In turn, this effect would decrease the relative amount of alkoxide and the rate of hydride transfer. Substitution of glutamate apparently has a larger effect than the substitution with a glutamine, since glutamate, with its negative charge, would more effectively oppose the development of a negative charge on the alcohol bound to the zinc.

These interpretations are supported by pH dependency studies (Figure 3, Table 6). For wild type enzyme there is a group with a pK of 7.7 that must be unprotonated for maximum catalytic activity. For the glutamine enzyme the pH dependency is altered so that the pK of 7.7 for wild-type enzyme is no longer visible, and instead a wavy function that can be described by mechanism with two ionizations, with pK values of 6.8 and 8.7, is obtained. For the glutamate enzyme, pK values of 6.5 and 8.9 are required to fit the data.

These results are significant for several reasons. First, they show that substitution of the histidine with glutamine or glutamate removes the pK value 7.7 and decreases the catalytic efficiency by factors of 13 to 70 at pH 7.3. These results are consistent with the role for His-51 in base catalysis. On the other hand, the results also indicate that these mutant enzymes still exhibit a pH dependency. From examination of the three-dimensional model for this enzyme, we conclude that these effects of pH must be mediated by amino acid residues that do not directly participate in the catalytic mechanism. The pK value near 7 could be due to the indirect effect of the deprotonation of His-47, and the pK at about 9 could be due to the deprotonation of Lys-228. These interpretations, however, are not reasonable since both of these residues should be in the protonated form for the best interactions with coenzyme.

It is also possible that one of the pK values, 7 or 9, could be due to ionization of the water bound to zinc. Since the Gln-68 and Asn-49 mutants exhibit pK values of 6.7 and 6.1, respectively, it would be reasonable to assign the lower pK to the ionization of the water bound zinc in the enzyme-NAD$^+$ complex. The reduction of pK from 7.7 to 6.1 in the Asn-49 enzyme would be consistent with the more electropositive environment of the zinc. If this assignment can be made, then one still should like to explain the pK value of 9. At this time no assignment seems reasonable, and we conclude that this part of the pH dependency is indeed controlled by ionization of one or more groups that are not within the

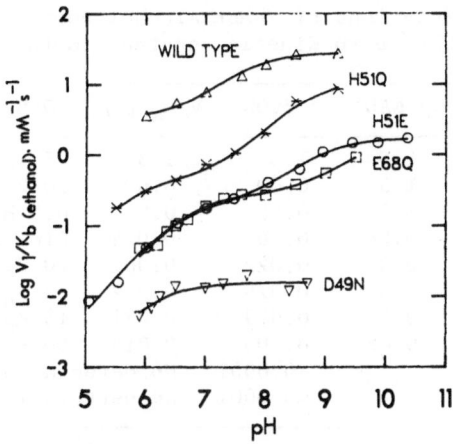

Figure 3. pH Dependence of Active Site Mutants. The catalytic efficiency with ethanol was determined in 20 mM $Na_4P_2O_7$ adjusted to the desired pH and 0.1 ionic strength with sodium phosphate buffer (Ganzhorn and Plapp, 1988; Gould and Plapp, 1990).

Table 6. pH Dependence of ADH Mutants Acting on Ethanol

The data from Figure 3 were fitted to an equation below, or to a simpler form that is derived for the general mechanism.

H_2E-NAD^+

$\mid$ pK_a

$\mid$ k_2

HE-NAD^+ --->

$\mid$ pK_b

$\mid$ k_3

E-NAD^+ --->

$$V/K_{obs} = \frac{k_2 + k_3 K_b/[H^+]}{1 + K_b/[H^+]}$$

$$V/K_{obs} = \frac{k_2 + k_3 K_a/[H^+]}{1 + K_a/[H^+] + [H^+]/K_b}$$

	WT	H51E	H51Q	D49N	E68Q
pK_a	-	6.5	6.8	-	6.7
pK_b	7.7	8.9	8.7	6.1	9.6
k_2	3.2	0.2	1.0	-	0.27
k_3	25	2.1	12	0.016	2

active site. These results also illustrate the difficulties in interpreting pH dependency studies in the wild type enzyme. If the pK values of distant groups can affect activity at the active site, then the pH dependency curve must represent the net result of ionization of many groups. In this case, we would suggest that activity of wild type enzyme is controlled mostly by the ionization of His-51. Then, how do ionizations of other, distant groups affect activity? Furthermore, what mechanisms for base catalysis operate in those enzymes that have tyrosine or threonine at residue 51?

Further evidence for the importance of the proton relay system comes from studies in which Thr-48 was replaced by alanine or cysteine. Both of these enzymes had no detectable activity. These mutations could block the transfer of proton from the alcohol to the solvent or, of course, could perturb the structure of the transition state.

Table 7. Mutations of Yeast Alcohol Dehydrogenase I
Relative Change in Kinetic Constant, Mutant:WT

Mutation	K_d NAD$^+$	V EtOH	V/K_m EtOH	Other
Met294Leu	1.5	1.5	1.3	8 V/$K_{Butanol}$
Thr48Ser	1.3	0.60	0.60	10 2-Butanol $S>R$
Trp57Met	3.0	0.65	0.24	0.5 Hexanol
His47Arg	0.26	0.16	0.034	Slow E-NAD$^+$ isom
Asp49Asn	6.3	0.022	0.001	20 K_d CF$_3$CH$_2$OH
Glu68Gln	3.8	0.029	0.012	9 K_d CF$_3$CH$_2$OH
His51Gln	0.5	0.079	0.071	13 K_d CF$_3$CH$_2$OH
His51Glu	0.08	0.005	0.015	50 K_d CF$_3$CH$_2$OH
Thr48Cys		<0.0001	No detectable activity	
Thr48Ala		<0.0001	No detectable activity	

SUMMARY

Table 7 presents a brief summary of the effects of various mutations on some of the relevant kinetic constants. The results illustrate several important features of the use of site-directed mutagenesis in exploring structure and function of enzymes. Note that most of the mutations affect a given step or kinetic parameter in the mechanism, such as the binding of NAD$^+$ or the turnover number with ethanol. Furthermore, one mutation can affect many steps in the mechanism. Thus, it is difficult to ascribe a particular role to an amino acid residue. It is also difficult to quantify the function of a residue, since the magnitudes of the effects on kinetic parameters will be modulated by the other amino acid residues that participate in the reaction. Comprehensive and quantitative kinetic studies of many mutant enzymes are required if we are to understand catalysis and specificity.

We are reluctant to describe any residue as "essential" for activity, since substitution with some amino acid can probably produce an enzyme with some residual activity. (Maybe the Thr48Gly enzyme would be active, as a water molecule could substitute for the hydroxyl of the threonine.) Likewise, when substitution of a residue partially, but not totally, decreases activity, it does not necessarily mean that the residue is "not essential". The change in activity can reflect the contribution of that residue to catalysis. On the other hand, if various substitutions of a residue do not change activity, it would be reasonable to conclude that the residue is not essential (Plapp et al., 1971). Most of the amino acid residues at the active site are involved in the catalytic mechanism, either by contacting the substrates directly or by participating in the chemistry. Some of the residues that are outside of the active site are indirectly involved, by affecting the structure of the protein. Substitution of an important amino acid residue should significantly affect activity, and studies on the kinetics and structure should allow one to distinguish among the various explanations.

ACKNOWLEDGMENTS

This work was supported by grant AA06223 from the National Institutes of Alcohol Abuse and Alcoholism. We also thank Drs. E. T. Young and B. D. Hall for providing the cloned gene for the alcohol dehydrogenase and the appropriate host yeast strains.

REFERENCES

Brändén, C.-I., Jörnvall, H., Eklund, H., and Furugren, B. (1975) Alcohol Dehydrogenases, *The Enzymes*, 3rd Ed., **11**, 103-190.

Bennetzen, J. L., and Hall, B. D. (1982) The primary structure of the *Saccharomyces cerevisiae* gene for alcohol dehydrogenase, *J. Biol. Chem.* **157**, 3018-3025.

Bosron, W. F., Magnes, L. J., and Li, T.-K. (1983) Kinetic and electrophoretic properties of native and recombined isoenzymes of human liver alcohol dehydrogenase, *Biochemistry* **22**, 1852-1857.

Cornell, N. W. (1983) Properties of alcohol dehydrogenase and ethanol oxidation *in vivo* and in hepatocytes, *Pharmacol. Biochem. Behav.* **18**, Suppl. 1, 215-221.

Dickenson, C. J., and Dickinson, F. M. (1975) A study of the pH and temperature dependence of the reactions of yeast alcohol dehydrogenase with ethanol, acetaldehyde and butyraldehyde as substrates, *Biochem. J.* **147**, 303-311.

Dickinson, F. M., and Monger, G. P. (1973) A Study of the kinetics and mechanism of yeast alcohol dehydrogenase with a variety of substrates, *Biochem. J.* **131**, 261-270.

Eklund, H., Samama, J.-P., Wallén, L., Brändén, C.-I., and Åkeson, Å. (1981) Structure of a triclinic ternary complex of horse liver alcohol dehydrogenase at 2.9 Å resolution, *J. Mol. Biol.* **146**, 561-587.

Eklund, H., Plapp, B. V., Samama, J.-P., and Brändén, C.-I. (1982) Binding of substrate in a ternary complex of horse liver alcohol dehydrogenase, *J. Biol. Chem.* **257**, 14349-14358.

Ganzhorn, A. J., Green, D. W., Hershey, A. D., Gould, R. M., and Plapp, B. V. (1987) Kinetic characterization of yeast alcohol dehydrogenases. Amino acid residue 294 and substrate specificity, *J. Biol. Chem.* **262**, 3754-3761.

Ganzhorn, A. J., and Plapp, B. V. (1988) Carboxyl groups near the active site zinc contribute to catalysis in yeast alcohol dehydrogenase, *J. Biol. Chem.* **263**, 5446-5454.

Gould, R. M., and Plapp, B. V. (1990) Substitution of arginine for histidine-47 in the coenzyme binding site of yeast alcohol dehydrogenase I, *Biochemistry* **29**, 5463-5468.

Klinman, J. P. (1981) Probes of mechanism and transition-state structure in the alcohol dehydrogenase reaction, *Crit. Rev. Biochem.* **10**, 39-78.

Pettersson, G. (1987) Liver alcohol dehydrogenase, *Crit. Rev. Biochem.* **21**, 349-389.

Plapp, B. V., Moore, S., and Stein, W. H. (1971) Activity of bovine pancreatic deoxyribonuclease A with modified amino groups, *J. Biol. Chem.* **246**, 939-945.

Russell, D. W., Smith, M., Williamson, V. M., and Young, E. T. (1983) Nucleotide sequence of the yeast alcohol dehydrogenase II gene, *J. Biol. Chem.* **258**, 2674-2682.

Stone, C. L., Li, T.-K., and Bosron, W. F. (1989) Stereospecific oxidation of secondary alcohols by human alcohol dehydrogenases, *J. Biol. Chem.* **264**, 11112-11116.

Wratten, C. C., and Cleland, W. W. (1963) Product inhibition studies on yeast and liver alcohol dehydrogenases, *Biochemistry* **2**, 935-941.

Yin, S.-J., Bosron, W. F., Magnes, L. J., and Li, T.-K. (1984) Human liver alcohol dehydrogenase: Purification and kinetic characterization of the $\beta_2\beta_2$, $\beta_2\beta_1$, $\alpha\beta_2$, and $\beta_2\gamma_1$ "Oriental" isoenzymes, *Biochemistry* **23**, 5847-5863.

ALCOHOL DEHYDROGENASE GENE EXPRESSION
and CLONING of the MOUSE χ-LIKE ADH

Howard J. Edenberg, Celeste J. Brown, Lucinda G. Carr, Wei-Hsien Ho, and Man-Wook Hur

Department of Biochemistry and Molecular Biology
Indiana University School of Medicine
Indianapolis, IN 46202-5122, U.S.A.

INTRODUCTION

Some of the differences between individuals in the metabolism of alcohol and in their susceptibility to alcoholism or its complications may be due to differences in the metabolism of ethanol. The amount of each different isozyme expressed will influence the overall alcohol dehydrogenase activity in a given tissue, and could play an important role in mediating or modulating the effects of ethanol. For these reasons, we have been studying the structure and expression of the mammalian alcohol dehydrogenases.

Humans have three closely related alcohol dehydrogenase genes, *ADH1*, *ADH2*, and *ADH3*, that catalyze ethanol metabolism in the liver (Bosron and Li, 1986; Smith, 1986). We have cloned the three human class I *ADH* genes (Carr *et al.*, 1989b; and unpublished data) and the closely related mouse *Adh-1* gene (Zhang *et al.*, 1987), and are analyzing their transcriptional regulation. We are studying specific DNA-protein interactions important to transcriptional control, using the techniques of gel retardation, methylation interference, and *in vitro* transcription. Two sites of strong interaction between the proximal promoter region of *ADH2* and nuclear proteins from mouse liver or rat hepatoma cells were localized to regions homologous to the mouse *Adh-1* gene (Zhang *et al.*, 1987; Carr *et al.*, 1989a; Carr and Edenberg, 1990). These sequences lie between nt -94 and -84 and between nt -72 and -64. Both sequences were found to be important to the initiation of transcription (Carr *et al.*, 1989a; Carr and Edenberg, 1990).

We have also cloned a cDNA for the mouse class III (χ-like) alcohol dehydrogenase Adh-B_2 from a mouse liver cDNA library, and compared it to the human (Kaiser *et al.*, 1988; Giri *et al.*, 1989; Sharma *et al.*, 1989), rat (Julia *et al.*, 1988) and horse (Kaiser *et al.*, 1989) χ-ADHs.

Enzymology and Molecular Biology of Carbonyl Metabolism 3
Edited by H. Weiner *et al.*, Plenum Press, New York, 1990

RESULTS AND DISCUSSION

IMPORTANT PROMOTER ELEMENTS IN THE HUMAN *ADH2* GENE

Gene expression in mammalian cells is primarily regulated at the initiation of transcription. The decisions about whether to transcribe a particular gene in a given cell, and if so at what rate, are determined by a complex interplay among sets of DNA sequences and the transcriptional regulatory proteins that bind to them. Many important regulatory sequences are located in the region just 5' (upstream) of the transcriptional start site. We are analyzing the 5' sequences of the human (and mouse) alcohol dehydrogenase genes.

We have demonstrated that the 5' region of the human *ADH2* gene (extending out to -1.3 kb) is capable of directing accurate *in vitro* transcription in nuclear extracts from hepatoma cells (Carr and Edenberg, 1990). Deletions that remove the most distal portions of this region actually increase the extent of transcription, suggesting the presence of inhibitory sequences (sequences to which transcriptional repressors bind). The highest level of accurate *in vitro* transcription is found with a fragment extending to nt -93. A slightly more truncated fragment, extending only to nt -55 but still containing the TATA box and potential CCAAT sequence, directs only a very low level of transcription (Carr and Edenberg, 1990). This demonstrates that one or more critical positively-acting sequences are located between nt -93 and -55.

The region between nt -93 and -55 contains two regions nearly identical to important control sequences in the mouse *Adh-1* gene (Zhang *et al.*, 1987; Carr *et al.*, 1989a; Carr and Edenberg, 1990) (Figure 1). Gel mobility shift (gel retardation) assays (Fried and Crothers, 1981; Garner and Revzin, 1981) allow us to detect sequence-specific binding of proteins to these DNA sequences. We synthesized oligonucleotides to make short double-stranded segments of DNA that span each of the two conserved sequences, and could clearly demonstrate strong specific binding of proteins to both of them. To pinpoint those nucleotides whose methylation prevents binding of the sequence-specific proteins, we used these oligos in methylation interference assays (Siebenlist *et al.*, 1980). The results of these experiments identified two core sequences (Carr *et al.*, 1989a; Carr and Edenberg, 1990), as depicted in Figure 1.

```
                 •  •••  •  •••  -80              •   •  ••  •   -60
human ADH2   AGTGGGTGTGGCTTAAAGACATAGATCACGTGTGGA
mouse Adh-1  GAAGGGTGTGTCGTAAAGGGCCAGATCACGTGTGGG
             ---G3T-----              ----CACGTG---
```

Figure 1. Important regulatory sequences in the *ADH2* gene.

The dots • above the human ADH2 sequence represent nucleotides whose methylation prevents protein binding. The TATA box (not shown) extends from nt -27 to nt -21 of the human sequence (Carr *et al.*, 1989b).

We used *in vitro* transcription competition assays to confirm the importance of both of these sequences independently. The promoter fragment extending to nt -93 directed *in vitro* transcription at a relatively high rate, compared to the fragment extending only to nt -55 (Carr and Edenberg, 1990). When a double-stranded oligonucleotide containing the G3T sequence is added to the *in vitro* transcription assay in 20- or 50- fold molar excess, the protein that normally binds to the G3T sequence in the promoter is expected to bind to the oligonucleotide instead. This effectively sequesters the protein and renders it unavailable for binding to the promoter. When that was done, transcription from the ADH2 promoter was greatly reduced, demonstrating that the protein that binds to the G3T sequence is a transcriptional stimulatory factor (Carr and Edenberg, 1990). A nearly identical oligonucleotide, in which nucleotides identified as contacting the protein were altered, did not bind to a protein in the gel mobility shift assays and did not affect *in vitro* transcription. This confirms the importance of the G3T sequence. The same experiments were performed with an excess of oligonucleotide containing the CACGTG sequence, with similar (but even more dramatic) results: transcription was reduced to nearly undetectable levels (Carr and Edenberg, 1990). Thus both of these sequences bind to transcriptional activators, and both are important in directing transcription of the *ADH2* gene.

We have evidence that an oligonucleotide containing the Sp1 consensus sequence (Carr and Edenberg, 1990; and unpublished data) can compete with the G3T oligonucleotide for binding to the liver extract. Thus the protein binding to the G3T sequence may be related to Sp1. The G3T sequence is nearly identical to a sequence originally shown to be important for β-globin gene expression (Myers *et al.*, 1986). A similar sequence has been reported in the tryptophan oxygenase gene (Schüle *et al.*, 1988). A sequence related to the G3T has been described in the SV40 enhancer, where it is called the GT-I element (Fromental *et al.*, 1988; Xiao *et al.*, 1987).

A CACGTG sequence is found in the Adenovirus major late promoter (Yu and Manley, 1984; Miyamoto *et al.*, 1984), where it binds to a transcriptional factor (Carthew *et al.*, 1985; Sawadogo and Roeder, 1985). There is also homology (5/6 identities) to sequences important in the expression of several immunoglobulin genes (Ephrussi *et al.*, 1985; Sen and Baltimore, 1986) and the γ-fibrinogen promoter (Chodosh *et al.*, 1987). We do not yet know whether the same proteins interact with these different but related sequences. There are families of several regulatory proteins that bind to families of related sequences. Expression of different members of a regulatory family in different tissues may be important.

The *ADH1*, *ADH2* and *ADH3* genes are very closely related. Pairwise comparisons show 94% identity in the amino acid sequences they encode (Ikuta *et al.*, 1986; Hoog *et al.*, 1986; Heden *et al.*, 1986; von Bahr-Lindstrom *et al.*, 1986; Edenberg *et al.*, 1989; Duester *et al.*, 1984). There are also significant regions of identity in the 5' region (unpublished data, and see Figure 2). There are small differences among the *ADH1*, *ADH2* and *ADH3* genes in these important cis-acting regulatory sequences (Figure 2). We are interested in the possibility that these differences might affect gene expression. We are beginning to compare the binding of nuclear proteins to comparable regions of these three genes, to determine whether the same factor may bind with different affinity, or whether a different factor may bind. Preliminary evidence indicates that one of the transcription factors has a higher affinity for a sequence in the *ADH2* gene than for the corresponding sequence in the *ADH3* gene.

```
           • ••• • •••  -80         •  • •• •  -60
 ADH2    AGTGGGTGTGGCTTAAAGACATAGATCACGTGTGGA
 ADH3    GAC-------------A--C-------C-----T-
 ADH1    AAT-------------G--C-------T-----G-
         ^^^ G3T           ^    ^    CACGTG  ^
```

Figure 2. Differences among the human *ADH* proximal promoters

The dots above the human *ADH2* sequence represent nucleotides whose methylation prevents protein binding. Exact matches of both *ADH1* and *ADH3* to the *ADH2* sequence are depicted as dashes; the nucleotides are shown at positions where any of the *ADHs* differ. [Data from Duester *et al.*, 1986; Carr *et al.*, 1989b; and unpublished.]

We are continuing our mapping of the *ADH2* promoter, to discover other important DNA-protein complexes. We have preliminary evidence that many different regions of the *ADH2* promoter specifically bind to liver nuclear proteins, and are mapping the location of these binding sites.

Despite their relatively recent divergence, the three human Class I *ADH* genes exhibit different patterns of tissue-specificity (Smith, 1986). We are, therefore, also examining the tissue-specificity of binding to different regions of the *ADH2* promoter. Preliminary evidence indicates that some fragments bind differently to extracts made from different mouse tissues.

We believe that detailed examination of the regulation of human *ADH* genes will illuminate differences in regulation that affect alcohol metabolism and susceptibility to alcohol-related problems.

CLONING OF THE MOUSE χ-ADH cDNA

As part of our analysis of ADH structure and function, we cloned a cDNA for the mouse class III (χ-like) alcohol dehydrogenase Adh-B_2. We cloned this from a cDNA library we created from the liver of a C57BL/6J mouse. The nucleotide sequence of the mouse Adh-B_2 cDNA will be reported elsewhere. The nucleotide sequence within the coding region of the mouse Adh-B_2

cDNA shares 87% identity with the human χ-ADH cDNA (Giri *et al.*, 1989; Sharma *et al.*, 1989).

One interesting difference between the mouse Adh-B$_2$ cDNA and the human χ-ADH cDNA reported by Giri et al. (1989) is that the mouse sequence does not encode the additional 19 amino acids reported at the N-terminal coding region. In fact, the two sequences diverge substantially in the 5' non-coding region. The significance of this difference is not clear.

The protein encoded by the Adh-B$_2$ cDNA (see Figure 3) is 92.8% identical in amino acid sequence to the human χ-ADH (Kaiser *et al.*, 1988), 96.5% identical in amino acid sequence to the rat ADH-2 (Julia *et al.*, 1988), and 91.2% identical to the horse ADH-BB (Kaiser *et al.*,1989). All of the class III ADHs begin at the position of the third amino acid of the class I ADHs, and have an amino acid inserted after position 60 of the class I ADHs.

There are 41 positions at which at least one of the class III ADHs differ from the others (out of 373 amino acids), thus at 89% of the positions there is complete identity among the enzymes from four species. By comparison, when the seven class I enzymes from these same species are compared, there are 103 sites with differences, corresponding to 72.5% identity. The relative conservation of the class III ADHs has been pointed out previously, based upon comparisons among three species (Kaiser *et al.*,1989). At 28 of the 41 sites at which class III ADHs vary, all 8 known mammalian class I ADHs are identical; this is in excellent agreement with the expectation on a random basis (29 sites are expected, since 71.7% of all residues are identical among the class I ADHs).

Among the 41 positions at which at least one of the class III ADHs differs, the horse ADH-BB has the largest number of unique residues (13), followed by the mouse (10), human (5) and rat (4). At 12 of these 41 sites, the horse and human enzymes share one amino acid and the mouse and rat enzymes share another amino acid; no other pairwise combination occurs more than twice. This indicates that the horse and human enzymes form one branch of an evolutionary tree, and the mouse and rat enzymes another.

The location of the 41 sites at which there is at least one difference among the class III enzymes is interesting. The regions near both ends of the polypeptide are more variable than the remainder. There are 3 sites in the 4 amino acid long region encoded by exon 1 (the initiating Met is not shown in Figure 3). There is a cluster of 9 sites from amino acid 348 to the end of the polypeptide (aa374), including 2 in the 7 amino acid long region encoded by exon 9. There are many differences among class I ADHs in the region between amino acids 108 and 125 (Edenberg *et al.*, 1985; Crabb and Edenberg, 1986), which flanks the fourth intron. This region of the class III genes is the least variable in the protein, with only two sites of variation between amino acids 61 and 135 (following the class I numbering system). There are 4 variable sites out of 5 just beyond this region, from amino acid 136 to 140.

```
              3      10        20        30        40
Mouse B₂:   ANQVIRCKAAVAWEAGKPLSIEEIEVAPPKAHEVRIKI
Rat ADH-2:  ANQVIRCKAAVAWEAGKPLSIEEIEVAPPQAHEVRIKI
Human χ:    ANEVIKCKAAVAWEAGKPLSIEEIEVAPPKAHEVRIKI
Horse BB:   SAEVIKCKAAVAWEAGKPVSIEEVEVAPPKAHEVRIKI
            ^^^   ^              ^      ^        ^

                    50        60·       70        80       89
Mouse B₂:   LATAVCHTDAYTLSGRDPEGCFPVILGHEGAGIVESVGEGVTKLKAGDTV
Rat ADH-2:  IATAVCHTDAYTLSGADPEGCFPVILGHEGAGIVESVGEGVTKLKAGDTV
Human χ:    IATAVCHTDAYTLSGADPEGCFPVILGHEGAGIVESVGEGVTKLKAGDTV
Horse BB:   IATAVCHTDAYTLSGADPEGSFPVILGHEGAGIVESVGEGVTKLKAGDTV
            ^              ^     ^

                   100       110       120       130      139
Mouse B₂:   IPLYIPQCGECKFCLNPKTNLCQKIRVTQGKGLMPDGTSRFTCKGKSVFH
Rat ADH-2:  IPLYIPQCGECKFCLNPKTNLCQKIRVTQGKGLMPDGTSRFTCKGKPILH
Human χ:    IPLYIPQCGECKFCLNPKTNLCQKIRVTQGKGLMPDGTSRFTCKGKTILH
Horse BB:   IPLYIPQCGECKFCLNPQTNLCQKIRTTQGKGLMPDGTSRFTCKGKTILH
                             ^        ^                  ^^^

                   150       160       170       180      189
Mouse B₂:   FMGTSTFSEYTVVADISVAKIDPSAPLDKVCLLGCGISTGYGAAVNTAKV
Rat ADH-2:  FMGTSTFSEYTVVADISVAKIDPSAPLDKVCLLGCGISTGYGAAVNTAKV
Human χ:    YMGTSTFSEYTVVADISVAKIDPLAPLDKVCLLGCGISTGYGAAVNTAKL
Horse BB:   YMGTSTFSEYTVVADISVAKIDPLAPLDKVCLLGCGVSTGYGAAVNTAKV
            ^                      ^              ^           ^

                   200       210       220       230      239
Mouse B₂:   EPGSTCAVFGLGGVGLAVIMGCKVAGASRIIGIDINKDKFAKAKEFGASE
Rat ADH-2:  EPGSTCAVFGLGGVGLAVIMGCKVAGASRIIGIDINKDKFAKAKEFGATE
Human χ:    EPGSVCAVFGLGGVGLAVIMGCKVAGASRIIGVDINKDKFARAKEFGATE
Horse BB:   EPGSTCAIFGLGGVGLAVIMGCKVAGASRIIGVDINKDKFAKAKEFGASE
                ^   ^                         ^      ^     ^

                   250       260       270       280      289
Mouse B₂:   CISPQDFSKSIQEVLVEMTDGGVDYSFECIGNVKVMRSALEAAHKGWGVS
Rat ADH-2:  CINPQDFSKSIQEVLIEMTDGGVDFSFECIGNVKVMRSALEAAHKGWGVS
Human χ:    CINPQDFSKPIQEVLIEMTDGGVDYSFECIGNVKVMRAALEACHKGWGVS
Horse BB:   CINPQDFSKPIQEVLIEMTDGGVDYSFECIGNVKVMRAALEACHKGWGVS
              ^     ^      ^         ^              ^   ^

                   300       310       320       330      339
Mouse B₂:   VVVGVAASGEEISTRPFQLVTGRTWKGTAFGGWKSVESVPKLVSEYMSKK
Rat ADH-2:  VVVGVAASGEEISTRPFQLVTGRTWKGTAFGGWKSVESVPKLVSEYMSKK
Human χ:    VVVGVAASGEEIATRPFQLVTGRTWKGTAFGGWKSVESVPKLVSEYMSKK
Horse BB:   VVVGVAASGEEIATRPFQLVTGRTWKGTAFGGWKSVESIPKLVSEYMSKK
                        ^                             ^

                   350       360       370
Mouse B₂:   IKVDEFVTGNLSFDQINQAFDLMHSGDSIRTVLKM*
Rat ADH-2:  IKVDEFVTGNLSFDQINKAFDLMHSGNSIRTVLKL*
Human χ:    IKVDEFVTHNLSFDEINKAFELMHSGKSIRTVVKI*
Horse BB:   IKVDEFVTHSLSFDQINEAFELMHAGKSIRTVVKL*
            ^^    ^  ^  ^  ^   ^ ^      ^ ^
```

Figure 3. Amino Acid Sequences of Class III ADHs

The ADH-B$_2$ mRNA was most abundant in the mouse liver. We detected (by Northern and dot-blot analyses) the presence of Adh-B$_2$ mRNA in many mouse tissues, including brain, intestine, kidney, spleen and stomach. This is in agreement with the previously reported tissue distribution of the mouse Adh-B$_2$ enzymatic activity (Holmes, 1978; Holmes *et al.*, 1986).

Recently, Koivusalo *et al.* (1989 and this volume) have shown that the sequence of 57 amino acids (from 5 tryptic peptides) of the rat liver formaldehyde dehydrogenase (FDH) exactly matches the sequence of the rat liver class III ADH, ADH-2. They have also shown that FDH and class III ADH copurify and are similarly distributed in all tissues examined. This indicates that the class III ADH and the FDH are the same protein. Since the class III ADHs are very closely related in amino acid sequence to the class I and class II ADHs (about 60% identity), and have clearly evolved from a common precursor, it appears reasonable to retain the name alcohol dehydrogenase for these enzymes.

ACKNOWLEDGMENTS

The research described here was supported by USPHS research grant AA06460 from the N.I.A.A.A.; C.J.B. and L.G.C. were supported by a training grant T32 AA07463, also from the N.I.A.A.A.

We thank Kathyrn A. Baltz and Ronald E. Jerome for excellent technical assistance.

Figure 3 Legend. Numbering follows the class I mouse Adh-A$_2$ sequence (Edenberg *et al.*, 1985); the initiating Met is not shown, since it is removed and the N-terminus blocked in the ADH proteins analyzed to date (Fairwell *et al.*, 1987). The class III ADHs have an amino acid inserted after aa60 of the class I proteins, indicated by a •. The termination codon is indicated by a *. Sites where there are differences among the four class III ADHs are indicated by carats ^ beneath the horse-BB sequence. The eight introns in the mouse *Adh-1* gene (encoding Adh-A$_2$) are located after the codons for K$_5$, K$_{39}$, within the codons for G$_{86}$, D$_{115}$, after the codons for K$_{188}$, M$_{275}$, and within the codons for G$_{321}$, S$_{367}$ (Zhang *et al.*, 1987); other class I genes have introns in the same locations. Source of sequences: Mouse B$_2$: this manuscript. Rat ADH-2: protein (Julia *et al.*, 1988). Human χ: protein (Kaiser *et al.*, 1988). Horse BB: protein, (Kaiser *et al.*, 1989).

REFERENCES

Bosron, W.F., and Li, T.-K., 1986, Genetic polymorphism of human liver alcohol and aldehyde dehydrogenases, and their relationship to alcohol metabolism and alcoholism. *Hepatology* 6:502-510.

Carr, L.G., and Edenberg, H.J., 1990, Cis-acting sequences involved in protein binding and in vitro transcription of the human alcohol dehydrogenase gene *ADH2*. *J. Biol. Chem.* 265:1658-1664.

Carr, L.G., Zhang, K., and Edenberg, H.J., 1989a, Protein-DNA interactions in the 5' region of the mouse alcohol dehydrogenase gene *Adh-1*. *Gene* 78:277-285.

Carr, L.G., Xu, Y., Ho, W.-H., and Edenberg, H.J., 1989b, Nucleotide Sequence of the *ADH2³* Gene Encoding the Human Alcohol Dehydrogenase ß3 Subunit. *Alcohol. Clin. Exp. Res.* 13:594-586.

Carthew, R.W., L.A. Chodosh, and P.A. Sharp (1985). An RNA polymerase II transcription factor binds to an upstream element in the adenovirus major late promoter. *Cell* 43:439-448.

Chodosh, L.A., R.W. Carthew, J.G. Morgan, G.R. Crabtree, and P.A. Sharp (1987). The adenovirus major late transcription factor activates the rat gamma-fibrinogen promoter. *Science* 238:684-688.

Crabb, D.W., and Edenberg, H.J., 1986, Complete amino acid sequence of rat liver alcohol dehydrogenase deduced from the cDNA sequence. *Gene* 48:287-291.

Duester, G., Hatfield, G.W., Buhler, R., Hempel, J., Jornvall, H., and Smith, M., 1984, Molecular cloning and characterization of a cDNA for the beta subunit of human alcohol dehydrogenase. *Proc. Natl. Acad. Sci. USA.* 81:4055-4059.

Duester, G., Smith, M., Bilanchone, V., and Hatfield, G., 1986, Molecular analysis of the human class I alcohol dehydrogenase gene family and nucleotide sequence of the gene encoding the β subunit. *J. Biol. Chem.* 261:2027-2033.

Edenberg, H.J., Zhang, K., Fong, K., Bosron, W.F., and Li, T.-K., 1985, Cloning and sequencing of cDNA encoding the complete mouse liver alcohol dehydrogenase. *Proc. Natl. Acad. Sci. USA.* 82:2262-2266.

Edenberg, H.J., Dailey, T.L., and Zhang, K., 1989, Human alcohol dehydrogenase cDNAs: structure and expression. *Prog. Clin. Biol. Res.* 290: 181-192.

Ephrussi, A., G.M. Church, S. Tonegawa, and W. Gilbert (1985). B lineage-specific interactions of an immunoglobulin enhancer with cellular factors in vivo. *Science* 227, 134-140.

Fairwell, T., Julia, P., Kaiser, R., Holmquist, B., Pares, X., Vallee, B.L., and Jornvall, H., 1987, Acetylated N-terminal structures of class III alcohol dehydrogenases Differences among the three enzyme classes. *FEBS. Lett.* 222:99-103.

Fried, M., and D.M.Crothers (1981). Equilibria kinetics of lac repressor-operator interactions by polyacrylamide gel electrophoresis. *Nucleic Acids Res.* 9: 6506-6525.

Fromental, C., M. Kanno, H. Nomiyama, and P. Chambon (1988). Cooperativity and heirarchical levels of functional organization in the SV40 enhancer. *Cell* **54**:943-953.

Garner, M.M. and A. Revzin (1981). A gel electrophoresis method for quantifying the binding of proteins to specific DNA regions: application to components of the Escherichia coli lactose operon regulatory system. *Nucleic Acids Res.* **9**:3047-3060.

Giri, P.R., Krug, J.F., Kozak, C., Moretti, T., O'Brien, S.J., Seuanez, H.N., and Goldman, D., 1989, Cloning and comparative mapping of a human class III (χ) alcohol dehydrogenase cDNA. *Biochem. Biophys. Res. Commun.* **164**:453-460.

Heden, L.O., Hoog, J.O., Larsson, K., Lake, M., Lagerholm, E., Holmgren, A., Vallee, B.L., Jornvall, H., and von Bahr-Lindstrom, H., 1986, cDNA clones coding for the beta-subunit of human liver alcohol dehydrogenase have differently sized 3'-non-coding regions. *FEBS. Lett.* **194**:327-332.

Holmes, R.S., 1978, Electrophoretic analyses of alcohol dehydrogenase, aldehyde dehydrogenase, aldehyde oxidase, sorbitol dehydrogenase and xanthine oxidase from mouse tissues. *Comp. Biochem. Physiol. [B]* **61**:339-346.

Holmes, R.S., Duley, J.A., Algar, E.M., Mather, P.B., and Rout, U.K., 1986, Biochemical and genetic studies on enzymes of alcohol metabolism: the mouse as a model organism for human studies. *Alcohol. Alcohol.* **21**:41-56.

Hoog, J.O., Heden, L.O., Larsson, K., Jornvall, H., and von Bahr-Lindstrom, H., 1986, The gamma 1 and gamma 2 subunits of human liver alcohol dehydrogenase. cDNA structures, two amino acid replacements, and compatibility with changes in the enzymatic properties. *Eur. J. Biochem.* **159**:215-218.

Ikuta, T., Szeto, S., and Yoshida, A., 1986, Three human alcohol dehydrogenase subunits: cDNA structure and molecular and evolutionary divergence. *Proc. Natl. Acad. Sci. USA.* **83**:634-638.

Julia, P., Pares, X., and Jornvall, H., 1988, Rat liver alcohol dehydrogenase of class III. Primary structure, functional consequences and relationships to other alcohol dehydrogenases. *Eur. J. Biochem.* **172**:73-83.

Kaiser, R., Holmquist, B., Hempel, J., Vallee, B.L., and Jornvall, H., 1988, Class III human liver alcohol dehydrogenase: a novel structural type equidistantly related to the class I and class II enzymes. *Biochemistry* **27**:1132-1140.

Kaiser, R., Holmquist, B., Vallee, B.L., and Jornvall, H., 1989, Characteristics of mammalian class III alcohol dehydrogenases, an enzyme less variable than the traditional liver enzyme of class I. *Biochemistry* **28**:8432-8438.

Koivusalo, M., Baumann, M., and Uotila, L., 1989, Evidence for the identity of glutathione-dependent formaldehyde dehydrogenase and class III alcohol dehydrogenase. *FEBS. Lett.* **257**:105-109.

Miyamoto, N.G., V. Moncollin, R. Hen, J.M. Egly, and P. Chambon (1984). Stimulation of in vitro transcription by the upstream element of the adenovirus-2 major late promoter involves a specific factor. *Nucleic Acids Res.* **12**: 8779-8799.

Myers, R.M., K. Tilly, and T. Maniatis (1986). Fine structure genetic analysis of a β-globin promoter. *Science* **232**:613-618.

Sawadogo, M., and R.G. Roeder (1985). Interaction of a gene-specific transcription factor with the adenovirus major late promoter upstream of the TATA box region. *Cell* **43**:165-175.

Schüle, R., M. Muller, H. Otsuka-Murakami and R. Renkowitz (1988). Cooperativity of the glucocorticoid receptor and the CACCC-box binding factor. *Nature* **332**:87-90.

Sen, R., and D.Baltimore (1986). Multiple nuclear factors interact with the immunoglobin enhancer sequences. *Cell* <u>46</u>, 705-716.

Sharma, C.P., Fox, E.A., Holmquist, B., Jornvall, H., and Vallee, B.L., 1989, cDNA sequence of human class III alcohol dehydrogenase. *Biochem. Biophys. Res. Commun.* **164**:631-637.

Siebenlist, U., R. Simpson, and W. Gilbert (1980). E. coli RNA polymerase interacts homologously with two different promoters. *Cell* <u>20</u>, 269-281.

Smith, M., 1986, Genetics of human alcohol and aldehyde dehydrogenases. *Adv. Human. Genet.* **15**:249-290.

von Bahr-Lindstrom, H., Hoog, J.O., Heden, L.O., Kaiser, R., Fleetwood, L., Larsson, K., Lake, M., Holmquist, B., Holmgren, A., Hempel, J., Vallee, B.L., and Jornvall, H., 1986, cDNA and protein structure for the α subunit of human liver alcohol dehydrogenase. *Biochemistry* **25**:2465-2470.

Xiao, J.H., I. Davidson, M. Macchi, R. Rosales, M. Vigneron, A. Staub, and P. Chambon (1987). In vitro binding of several cell-specific and ubiquitous nuclear proteins to the GT-I motif of the SV40 enhancer. *Genes Dev.* **1**:794-807.

Yu, Y.T., and J.L. Manley (1984). Generation and functional analyses for base-substitution mutants of the adenovirus 2 major late promoter. *Nucleic Acids Res.* **12**:9309-9321.

Zhang, K., Bosron, W.F., and Edenberg, H.J., 1987, Structure of the mouse *Adh-1* gene and identification of a deletion in a long alternating purine-pyrimidine sequence in the first intron of strains expressing low alcohol dehydrogenase activity. *Gene* **57**:27-36.

ROLE OF ALCOHOL DEHYDROGENASE POLYMORPHISM

IN ETHANOL METABOLISM AND ALCOHOL-RELATED DISEASES

Patrice Couzigou, Benoît Fleury, Alexis Groppi, Albert Iron, Christiane Coutelle,
André Cassaigne, Joël Bégueret

Departement of Hepatogastroenterology, Hopital Haut-Lévêque, 33604 Pessac
Department of Genetics and Biochemistry, University Bordeaux II, 33076 Bordeaux
France

INTRODUCTION

In alcohol toxicity, a lot of data point to genetic factors associated with environmental factors. These genetic factors can intervene at different levels : genetic variability in the distribution of polymorphic alcohol and aldehyde dehydrogenase isozymes (Bosron, et al., 1988) could play a role in determining individual difference in ethanol metabolism and toxicity. In the future, other polymorphisms in alcohol metabolism may be described. Polymorphism is also to be considered in detoxication and/or target organ metabolism (as collagen polymorphism in alcoholic cirrhosis)(Weiner, et al., 1988).

Alcohol dehydrogenase (ADH) polymorphism is well known, especially in the class I at the loci ADH_2 and ADH_3 (Bosron, et al., 1988). Aldehyde dehydrogenase (ALDH) genetic polymorphism has been observed for $ALDH_2$ but not in Caucasian populations. Only a few cases of variation in $ALDH_1$ has been reported (Goedde, et al., 1989). In view of these data, we have chosen to focus on ADH polymorphism in France, especially ADH_3 polymorphism. Indeed, phenotypic studies show a rather well balance distribution about ADH_3 with a little tendency in favor of γ_1 sub-unit (ADH_3^1 allele). In the opposite, in France, the frequency of β_2 sub-unit (ADH_2^2 allele) is very low (Coutelle, et al., 1989), in comparaison of the previously published data in others Caucasian populations (Agarwal, et al., 1989).

Genotype determination of human ADH using direct mutagenesis by Polymerase chain reaction (P.C.R.) (Groppi, et al., in press.) allow us to study the role of ADH polymorphism in variations of alcohol metabolism, in the development of alcoholic liver disease and alcoholic chronic pancreatitis.

SUBJECTS AND METHODS

SUBJECTS

1) For alcohol metabolism study, sixteen caucasian non alcoholic male students were studied (median age : 22; 19-32 years, weight : 73.3 ± 8.6 kg). All subjects were social drinkers and on no medication. 8 subjects had ADH_3^1 ADH_3^1 genotype and 8 subjects ADH_3^2 ADH_3^2 genotype. They were picked out after screening of 34 subjects. Age and weight were similar between the two groups. Ethanol was administered as a 5 % sterile solution with isotonic glucose solution at a rate of 0,6 g.kg $^{-1}$ over a total infusion time of 90 minutes. Blood samples were taken every 15 min during 1 h, then every 30 min during 5 hours. Blood alcohol level was measured by enzymatic technique (radiative energy method) on a TDX system.

2) For cirrhosis study, in the first part of the study, in Bordeaux, 46 caucasian alcoholic (>100 gr pure alcohol per day for more 5 years ; 31 males, 15 females; median age 57 : 35-87) patients with cirrhosis were studied. They were Ag HbS negative, with ascite and/or persistent oesophageal varices treated by sclerotherapy. 39 caucasian medical students, (median age : 23 ; 19-32) in good health and with alcohol consumption < 40 gr/d represented the control group.
In the second part of the cirrhosis study, were included 19 alcoholic (>100 gr pure alcohol per day for more 5 years) French West Indian patients (16 males, 3 females) (age median 51 ; 40-68) with ascite and/or oesophageal varices. 24 French West Indian subjects without clinical or biological liver disease (19 males, 5 females) (age median 50 ; 35-69) were studied as controls.This part of the study has been realized with the collaboration of Doctors T. Jasawank and R. Bonnet (Hôpital de Pointe à Pitre , Guadeloupe, France).

3) For alcoholic chronic pancreatitis study, were included 15 caucasian alcoholic (>100 gr pure alcohol per day for more 5 years) patients (13 males, 2 females) (age median 45 ; 33-69) with chronic pancreatitis (pancreatitis calcifications and/or pseudocyst). 39 caucasian medical students (age median 23 ; 19 -32) with alcohol consumption < 40 gr/d represent the control group.This part of the study has been performed with collaboration of Professor Gérolami (Marseille, France).

4) Statistical analysis was performed using Hardy-Weinberg law with X^2 test and Odds ratio (Mantel-Haenszel method).

5) Genotyping of human ADH at ADH_2 and ADH_3 loci was carried out on blood microsample collected on blotting paper, very convenient for storage and handling and so for epidemiological studies. This method is not radioactive and use an internal control for digestion. $ADH_2{}^2$ differs from both $ADH_2{}^1$ and $ADH_2{}^3$ by a Mae III restriction site present in coding sequence of exon 3, corresponding to an aminoacid substitution at position 47. Thirty five base pairs downstream, another Mae III site present in all ADH genes was included in the amplified fragment as an internal control for digestion.(fig 1) For $ADH_2{}^3$, the sub-unit β_3 differ from both β_1 and β_2 at position 369 due to cytosine/thymine substitution in the coding sequence of exon 9 but without creation of restriction fragments length polymorphism (R.F.L.P.) The region surrounding the codon for aminoacid 369 was amplified. After, using direct mutagenesis by P.C.R., a restriction site was created, specific of $ADH_2{}^3$, and recognized by Alu I (with a sequence AGCT for $ADH_2{}^3$ and AGCC for $ADH_2{}^1$ and $ADH_2{}^2$. Moreover, two other Alu I restriction sites in the amplified fragment provide an internal control for digestion, one of these only present in the ADH_2 genes (fig2). For differentiation between $ADH_3{}^1$ and $ADH_3{}^2$, it was chosen to amplify a 145 bp segment of the exon 8 surrounding the codon for aminoacid 349 (one of the two aminoacid differences between γ_1 and γ_2 subunits). The substitution of guanine ($ADH_3{}^1$) by adenine ($ADH_3{}^2$) creates a Ssp I restriction site in $ADH_3{}^1$. For internal control of digestion, was created a new Ssp I site using mutagenesis by P.C.R. (fig 3) .(Groppi, et al., in press).

ADH 2 Exon III

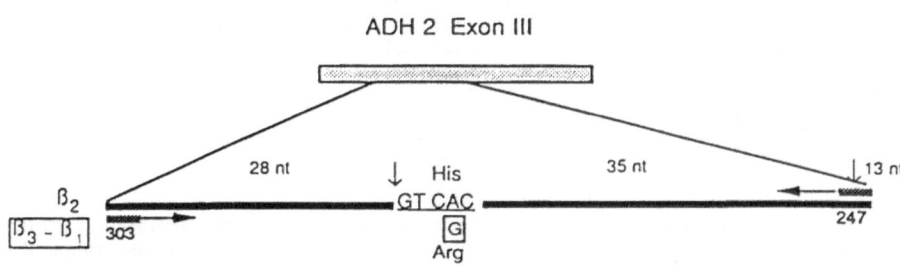

303/247:primers

303 : ATTCTGTAGATGGTGGCTGT

247 : GAAGGGGGGTCACCAGGTTG

↓ : Mae III cutting site, GTNAC : Mae III recognition sequence

Figure 1. STRATEGY FOR THE DETECTION OF $ADH_2{}^2$ ALLELE

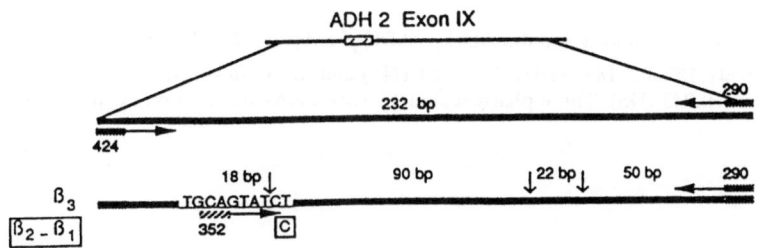

424, 290 et 352 : primers

424 : TGGACTCTCACAACAAGCATGGT

290 : TTTCTTTGGAAAGCCCCCAT

352 : TCTTTCCTATTGCAGTAGC

(*)represents a mismatch

↓ : Alu I cutting site, <u>AGCT</u> : Alu I recognition sequence

Figure 2. STRATEGY FOR THE DETECTION OF ADH$_2$^{3} ALLELE

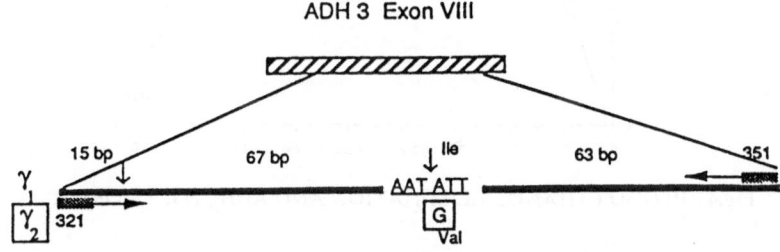

321/351 : primers

321 : GCTTTAAGAGTA<u>AATATT</u>CTGTCCCC

351 : AATCTACCTCTTTCCGAAGC

(*) represents a mismatch

↓ : SspI cutting site, <u>AATATT</u> : SspI recognition sequence

Figure 3. STRATEGY FOR THE DIFFERENTIATION OF ADH$_3$^{1}/ADH$_3$^{2} ALLELES

RESULTS

ETHANOL ELIMINATION RATE IN HEALTHY VOLUNTEERS WITH $ADH_3{}^1$-$ADH_3{}^1$ OR $ADH_3{}^2$-$ADH_3{}^2$ GENOTYPES

For the $ADH_3{}^1$-$ADH_3{}^1$ subjects and the $ADH_3{}^2$-$ADH_3{}^2$ subjects, the ethanol elimination rate ($mg.Kg^{-1}.h^{-1}$) was respectively 137.4 ± 14.6 and 149 ± 23.1 (NS) and the Widmark β factor ($mgr.ml^{-1}.h^{-1}$) 0.129 ± 0.039 and 0.150 ± 0.043 (NS). These pharmacokinetic data were calculated on the pseudo linear part of the curve (Fig. 4).

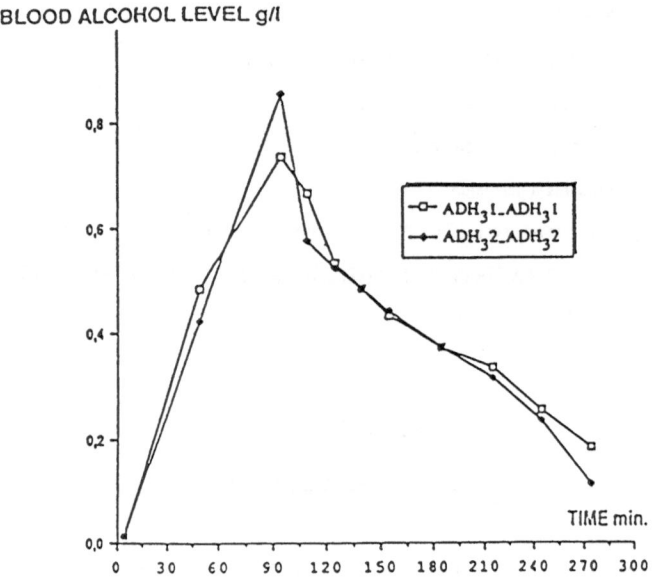

Fig.4. BLOOD ETHANOL ELIMINATION AND ADH_3 POLYMORPHISM

ADH POLYMORPHISM AND ALCOHOLIC CIRRHOSIS

The results about cirrhosis are presented in the Tables I and II for the Caucasian patients in Bordeaux and in the Tables III and IV for the study realized on French West Indian population. The Bordeaux study doesn't show any difference for ADH_3 polymorphism between cirrhosis and controls and the very low frequency of $ADH_2{}^2$ allele doesn't allow us to test an eventual role of ADH_2 polymorphism in the occurrence of alcoholic cirrhosis in France. No $ADH_2{}^2$ allele was found in this population sample. No difference between controls and cirrhosis was found for $ADH_2{}^3$ allele.

TABLE I

ADH2 POLYMORPHISM AND CIRRHOSIS (Study I)

	ADH_2^1-ADH_2^1	ADH_2^1-ADH_2^1	ADH_2^2-ADH_2^2
Controls (n=39) Observed Calculated	38 (97,5 %) 37,993(97,5 %)	1 (2,5 %) 1 (2,5 %)	0 (0 %) 0,007 (10^{-4} %)
Cirrhosis (n=46) Observed Calculated	44 (95,7%) 44,8	2 (4,3%) 1,19	0 (0%) 0,01

TABLE II

ADH3 POLYMORPHISM AND CIRRHOSIS (Study I)

	ADH_3^1-ADH_3^1	ADH_3^1-ADH_3^2	ADH_3^2-ADH_3^2
Controls (n=39) Observed Calculated	14 (35,9 %) 13 (33,3 %)	17 (43,6 %) 19 (48,8 %)	8 (20,5 %) 7 (17,9 %)
Cirrhosis (n=46) Observed Calculated	13 (28,3 %) 15,3	26 (56,5 %) 22,5	7 (15,2 %) 8,2

$X^2 = 1,767$ (NS)   Odds ratio : 0,7 (0,32 - 1,51)

TABLE III

ADH2 POLYMORPHISM AND CIRRHOSIS (Study II)

	$ADH_2{}^1\text{-}ADH_2{}^1$	$ADH_2{}^1\text{-}ADH_2{}^3$	$ADH_2{}^3\text{-}ADH_2{}^3$
Controls (n=24)			
Observed	16 (66,6 %)	8 (33,3 %)	0 (%)
Calculated	16,7 (69,6 %)	6,6 (27,4 %)	0,7 (3 %)
Cirrhosis (n=19)			
Observed	13 (68,4%)	6 (31,6%)	0 (%)
Calculated	13,3	5,2	0,6

X^2 : 0,197   odds ratio 0,92 (0,64-1,31)
Absence of $ADH_2{}^2$ allele $ADH_2{}^3$ frequency : 0,166

TABLE IV

ADH3 POLYMORPHISM AND CIRRHOSIS (Study II)

	$ADH_3{}^1\text{-}ADH_3{}^1$	$ADH_3{}^1\text{-}ADH_3{}^2$	$ADH_3{}^2\text{-}ADH_3{}^2$
Controls (n=23)			
Observed	13 (56,5 %)	9 (39,1 %)	1 (4,4 %)
Calculated	13,3 (57,8 %)	8,4 (36,5 %)	1,3 (5,7 %)
Cirrhosis (n=19)			
Observed	15 (78,9 %)	4 (21,1 %)	0 (0 %)
Calculated	11	6,9	1,1

X^2 : 1,45   odds ratio 0,35 (0,11-1,07)

The results about alcoholic chronic pancreatitis are presented in the Table V. No significant difference is observed between controls and alcoholic chronic pancreatitis for ADH3 polymorphism.

TABLE V

ADH3 POLYMORPHISM AND
ALCOHOLIC CHRONIC PANCREATITIS

	$ADH_3{}^2\text{-}ADH_3{}^2$	$ADH_3{}^1\text{-}ADH_3{}^2$	$ADH_3{}^2\text{-}ADH_3{}^2$
Controls(n = 39) Observed Calculated	14 (35,9 %) 13 (33,3 %)	17 (43,6 %) 19 (48,8 %)	8 (20,5 %) 7 (17,9 %)
Pancreatitis(n = 15) Observed Calculated	9 (60 %) 5	4 (26,7 %) 7,3	2 (13,3 %) 2,7

$\chi2$: 0,5 : NS, Odds ratio : 0,77 (0,37 - 1,59)

DISCUSSION

About genotype determination of ADH_2 and ADH_3 loci, the methodology used in this work presents advantages in comparison with previous methods (Gennari, *et al.*, 1988, Xu, *et al.*, 1988). The use of an internal control for digestion avoids all the confusions due to incomplete digestions. It requires only a blood microsample collected on blotting paper. This method of sampling offers great advantages in storage or handling. The use of direct mutagenesis by P.C.R. allows us to determine the genotype unambiguously without needing allele specific oligonucleotidic probes which requires more manipulations and are less easy to use than R.F.L.P.

Measurement of alcohol elimination rate in humans reveals a large, two-three fold, variation between individuals. When elimination rate was measured in twins, it was found that about half of the variation could be accounted for by genetic factor (Kopun, *et al.*, 1977, Martin, *et al.*, 1985). Moreover, preliminary results point to no significant difference in the rate of ethanol metabolism between normal and atypical ADH phenotype carriers (Edwards, *et al.*, 1967). Our study is the first using genotype determination of the subjects. The first analysis shows no difference but it is possible that significant difference using Michaelis-Menten model at lower concentrations gives different results and new look about this point is in progress. In France, it is difficult to study an eventual role of ADH_2 polymorphism in ethanol metabolism in view of the very low frequency of $ADH_2{}^2$ allele. In French West Indies, the frequency of $ADH_2{}^3$ is similar to these of black American or African populations (Iron, *et al.*, 1989). In view of the rather low frequency of $ADH_2{}^3$ allele in black populations, a lot of subjects are to be test to pick out $ADH_2{}^1$ homozygous subjects and $ADH_2{}^3$ homozygous subjects and after perform comparative alcohol pharmacokinetic studies.

20-30 % of chronic alcohol abusers develop cirrhosis. A twin study favors the existence of a genetic factor in cirrhosis (Hrubec. *et al.*,1981) but a lot of other factors could contribute to individual susceptibility for alcoholic cirrhosis, especially nutrition, virus B and/or C, congeneres in alcoholic beverages. The present study shows no significant difference between controls and alcoholic cirrhosis patients for $ADH_2{}^3$ allele and for ADH_3 alleles. However, it's possible that, at moderate alcohol consumption, a particular genotype could play a significant role in the development of alcoholic cirrhosis. Moreover, at the expression level, a polymorphism could play a significant role. In a recent genotyping study in Japanese, (Shibuya, *et al.*, 1988) no difference between patients with alcoholic liver disease and control group was found at the ADH_2 locus.

About alcoholic chronic pancreatitis, some data also point out to the existence of a genetic factor in the development of alcoholic chronic pancreatitis, especially about deficiency of pancreatic stone protein, but variations in alcohol metabolism could also intervene (Sarles, *et al.*, 1990). The present results are too small to conclude about ADH_3 polymorphism even if it is questionnable that a larger study would show a significant difference in view of the Odds ratio confidence interval. It is possible that, as for cirrhosis, a particular genotype could intervene in the development of alcoholic chronic pancreatitis, only at moderate alcohol consumption. A role of a genetic polymorphism in expression of ADH is also possible.

Acknowledgements -- The authors express their gratitude to Mrs O. Rus for her valuable contribution. This work has been supported by two research grants (n° 87-023, 1987 and n° 887-006, 1988) from Institut National de la Recherche Médicale (INSERM).

REFERENCES

Agarwal, D.P., Goedde, H.W.,1989, Enzymology of alcohol degradation. In *Alcoholism Biomedical and Genetic Aspects*, Goedde, H.W. and Agarwal, D.P. eds, pp. 3-20. Pergamon Press, New York.

Bosron, W.F., Lumeng, L., Li, T.K., 1988, Genetic polymorphism of enzymes of alcohol metabolism and susceptibility to alcoholic liver disease. *Molec. Aspects Med.,* **10** : 147-158.

Coutelle, C., Fleury, B., Couzigou, P., Poupon, R.E., Nalpas, B., Iron, A., Higueret, D., Saric, J., Masson, B., Béraud, C., Cassaigne, A., 1989, Distribution of β and γ isozymes of hepatic alcohol dehydrogenase (ADH) in France (Abstract), *Alcohol Alcoholism ,* **24** : 369.

Edwards, J.A., Price Evans, D.A., 1967, Ethanol metabolism in subjects possessing typical and atypical liver alcohol dehydrogenase, *Clin. Pharmacol. Ther.,* **8** : 824-829.

Gennari, K., Wermuth, B., Muellener, D.,Ehrig, E., and von Wartburg, J.P., 1988, Genotyping of human class I alcohol dehydrogenase. Analysis of enzymatically amplified DNA with allele-specific oligonucleotides, *FEBS Letters ,* **228**: 305-309.

Goedde, H.W., Agarwal, D.P., 1989, Acetaldehyde metabolism : genetic variation and physiological implications, In *Alcoholism Biomedical and Genetic Aspects*, Goedde, H.W. and Agarwal, D.P. eds, pp. 21-56. Pergamon Press, New York.

Groppi, A., Bégueret, J., Iron, A., Advanced methods in the genotype determination of the human alcohol dehydrogenase at the ADH 2 and ADH 3 loci using directed mutagenesis by P.C.R., *Clin. Chem.,* in press.

Hrubec, Z., Ommen, G.S.,1981, Evidence of genetic predisposition to alcoholic cirrhosis and psychosis : twin concordances for alcoholism and its biological end-points by zygotisis among male veterans. *Alcoholism Clin. Exp. Res.,* **5** : 207-215.

Iron, A., Groppi, A., Fleury, B., Couzigou, P., Bégueret, J., Cassaigne, A., 1989, A procedure for complete genotyping human class I alcohol dehydrogenase (ADH) with the polymerase chain reaction (P.C.R.) on a micro-blood sample (Abstract). *Alcohol Alcoholism ,* **24** : 377.

Kopun, M., Propping, P., 1977, The kinetics of ethanol absorption and elimination in twins and supplementary repetitive experiments in singleton subjects, *Eur. J. Clin. Pharmacol.,***11** : 337-344

Martin, N.G., Perl,J., Oakeshott, J.G., Gibson, J.B., Starmer, G.A., Wilks, A.V., 1985, A twin study of ethanol metabolism , *Behav. Genet.,* **15** : 93-109.

Sarles, H., Bernard, J.P., Gullo, L., 1990, Pathogenesis of chronic pancreatitis, *Gut,* **31** : 629-632.

Shibuya, A., Yoshida, A., 1988, Genotypes of alcohol metabolizing enzymes in japanese with alcohol liver diseases : a strong association of the usual caucasian-type aldehyde dehydrogenase gene ($ALDH_2{}^1$) with the disease,*Am. J. Hum. Genet.,* **43**: 744-748.

Weiner, F.R., Eskreis, D.S., Compton, K.V., Orrego, H., Zern, M.A., 1988, Haplotype analysis of a type I collagen genetic and its association with alcoholic cirrhosis in man, *Molec.Aspects Med.,* **10** : 159-168.

Xu, Y., Carr, L.G., ßosron, W.F., Li, T.K., Edenberg, H.J., 1988, Genotyping of human alcohol dehydrogenases at the ADH2 and ADH3 loci following sequence amplification, *Genomics ,* **2**: 209-214.

CHARACTERIZATION OF HUMAN ALCOHOL DEHYDROGENASES

CONTAINING SUBSTITUTIONS AT AMINO ACIDS 47 AND 51

Thomas D. Hurley, Torsten Ehrig, Howard J. Edenberg, and William F. Bosron

Department of Biochemistry and Molecular Biology, Indiana University School of Medicine, Indianapolis, IN 46202

INTRODUCTION

The kinetic properties of multiple molecular forms of human liver alcohol dehydrogenase (ADH) have been examined in an effort to gain a better understanding of the structure-function relationships of this enzyme, which catalyzes the rate-limiting step during ethanol metabolism in the liver (Ehrig et al., 1990). In order to accomplish this, a system was developed to express the cloned human $\beta_1\beta_1$ enzyme and mutants at specific amino acid positions which were prepared by site-directed mutagenesis in *E. coli* (Hurley et al., 1990).

The expressed enzymes were purified to homogeneity and their steady-state kinetic and coenzyme binding properties compared. X-ray data on crystals of the recombinant $\beta_1\beta_1$ enzyme grown in the presence of NAD^+ was collected. The structure was solved by molecular replacement with the horse EE enzyme as the search structure, and it is being refined to 3.3 angstroms resolution.

THE ROLE OF ARGININE-47 IN COENZYME BINDING AND THE V_{MAX} OF ETHANOL OXIDATION

Arg-47 has been suggested to play an important role in coenzyme binding, based on the structure of the horse EE enzyme (Eklund et al., 1981). In the horse structure, the guanidino nitrogen of this amino acid hydrogen bonds with an adenosine phosphate oxygen of NAD(H) (Eklund et al., 1984). The preliminary x-ray structure of the $\beta_1\beta_1$ enzyme indicates that Arg-47 forms a hydrogen bond with the 3' oxygen of the adenosine ribose and a hydrogen bond/ion pair with one of the adenosine phosphate oxygens of NAD(H). Therefore, the substitution of this amino acid in the α (Gly for Arg) and β_2 (His for Arg) enzymes would be expected to have substantial effects on the kinetics of coenzyme binding and catalysis. The human $\beta_2\beta_2$ enzyme does display decreased affinity for NAD^+ (a higher K_i for NAD^+ than $\beta_1\beta_1$ in Table 1). Surprisingly, the human $\alpha\alpha$ enzyme with a Gly at position 47 displays an increased affinity for coenzyme (a lower K_i for NAD^+ than $\beta_1\beta_1$ in Table 1).

In order to systematically assess the role of Arg-47 in coenzyme binding and catalysis, a series of enzymes which possess amino acids with decreasing basic character at position 47 [Lys (K), His (H), Gln (Q), and Gly (G)] were prepared (Hurley et al. 1990). The recombinant enzymes were expressed in *E. coli*, purified, and their interaction with NAD(H) was characterized using steady-state kinetics and equilibrium binding studies.

TABLE 1. Kinetic Constants of Liver ADH Enzymes

	$\beta_1\beta_1$	$\beta_2\beta_2$	$\alpha\alpha$
Amino acid 47	Arg	His	Gly
V_{max} (min^{-1})	3.4	400	27
K_i(NAD$^+$) (μ M)	90	340	32

K_i values are from (Bosron et al., 1983) and V_{max} values are from (Burnell et al., 1989).

The human $\beta_1\beta_1$ and $\beta_2\beta_2$ enzymes have been shown to obey the ordered Bi Bi mechanism of substrate addition (Yin et al., 1984). Product inhibition studies of the β47Q and β47G enzymes were consistent with a change in the mechanism of substrate addition. The product inhibition patterns found for the β47Q and β47G enzymes are listed in Table 2, along with the expected patterns for an ordered Bi Bi, Theorell-Chance ordered Bi Bi, and rapid equilibrium random Bi Bi mechanisms. The pattern exhibited by β47Q is consistent with the Theorell-Chance ordered Bi Bi mechanism and the pattern exhibited by β47G is consistent with the rapid equilibrium random Bi Bi mechanism.

Table 2. Product Inhibition Patterns of Bi Bi Enzymes

Mechanism	ALD vs. NAD$^+$ (Subsat. EtOH)	ALD vs. EtOH (Subsat. NAD$^+$)
Ordered Bi Bi (ternary complex)	Noncompetitive	Noncompetitive
Ordered Bi Bi (Theorell-Chance)	Noncompetitive	Competitive
Rapid Equilibrium Random Bi Bi	Competitive	Competitive
Enzyme		
β47Q	Noncompetitive	Competitive
β47G	Competitive	Competitive

For all mechanisms, an examination of the effects of amino acid substitutions on K_i(NAD$^+$) and K_d(NADH) provides the best indication of the role of the amino acid 47 side chain in coenzyme binding. If we exclude β47G from the discussion for the moment, it is obvious from Table 3 that the less basic amino acids substituted at position 47 show lower affinity than β47R (higher K_i for NAD$^+$ and higher K_d for NADH). In addition, each of these enzymes also exhibits a higher V_{max} value than β47R. The increased V_{max} values are consistent with an increase in the rate of NADH dissociation, which has been shown to be partially rate-limiting for ethanol oxidation with $\beta_1\beta_1$ (β47R) and $\beta_2\beta_2$ (β47H) (Yin et al., 1984). Interestingly, the β47Q enzyme exhibits the highest V_{max} of any mammalian ADH. The data obtained from this series of enzymes clearly confirms that Arg-47 contributes to coenzyme binding and kinetics, since substitution of less basic amino acids result in decreased affinity for NAD(H) and increased V_{max}.

Table 3. Kinetic Constants of Expressed ADH Variants

	$\beta 47R$	$\beta 47K$	$\beta 47H$	$\beta 47Q$	$\beta 47G$
Amino acid 47	Arg	Lys	His	Gln	Gly
pK_a of AA-47	12.5	10.5	6.0	-	-
V_{max} (min^{-1})	3.8	10	370	840	2.1
K_i(NAD$^+$) (μM)	50	200	360	600	390
K_d(NADH) (μM)	0.16	0.34	3.7	2.6	0.79

The substitution of Gly for Arg-47 in $\beta_1\beta_1$ does not alter the enzyme in the same manner as the other substitutions. $\beta 47G$ exhibits affinities for NAD$^+$ and NADH which are intermediate between $\beta 47K$ and $\beta 47Q$, and a V_{max} lower than that of $\beta 47R$. That this enzyme should show moderately high affinity for NAD(H) is surprising, since the substitution of Gly at this position, unlike the other substitutions, completely removes a potential hydrogen-bonding donor to the NAD(H) molecule. A possible explanation for this anomalous behavior is that the substitution of Gly at position 47 results in a conformational change in the enzyme which somehow both compensates for the loss of a hydrogen bond donor and alters the mechanism of substrate addition (Table 2). The kinetic properties of $\beta 47G$ (Table 3) are also quite different from those of $\alpha\alpha$ (a lower V_{max} and higher K_i for NAD$^+$). One or more of the additional 22 substitutions present in $\alpha\alpha$ versus $\beta 47G$ appear to alter coenzyme binding and kinetics.

THE ROLE OF HISTIDINE-51 AS A GENERAL BASE CATALYST

The rate of alcohol oxidation in the EE isoenzyme of horse liver ADH shows a pH dependence which is related to the pK_a of the alcohol substrate. This suggests that during catalysis the alcohol deprotonates to form the alcoxide ion (Kvassman et al., 1981) which then undergoes hydride transfer to yield the aldehyde. It has been proposed that during catalysis the proton is shuttled from the active center through a so-called "proton-relay system" involving the hydroxyl group of Ser-48, the 2' hydroxyl of the coenzyme nicotinamide ribose, and the imidazole nitrogens of His-51. These residues form a hydrogen-bonded chain extending from the alcohol to the solvent through the exposed imidazole of His-51 (Eklund et al., 1982). It is conceivable that the imidazole of His-51, with it's pK_a close to 7.0, is the actual proton acceptor during catalysis *in vivo*. We have mutated His-51 in the homologous $\beta_1\beta_1$ enzyme ($\beta 51H$) to Gln ($\beta 51Q$) to test the role of His-51 in proton transfer.

The steady-state kinetic parameter $V_{max}/K_{M(EtOH)}$ (V/K_b), representing the catalytic efficiency of alcohol oxidation, was approximately 6-fold lower in the mutant enzyme ($\beta 51Q$) compared to the wildtype ($\beta 51H$) in 10 mM ACES (Table 4). This drop could be restored by the addition of glycyl-glycine (Table 4). With 33 mM Na_2SO_4 no change in V/K_b is observed, demonstrating that the effect of glycyl-glycine is not due to an increase in the ionic strength of the buffer. These findings imply that, in the mutant enzyme, the loss of the imidazole of His-51 removes the proton acceptor and the catalytic efficiency is reduced. The missing proton acceptor can, however, be substituted for by a base present in the solvent (glycyl-glycine), thus raising V/K_b to the wild-type value.

Table 4. Buffer Effects on V/K_b at pH 7.0

Buffer	V/K_b [(Units/mg)/mM]	
	$\beta 51H$	$\beta 51Q$
10 mM ACES	3.8	0.6
10 mM ACES + 0.1 M Glycyl-glycine	3.7	4.0
10 mM ACES + 33 mM Na_2SO_4	4.1	0.6

Site-directed mutagenesis of a putative acid/base catalyst in conjunction with the restoration of activity by addition of a buffer base has been reported for other enzymes. In aspartate aminotransferase and carbonic anhydrase, Lys-258 and His-64, respectively, had been postulated to be responsible for proton transfer. When these residues were mutated to Ala, the mutants showed a decreased activity which could be restored by the presence of high concentrations of buffers (Toney and Kirsch, 1989; Tu et al. 1989). The authors pointed out that a reduced activity as a result of the mutation did not necessarily indicate a specific function for the mutated amino acid, because structural rearrangements affecting other active site residues could have occurred. Accordingly, the restoration of activity in the Gln for His-51 ADH mutant by glycyl-glycine strongly suggests that this residue functions as a general base during ethanol oxidation.

(Supported by R01-AA07117 and T32-AA07462)

REFERENCES

Bosron, W. F., Magnes, L. J., and Li, T.-K., 1983, Kinetic and Electrophoretic Properties of Native and Recombined Isoenzymes of Human Liver Alcohol Dehydrogenase, Biochemistry, 22:1852-1857.

Burnell, J. C., Li, T.-K., and Bosron, W. F., 1989, Purification and Steady-State Kinetic Characterization of Human Liver $\beta_3\beta_3$ Alcohol Dehydrogenase, Biochemistry, 28:6810-6815.

Ehrig, T., Bosron, W. F., and Li, T.-K., 1990, Alcohol and Aldehyde Dehydrogenase, Alcohol and Alcoholism, 25:105-116.

Eklund, H., Samama, J. P., Wallen, L., and Branden, C. I., 1981, Structure of a Triclinic Complex of Horse Liver Alcohol Dehydrogenase at 2.9 Angstrom Resolution, J. Mol. Biol., 146:561-587.

Eklund, H., Plapp, B. V., Samama, J. P., and Branden, C. I., 1982, Binding of Substrate in a Ternary Complex of Horse Liver Alcohol Dehydrogenase, J. Biol. Chem., 257:14349-14358.

Eklund, H., Samama, J. P., and Jones, T. A., 1984, Crystallographic Investigations of Nicotinamide Adenine Dinucleotide Binding to Horse Liver Alcohol Dehydrogenase, Biochemistry, 23:5982-5996.

Hurley, T. D., Edenberg, H. J., and Bosron, W. F., 1990, Expression and Kinetic Characterization of Variants of Human $\beta_1\beta_1$ Alcohol Dehydrogenase Containing Substitutions at Position 47, J. Biol. Chem., in press., 1990.

Kvassman, J., Larsson, A., and Petterson, G., 1981, Substituent Effects on the Ionization Step Regulating Desorption and Catalytic Oxidation of Alcohols Bound to Liver Alcohol Dehydrogenase, Eur. J. Biochem., 114:555-563.

Toney, M., and Kirsch, J. F., 1989, Direct Bronsted Analysis of the Restoration of Activity to a Mutant Enzyme by Exogenous Amines, Science, 243:1485-1488.

Tu, C., Forsman, C., Jonsson, B. H., Lindskog, S., and Silverman, D. N., 1989, Role of Histidine 64 in the Catalytic Mechanism of Human Carbonic Anhydrase II Studied with a Site-Specific Mutant, <u>Biochemistry</u>, 28:7913-7918.

Yin, S.-J., Bosron, W. F., Magnes, L. J., and Li, T.-K., 1984, Human Liver Alcohol Dehydrogenase: Purification and Kinetic Characterization of the $\beta_2\beta_2$, $\beta_2\beta_1$, $\alpha\beta_2$, and $\beta_2\gamma_1$ "Oriental" Isoenzymes, <u>Biochemistry</u>, 23:5847-5853.

ENDOCRINE REGULATION AND METHYLATION PATTERNS OF RAT

CLASS I ALCOHOL DEHYDROGENASE IN LIVER AND KIDNEY

David W. Crabb, Mona Qulali, and Katrina M. Dipple

Departments of Medicine and of Biochemistry
and Molecular Biology
Indiana University School of Medicine
Indianapolis, IN

INTRODUCTION

Alcohol dehydrogenase (ADH) is expressed in a tissue-specific fashion and its activity is modulated by several hormones. Class I ADH activity is largely confined to the liver, with considerably smaller levels in the kidney. Experiments with intact animals indicate that thyroid hormone and testosterone reduce the activity of the enzyme in liver (Mezey and Potter, 1981; Rachamin et al., 1980), and that estrogen and growth hormone increase the activity (Teschke et al., 1986; Mezey and Potter, 1979). In primary hepatocyte cultures, growth hormone produced an increase in ADH activity and mRNA levels, and dihydrotestosterone reduced the activity (Mezey et al., 1986a, 1986b; Potter et al., 1989). Certain hepatoma cell lines express low levels of ADH activity and mRNA, and these can be induced by exposure to glucocorticoids (Wolfla et al., 1988; Dong et al., 1988). The enzyme activity in the kidney is induced by androgens in the mouse (Felder et al., 1988) and by estrogen in the rat (Dembic and Sabolic, 1982). Although the effect of androgens in mouse kidney is known to be mediated by an increase in the transcription of the gene, the mechanism of the effect of estrogens in rat kidney has not been reported.

To elucidate the molecular mechanisms of these examples of steroid induction of ADH, we studied the effect of glucocorticoids on the expression of the rat class I ADH gene in gene transfection experiments. We also examined the effect of estrogen on ADH activity and mRNA expression in the kidney. Because the tissue-specific expression of some genes is controlled by or associated with changes in the methylation of the gene, we restriction mapped the rat ADH gene for Msp I sites and compared the patterns of methylation of the ADH gene in several tissues: liver, kidney, and spleen.

METHODS

cDNA and Genomic Cloning

The rat ADH cDNA was obtained by screening a liver cDNA library with the mouse ADH cDNA (Crabb and Edenberg, 1986). The rat ADH gene was cloned from a Charon 4A genomic library. Two non-overlapping clones were obtained. The remaining portion of the gene was obtained by amplifying genomic DNA using primers specific for the third exon and the 3' end of the third intron (Crabb et al., 1989). The clones were mapped for Msp I sites by standard methods.

Enzyme Assays

Liver and kidney specimens were rapidly homogenized in 50 mM Hepes buffer, pH 8.4, containing dithiothreitol and were centrifuged at 100,000 x g. The supernatants were then assayed for enzyme activity and protein content. The enzyme assay was carried out at 25° in 0.5 M Tris HCl, pH 7.2 and started with the addition of 10 mM ethanol (Crabb et al., 1986). Activity was normalized to the content of protein in the supernatant and of DNA in the homogenates. Chloramphenicol acetyltransferase was assayed as described elsewhere (Gorman et al., 1982; Crabb and Dixon, 1987). The radioactivity in the acetylated chloramphenicol derivatives was quantified using a beta scanner (AMBIS Radioanalytic Imaging Systems).

Plasmid Constructions

pTE2 is the parent plasmid and contains the chloramphenicol acetyltransferase (CAT) gene and splice signals under the control of the herpes simplex virus (HSV) thymidine kinase promoter. This promoter contains a TATA box, a CCAAT box, and two Spl sites (McKnight and Tjian, 1986). pTE2ADH contains a 3.5 kb fragment of the upstream region of the ADH gene from the Bam HI site in exon 1 to the upstream Eco RI site (Figure 1). pMSG-CAT contains the CAT gene under the control of the simian virus 40 (SV40) promoter and the mouse mammary tumor virus (MMTV) long terminal repeat (LTR) (Alton and Vapnek, 1979) and was obtained from Pharmacia. p6RGR contains the rat glucocorticoid receptor driven by the Rous sarcoma virus (RSV) LTR. It was kindly provided by Dr. Keith Yamamoto.

Tissue Culture Methods

H4IIE and COS-7 cells were grown in Dulbecco's minimal essential medium (DMEM) with 5% fetal bovine serum. The cells were transfected using the calcium phosphate precipitation technique (Graham and van der Eb, 1973) and glycerol shocked at 4 hours after transfection. The cells were harvested 48 h later for CAT assays (Gorman et al., 1982). This included heat treatment of the extract to inactivate enzymes which interfere with the assay (Crabb and Dixon, 1987).

Northern and Southern Blots

RNA was prepared from the kidneys by the guanidinium isothiocyanate acid phenol method (Chomzynski and Sacchi, 1987). This method was modified by the addition of a LiCl precipitation step to remove polysaccharides and by additional phenol-chloroform extractions (Meltzer et al., 1990). The RNA was fractionated in 1% formaldehyde agarose gels, blotted to charge-modified nylon membranes (Nytran), and hybridized with the ADH cDNA and a housekeeping cDNA (CHO-B) which served as an internal control for the amount of RNA loaded and transferred.

Genomic Southern blots were performed using DNA isolated from kidney, liver, and spleen by proteinase K digestion, phenol extraction, and spooling of the high molecular weight DNA. Ten micrograms of DNA were then digested with 20 U of Msp I or Hpa II overnight and electrophoresed in a 0.6% agarose gel. The DNA was fragmented by exposing the gel to UV light for 15 min before the denaturation and renaturation steps, and the DNA was blotted by capillary action to nitrocellulose. The blots were hybridized with probes obtained from the ADH genomic clones.

RESULTS

Glucocorticoid-Responsiveness of DNA Sequences 5' to the ADH gene

3.5 kb of DNA from the 5'-flanking region of the ADH gene were cloned in a vector (pTE2) containing the HSV thymidine kinase (tk) promoter driving the chloramphenicol acetyltransferase (CAT) gene. We had found in the past that the available hepatoma cell lines, such as the H4IIE or H4IIEC3 cells, that were used to observe the effect of dexamethasone on ADH mRNA and activity, transfect poorly when any number of different transfection protocols were followed. COS-7 cells transfect very well but they and the cells from which they were derived (CV-1) are reported to contain few or no glucocorticoid receptors. To study the interaction of glucocorticoid receptors and the ADH flanking sequences, we chose to cotransfect pTE2ADH with p6RGR, a plasmid directing high level expression of the rat glucocorticoid receptor, into the COS-7 cells. Thus we hoped to circumvent the difficulty in transfecting the hepatoma cells. Control experiments showed that expression of a MMTV promoter-CAT reporter gene (pMSG-CAT, that contains glucocorticoid response elements (GRE)) was increased 7-fold by dexamethasone after transfection into COS-7 cells. This may represent the effect of low levels of the glucocorticoid receptor in this cell line, or a non-receptor mediated effect. When the COS-7 cells were cotransfected with the receptor plasmid and the reporter plasmid, there was at least a 17-fold increase in CAT activity in the presence of dexamethasone. Hence, the receptor was expressed in a functional form in the COS-7 cells.

The low levels of expression of pTE2, containing only the tk promoter, were not altered by dexamethasone in the presence or absence of p6RGR. Dexamethasone increased the expression of pTE2ADH approximately 3.4-fold when p6RGR and the pTE2ADH plasmids were cotransfected, compared to cotransfection and culturing in the absence of dexamethasone. The magnitude of this increase was similar to that of the increase in ADH expression observed in hepatoma cells (Wolfla et al., 1988; Dong et al., 1988) and suggested that one or more GREs are present within the ADH 5'-flanking region included in this construct.

Mechanism for the Estrogen Induction of Rat Kidney ADH

Rat kidney cortex contains much lower activity of ADH than does liver (0.04 vs 1.1 U/gm tissue). It had been reported that the activity of kidney ADH was higher in the kidneys of female rats than of male rats (Buttner, 1965) and that estrogen injections induced this activity in female rats (Dembic amd Sabolic, 1982). Using the dose of estradiol previously shown to induce the maximum response (1 mg/kg body weight/d), we replicated the finding that estrogen induces ADH activity 2-3 fold in kidney of female rats. This was accompanied by an approximate 3-4 fold increase in the amount of ADH mRNA present on Northern blots, compared with the amount of CHO-B message. Of interest, the ADH activity increased only 3-4 fold in kidney cortex of male rats treated with estrogen, while mRNA appears to increase 8-10 fold. These results are being extended by performing Western blots of the kidney extracts to determine if the induction in activity is accompanied by a proportional induction in the ADH protein. It is conceivable that estrogen alters both the RNA level as well as its translatability or the stability of ADH protein in kidney. It is noteworthy that estrogen does not induce ADH activity in the liver nearly as much as in the kidney, if at all (Rachamin et al., 1980). We are performing S1 nuclease assays to determine if the kidney and hepatic RNAs are transcribed from the same promoter and have the same cap site. There is precedent for the use of alternative ADH promoters in different stages of development in Drosophila (Benyajati et al., 1983).

Class I ADH Does Not Reside Within a HTF Island

DNA of vertebrates can be methylated at the 5' cytosine of the dinucleotide CpG. The majority of CpGs are methylated in mammals, with the notable exception of CpGs within so-called HTF (Hpa tiny fragment) islands. HTF islands are rich in G+C content and the number of CpGs is similar to that of GpC, which is not subject to methylation. The depletion of CpG in the remainder of the genome is thought to have resulted from the spontaneous deamination of methylated cytosine to generate thymine. This form of mutation cannot be repaired by DNA repair mechanisms, and therefore these mutations accumulate in the genome. Genes which are associated (usually in their 5' region) with HTF islands are often (but not always) considered to have housekeeping functions (Bird, 1986). On the other hand, tissue-specific genes are commonly relatively undermethylated in the tissues in which the gene is transcribed, but fully methylated in non-expressing tissues. We therefore examined the sequences surrounding the first exon of the class I ADH gene for the characteristics of an HTF island and performed blotting experiments to determine the methylation status of the gene in liver, kidney, and spleen. Sequences of exons 2 through

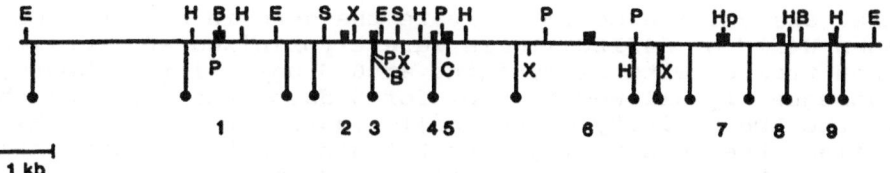

Figure 1. Msp I restriction map of the rat class I ADH gene. Msp I sites are indicated by closed circles. E = Eco RI, H = Hind III, P = Pst I, B = Bam HI, S = Sst I, X = Xba I, C = Cla I, Hp = Hpa I. Exons are shown as filled boxes and are numbered.

9, flanking introns, and the downstream flanking region, had roughly equal G+C and A+T contents but contained only 50% as many of the dinucleotide CpG compared with GpC. The 800 bp surrounding the cap site (-436 to +360 bp into the first intron) were A+T rich (58%) and very low in CpG (9 dinucleotides) compared with GpC (41 dinucleotides). This argues that the ADH gene is not associated with an HTF island, which is consistent with its tissue-specific expression in liver and, to a lesser extent, in kidney.

To determine which CpGs were methylated in different tissues, we digested total genomic DNA prepared from liver, kidney, and spleen with either Msp I or Hpa II, and then hybridized the DNA with different fragments of the ADH gene. Both enzymes recognize the sequence CCGG, but only Msp I can digest the site if the internal CpG is methylated. Msp I sites were mapped in the cloned fragments of the ADH gene (Figure 1). The genomic blots demonstrated several things. First, smears were observed when several genomic fragments were used as probes, indicating the presence of repetitive sequences. One lies within the first intron, probably the sequence of 26 consecutive TGs in this region (Crabb et al., 1989). This sequence is known to be a repetitive element in most eukaryotes from yeast to humans (Hamada et al., 1982). Additional repetitive sequences are probably present in the region surrounding the 5th and 6th exons. Second, most of the Msp I sites were methylated in all three tissues. One important exception was the presence of two unmethylated Msp I sites surrounding the first exon in DNA from liver. This suggests the possiblity that methylation of these sites (and presumably other CpGs which are not detected using this assay) is associated with inactivity of the gene in kidney and spleen. Our inability to detect unmethylated sites in kidney DNA may reflect the likelihood that only a small proportion of the cells in the kidney (most likely the proximal tubules (Goldstein and Maxwell, 1982)) express ADH.

In some instances, treatment of cells in tissue culture with the antineoplastic agent 5-azacytidine has been associated with increased expression of certain genes, e.g., globin or thyroglobulin (Avvedimento et al., 1989). Azacytidine inhibits methylation of cytosines. Since the ADH gene seems to be undermethylated in liver, we hypothesized that azacytidine could activate the gene in hepatoma cells. When H4IIE or H4IIEC3 cells were trypsinized and replated, then treated with azacytidine at 0.5 ug/ml for 24 hours, and subsequently allowed to grow for 6 days, there was no change in the ADH activity of the cells (data not shown). This may reflect the possibility that the gene is transcriptionally competent in the hepatoma cells (and appropriately undermethylated), but higher levels of expression are limited by some other factor, for instance the availability of certain liver specific transcription factors.

DISCUSSION

Several features of the regulation of expression of the class I ADH gene in the rat are becoming better understood. The ADH gene appears to be steroid-responsive. Although potential glucocorticoid response element have been seen in the proximal 5' flanking regions (about 400 bp upstream from the cap site) of human, mouse, and rat ADH genes, this region does not confer glucocorticoid-responsiveness in stably transfected cells (Crabb et al., 1989). In hepatoma cells, glucocorticoids induce ADH activity and mRNA. Although the effect in the hepatoma cells has not yet been proven to be due to a transcriptional effect, glucocorticoids also increase the expression of CAT from plasmid genes in which longer fragments (3500 bp) of the ADH upstream regions are linked to the CAT gene. The magnitude of the induction of CAT is relatively small; hence, the results must be repeated with internal controls to insure that this is a real phenomenon. It will also be important to find the specific regions involved in this induction by deletion analysis.

In the kidney, estradiol increases ADH activity by increasing the steady-state level of its mRNA. This may be due to increased rates of RNA transcription or to stabilization of the mRNA. This will be investigated by performing nuclear run-on assays. The induction of ADH mRNA is substantially greater than the induction of ADH activity in male rat kidney. This suggests that estradiol has other effects on the ADH message (e.g., on the translation rate) or ADH protein (e.g. on the degradation rate constant). One effect that is under consideration is that the estradiol treatment alters food intake and may thereby alter overall rates of protein degradation in a somewhat non-specific manner. Since both liver and kidney contain estrogen receptors, but liver ADH activity is modified to only a small extent by estrogen treatment, we will investigate whether the same promoter is being used in the two tissues.

Finally, the class I ADH gene appears to be a member of the class of tissue-specific genes that do not reside near an HTF island. From our preliminary experiments, undermethylation of Msp I sites near the 5' end of the gene is associated with high level expression of the gene (in the liver), and complete methylation is associated with low (kidney) or absent (spleen) expression.

282

ACKNOWLEDGMENTS

This work was supported by grants from the NIAAA (AA06434 and 00081 to DWC and AA 05310 to KMD). KMD was also supported by a Summer Research Fellowship from the American Diabetes Association, Indiana Affiliate, and MQ has been supported by a grant from the American Liver Foundation. We thank Ruth Ann Ross for excellent technical help in the determination of ADH activity in estrogen-treated rat kidney.

REFERENCES

Alton, N.K. and Vapnek, D., 1979, Nucleotide sequence analysis of the chloramphenicol resistance transposon Tn9, _Nature_, 282:864.

Avvedimento, E.V., Obici, S., Sanchez, M., Gallo, A., Musti, A., and Gottesman, M.E., 1989, Reactivation of thyroglobulin gene expression in transformed thyroid cells by 5-azacytidine, _Cell_, 58:1135.

Benyajati, C., Spoerel, N., Haymerle, H., and Ashburner, M., 1983, The messenger RNA for alcohol dehydrogenase in Drosophila melanogaster differs in its 5' end in different developmental stages, _Cell_, 33:125-133.

Bird, A.P., 1986, CpG rich islands and the function of DNA methylation, _Nature_, 321:209.

Buttner, H., 1965, Aldehyd- und Alkoholdehydrogenase-Aktivitat in Leber und Niere der Ratte, _Biochem. Z._, 341:300.

Chomzynski, P. and Sacchi, N., 1987, Single step method of RNA isolation by acid guanidinium thiocyanate-phenol-chloroform extraction, _Anal. Biochem._, 162:156.

Crabb, D.W., Bosron, W.F., and Li, T.-K., 1986, Role of the pituitary and neonatal androgenic imprinting in the hormonal regulation of liver alcohol dehydrogenase activity, _Biochem. Pharmacol._, 35:1527.

Crabb, D.W. and Dixon, J.E., 1987, A method for increasing the sensitivity of chloramphenicol acetyltransferase assays in extracts of transfected cultured cells, _Anal. Biochem._, 163:88.

Crabb, D.W. and Edenberg, H.J., 1986, Complete amino acid sequence of rat liver alcohol dehydrogenase deduced from the cDNA sequence, _Gene_, 48:287.

Crabb, D.W., Stein, P.M., Dipple, K.M., Hittle, J.B., Sidhu, R.S., Qulali, M., Zhang, K., and Edenberg, H.J., 1989, Structure and expression of the rat class I alcohol dehydrogenase gene, _Genomics_, 5:906-914.

Dembic, Z. and Sabolic, I., 1982, Alcohol dehydrogenase activity in rat kidney cortex stimulated by oestradiol, _Biochem. Biophys. Acta._, 714:331.

Dong, Y., Poellinger, L., Okret, S., Hoog, J., von Bahr-Lindstrom, H., Jornvall, H., and Gustafsson, J., 1988, Regulation of gene expression of class I alcohol dehydrogenase by glucocorticoids, _Proc. Natl. Acad. Sci. USA_, 85:767.

Felder, M.R., Watson, G., Huff, M.O., and Ceci, J.D., 1988, Mechanism of induction of mouse kidney alcohol dehydrogenase by androgen, _J. Biol. Chem._, 263:14531.

Goldstein, B. and Maxwell, D.S., 1982, The purification and immunocytochemical localization of rat liver and kidney alcohol dehydrogenase, _Alcoholism Clin. Exp. Res._, 6:142.

Gorman, C.M., Moffat, L.F., and Howard B.H., 1982, Recombinant genomes which express chloramphenicol acetyltransferase in mammalian cells, <u>Mol</u>. <u>Cell</u>. <u>Biol</u>., 2:1044.

Graham, F. and van der Eb, A., 1973, A new technique for the assay of infectivity of human adenovirus 5 DNA, <u>Virology</u>, 53:456.

Hamada, H., Petrino, M.G., and Kakunaga, T., 1982, A novel repeated element with Z-DNA-forming potential is widely found in evolutionarily diverse eukaryotic genomes. <u>Proc</u>. <u>Natl</u>. <u>Acad</u>. <u>Sci</u>. <u>USA</u>, 79:6465.

McKnight, S. and Tjian, R., 1986, Transcriptional selectivity of viral genes in mammalian cells, <u>Cell</u>, 46:795.

Meltzer, S.J., Mane, S.M., Wood, P.K., Johnson, L., and Needleman, S.W., 1990, An improvement of the single-step method of RNA isolation by acid guanidinium thiocyanate-phenol-chloroform extraction, <u>BioTechniques</u>, 8:148.

Mezey, E. and Potter, J.J., 1979, Rat liver alcohol dehydrogenase activity: effects of growth hormone and hypophysectomy, <u>Endocrinol</u>., 104:1667.

Mezey, E. and Potter, J.J., 1981, Effects of thyroidectomy and triiodothyronine administration on rat liver alcohol dehydrogenase, <u>Gastroenterol</u>., 80:566.

Mezey, E., Potter, J.J., and Diehl, A.M., 1986a, Depression of alcohol dehydrogenase activity in rat hepatocyte culture by dihydrotestosterone, <u>Biochem</u>. <u>Pharmacol</u>., 35:335.

Mezey, E., Potter, J.J., and Rhodes, D.L., 1986b, Effect of growth hormone on alcohol dehydrogenase activity in hepatocyte culture, <u>Gastroenterol</u>., 91:1271.

Potter J.J., Mezey, E., and Yang, V.W., 1989, Influence of growth hormone on the synthesis of rat liver alcohol dehydrogenase in primary hepatocyte culture, <u>Arch</u>. <u>Biochem</u>. <u>Biophys</u>. 274:548.

Rachamin, G., MacDonald, J.A., Wahid, S., Clapp, J.J., Khanna, J.M., and Israel, Y., 1980, Modulation of alcohol dehydrogenase and ethanol metabolism by sex hormones in the spontaneously hypertensive rat, <u>Biochem</u>. <u>J</u>., 186:483.

Teschke, R., Wannagat, F., Lowendorf, F., and Strohmeyer, G., 1986, Hepatic alcohol metabolizing enzymes after prolonged administration of sex hormones and alcohol in female rats, <u>Biochem</u>. <u>Pharmacol</u>., 35:521.

Wolfla, C.E., Ross, R.A., and Crabb, D.W., 1988, Induction of alcohol dehydrogenase activity and mRNA in hepatoma cells by dexamethasone, <u>Arch</u>. <u>Biochem</u>. <u>Biophys</u>., 263:69.

MAMMALIAN CLASS II ALCOHOL DEHYDROGENASE: SPECIES AND CLASS

COMPARISONS AT GENOMIC AND PROTEIN LEVELS

Jan-Olov Höög

Department of Chemistry I
Karolinska Institutet
S-104 01 Stockholm, Sweden

INTRODUCTION

Human class II alcohol dehydrogenase is formed of the π-subunit to an active dimer, and the enzyme belongs to the family of long-chain zinc-containing alcohol dehydrogenases (Jörnvall et al., 1987a). Including in the family is the two other mammalian alcohol dehydrogenases (ADH), class I and class III (Vallee and Bazzone, 1983) as the related sorbitol dehydrogenase (Jörnvall et al., 1987b). Class I ADH is the classical ADH that is responsible for the main metabolism of ethanol and the class III ADH has recently been shown to be identical to glutathione-dependent formaldehyde dehydrogenase (Koivusalo et al., 1989). The class II ADH, the least studied form of the different ADHs, shows a much higher Km for ethanol than the class I isozymes, 34 mM compared to 50 μM for the $\beta\beta$ isozyme of class I (Bosron et al., 1979). The $\pi\pi$ enzyme is also active towards norepinephrine metabolites (Mårdh et al., 1986). The class II ADH is encoded by its own gene, ADH4, that have been mapped to locus 4q22 in the human genome (McPhearson et el., 1989). This is in the same region where the other human ADH genes, ADH1-3 and ADH5, have been localized (Smith, 1986). The rat class II ADH, designated ADH-1, has been ascribed retinol dehydrogenase activity and this enzyme is also more anodic than the human class II form (Julià et al., 1986). This gives the class II enzyme other properties compared to the class I and III of mammalian ADH. For class I and III several structures are known at protein and DNA levels (Jörnvall et al., 1987b; Jörnvall et al., 1989; Kaiser et al., 1989). For the class II, however, only the human protein and cDNA structures have been determined (Höög et al., 1987).

Here the human class II genomic and the rat class II cDNA structures will be discussed in comparisons to earlier known mammalian ADH structures.

MATERIALS AND METHODS

A human fetal genomic library was provided by Dr. Lawn (Lawn et al., 1982) and a rat liver cDNA library was purchased from Clontech, CA. cDNA fragments about 500 bp in size from the coding region of an earlier isolated human class II cDNA-clone (Höög et al., 1987), were

isolated on agarose gels after restriction enzyme digestions. All
fragments were purified with the Gene-clean procedure (Bio 101, CA)
prior to nick-translation. Both cDNA-libraries were screened with
these probes under almost identical conditions and positive plaques
were isolated with the plate lysate method (Maniatis et al., 1982).
Isolated lambda DNA was sized on agarose gels and inserts were
liberated and subcloned into vector pEMBL8. DNA sequence analysis was
performed using T7 DNA polymerase (Pharmacia LKB Biotechnology) in the
dideoxy chain termination method (Sanger et al., 1979) both on single-
stranded and double-stranded DNA templates.

RESULTS AND DISCUSSION

Two independent class II ADH clones were isolated from the human
fetal genomic library. These two clones covered the entire ADH4 gene
and included about 6 kb upstream of the ATG initiator codon of the
gene. The class II ADH gene is divided into nine exons and eight
introns altogether covering 22 kb (von Bahr-Lindström et al., 1990).
All intron/exon borders show the consensus splicing signals GT and AG
(Mount, 1982). This is the same arrangement as for the other today
cloned mammalian ADH genes, and all the human genes show exactly the
same splicing points. For the class III, ADH5, gene is only a partial
sequence known indicating that this ADH gene also has the same
arrangement (Matsou & Yokoyama, 1990). However, the ADH4 gene covers a

Table 1. Intron sizes in different mammalian ADH genes.
ADH4 codes for the human class II π-subunit, ADH1 for the
α subunit and ADH2 for the β subunit of human class I ADH.
RAT and MOUSE are the class I genes from the two species.
All sizes are in kb. Σ is the total size (exons + introns)
from the ATG codon to the stop codon of the different
genes.

INTRON /	ADH4	ADH1	ADH2	RAT	MOUSE
1	1.6	3.2	2.8	2.0	2.1
2	0.8	0.6	0.6	0.6	0.4
3	2.9	1.8	1.7	0.9	2.0
4	2.7	0.1	0.1	0.1	0.1
5	6.1	1.9	2.0	2.3	3.6
6	3.6	2.2	2.2	1.8	1.8
7	0.6	0.6	0.6	0.9	0.7
8	2.6	2.8	2.8	0.7	0.9
Σ	22.0	14.3	13.9	10.4	12.7

Sizes from: ADH1 - Matsuo et al. 1989 FEBS Lett. 243, 57-60.
ADH2 - Duester et al. 1986 J. Biol. Chem. 261, 2027-2033.
RAT - Crabb et al. 1989 Genomics 5, 906-914. MOUSE - Zhang
et al. 1987 Gene 57, 27-36.

much larger region than the other analyzed class I genes. The ADH4 gene is 50% larger than the human class I genes and compared to the rat class I gene the ADH4 gene spans more than the double region (Table 1). The size differences are due to larger introns especially introns four and five. In the class II gene the sizes are 2.7 kb and 6.1 kb, respectively, versus 0.1 kb and 2.0 kb, respectively, in the human ADH2 gene.

The trancription start was determined to 61 bp upstream of the ATG initiator codon (von Bahr-Lindström et al., 1990) which is slightly less than for the human class I ADH gene (Duester et al., 1986). A TATA box and a CAAT elemnt are identified upstream of the determined transcription start point (Fig. 1). When the two promoter regions, ADH4 and ADH2, are aligned only 41% positionl identity is obtained when gaps are introduced to get the highest positional identity (Fig. 1). In the class I ADH promoter region glucocorticoid responsive elements have been identified by homology search to consensus sequences (Duester et al., 1986). These segments have later

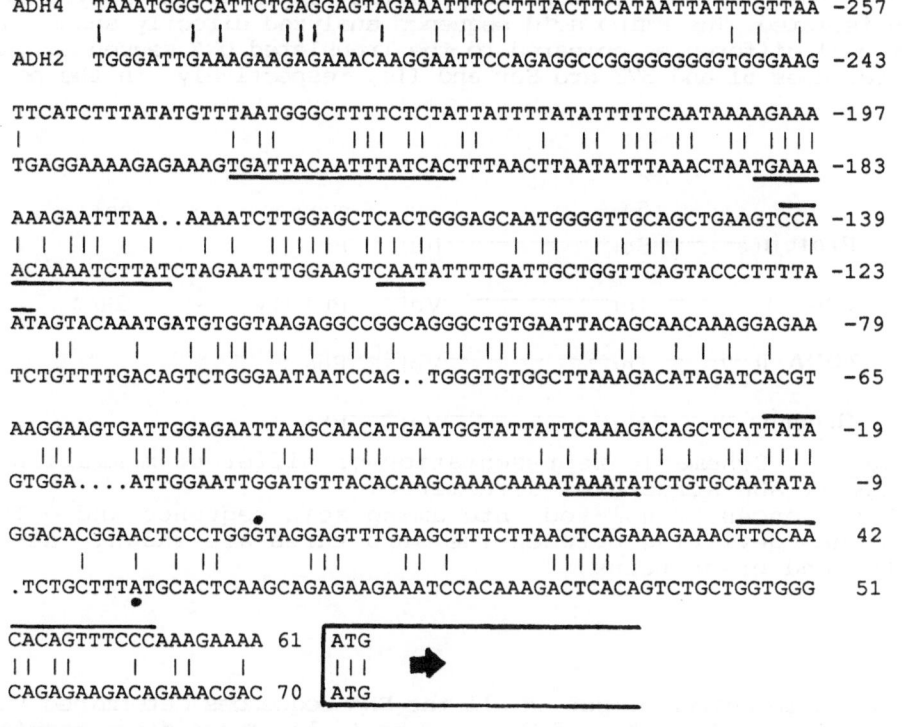

Figure 1. Comparison of 5' non-coding regions. ADH4 (top) and ADH2 (bottom) genes. Postulated glucocorticoid responsive elements in the ADH2 gene (Duester et al., 1986) are marked in the bottom row. Sequences with glucocorticoid responsive element homology found in the ADH4 gene are marked in the upper row. Transcription starts for the two genes are marked with dots. TATA-boxes and CAAT-elements are marked in the upper and bottom rows. Open box and arrow designate the start of the coding region - exon 1.

been verified by experimental data showing increased mRNA levels after induction with steroid-receptor complex (Dong et al., 1988). In footprinting studies the two identified segments have been shown to act in tandem (Duester et al., 1990). For the class II ADH gene two glucocorticoid responsive elements with 70% positional identity to consensus sequences were identified at non identical positions compared to the ADH2 gene. For the ADH4 gene the identified segments were not found in tandem, like for the ADH2 gene. One of these identified segments is located between transcription and translation starts and the other segment is located upstream of the responsive elements found in the ADH2 gene (Fig. 1). If these segments in the ADH4 gene will give increased mRNA levels has not been proven experimentally, but from the comparison studies the ADH4 gene seems to be regulated in a different way than the ADH2 gene.

Some deviations between the different structures determined for the human class II ADH were obtained. One cDNA-clone corresponded to a polypeptide chain elongated with twelve amino acid residues (Fig. 2). The expected stop codon was mutated to a GGA codon (Höög et al., 1987). For another cDNA clone and the genomic clone the stop codon was found in the position corresponding to the length of the polypeptide chain isolated. The amino acid sequence analyzed directly shows two positional differences compared to the translated DNA sequences. Amino acid residues 51 and 373 are Ser and Ile, respectively, in the poly-

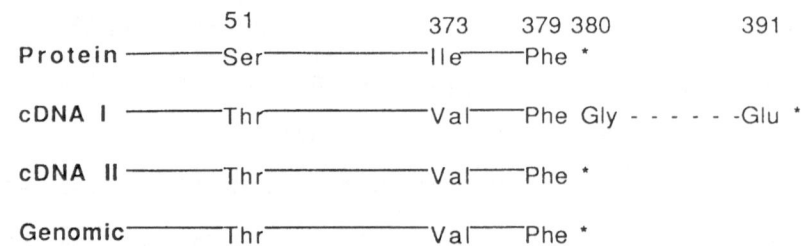

Figure 2. Schematic representation of different mammalian class II ADH structures determined.
DNA sequences translated into amino acid sequence and only residues deviating between the structures are shown. *denotes end of proteins.

peptide chain analyzed, but in all the DNA sequences determined the codons correspond to Thr and Val, respectively. These discrepancies might reflect alleles or individual sequence variations. The conservative Ile/Val difference at position 373 might not influence at all on the properties of the enzyme, but the Ser/Thr difference at position 51 will probably change properties. In the class I enzymes this residue is His and is one of the residues that is involved in a charge-relay system. In class III ADH there is a Tyr in position 51 and both these residues in class I and III ADH can form a hydrogen bond to the coenzyme NAD. Ser and Thr in the class II enzymes are much shorter residues and cannot form a direct hydrogen bond, but from computer modell building a water molecule can bridge hydrogen bonding from a Ser to the coenzyme (Eklund et al., 1990). These replacements in position 51 between the different ADH enzymes may also be one of

the explanations to the different responsivness to the competitive
inhibitor pyrazole, where the class I enzymes are inhibited strongly,
the class II enzyme poorly, and the class III enzyme hardly influenced
at all (Eklund et al., 1990).

To further elucidate the class II ADH relationship to the other
members in the family of long-chain alcohol dehydrogenases a class II
ADH cDNA-clone was isolated from a rat liver cDNA-library. DNA
sequence determination of this cDNA-clone gave several deviations from
the human class II enzyme. In the human class II ADH there are two
insertions of altogether five amino acid residues compared to the
human class I enzymes. One of these insertions are missing in the rat
class II ADH, the one at position 60, but the other insertion of four
amino acid residues at position 120 is conserved between the two
species. These insertions does not coincide with intron/exon borders
in the human ADH4 gene, they are localized in exon three and five,
respectively. Different lengths of human and rat ADH polypeptides was
also found for the class I enzymes (Crabb et al., 1988), where the rat
enzyme has one insertion, but here it coincided with an intron/exon
border. For the class III ADH, however, the rat and human polypeptides
have the same length.

The rat ADHs shows lower positional identities in any comparison
class I-II-III than the human enzymes. Noteworthy is that the
relationship between the three different classes is here stressed.

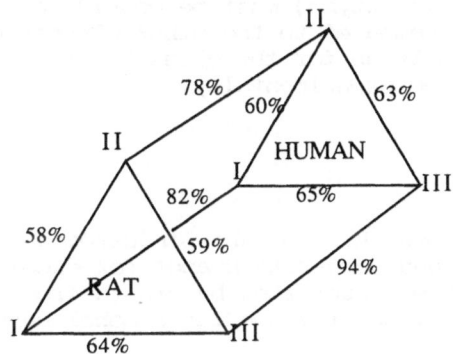

Figure 3. Comparisons of human and rat ADHs.
Positional identities at protein level are given in % of
compared amino acid residues. Rat sequences in the front
and human sequences in the back triangle. Human class I
ADH is represented of the γ-subunit.

Class I and III are the most closely related with the class II enzyme
as the most deviating form. The latter was also observed when an avian
ADH structure was compared to the human forms (Kaiser et al., 1990).
When interspecies comparisons between human and rat ADHs are performed
class I shows 82%, class II 78%, and class III 94% positional identity
(Fig. 3). The class III is the far most conserved structure which has
been shown earlier (Kaiser et al., 1989), and verifies that the class

III ADH/formaldehyde dehydrogenase has a more strict function (Koivusalo et al., 1989). Class I shows an intermediate positional identity and class II is the other extreme with the highest evolutionary rate. The functional meaning of the large structural variance of the class II enzymes is not known, but the enzymes can catalyze a broad range of hydroxyl-containing compounds (Bosron et al., 1979; Julià et al., 1986; Mårdh et al., 1986). Of the amino acid residues lining the substrate-binding cleft in the mammalian ADHs (Eklund et al., 1990) only 45% are strictly conserved between the two class II enzymes. As a functional response to this, a large amount of Km-values for alcohols differ widely between the human and rat class II ADHs, for ethanol the Km-values are 34 mM for the human enzyme and 5 M for the rat enzyme at pH 7.5 (Bosron et al., 1979; Julià et al., 1987). The extreme difference at this pH is mainly due to the low activity of the rat enzyme at pH 7.5. At pH 10 the difference in Km-values between the two enzymes is only three fold, but still the difference is large. All zinc-ligands, both to the active and to the structural zinc atom, are as expected conserved between the two class II enzymes.

In different species a class II ADH has been shown to appear in the stomach (Julià et al., 1987; Duley et al., 1985). A human enzyme of this type has been purified and a preliminary Km of 40 mM has been reported (Moreno et al., 1989), very close to the value for the enzyme purified from human liver. Furthermore, it was recently reported that a class II type ADH should be responsible for the differences in the rate of ethanol metabolism between men and women (Frezza et al., 1990). If this ADH has the same structure as the liver class II enzyme is not known.

The class II ADH ($\pi\pi$ enzyme) must be considered as a different enzyme, structurally, compared to the other classes of ADH and if this is true also functionally as for the class III ADH (Koivusalo et al., 1989) has to be verified experimentally.

ACKNOWLEDGEMENTS

I would like to thank Drs. H. von Bahr-Lindström and H. Jörnvall for valueable discussions and Margareta Brandt for excellent technical assistance. This study was supported by grants from the Swedish Medical Research Council and the Swedish Alcohol Research Foundation.

REFERENCES

von Bahr-Lindström, H., Jörnvall, H., & Höög, J.-O. (1990) Molecular cloning and characterization of the human ADH4 gene. Nucl. Acids Res. in press.
Bosron, W.F., Li, T.-K., Dafeldecker, W.P., & Vallee, B.L. (1979) Human liver π-alcohol dehydrogenase: Kinetic and molecular properties. Biochemisty 18, 1101-1105.
Crabb, D.W., Stein, P.M., Dipple, K.M., Hittle, J.B., Sidhu, R., Qulali, M., Zhang, K., & Edenberg, H.J. (1989) Structure and expression of the rat class I alcohol dehydrogenase gene. Genomics 5, 906-914.
Dong, Y., Poellinger, L., Okret, S., Höög, J.-O., von Bahr-Lindström, H., Jörnvall, H., & Gustafsson, J.-Å. (1988) Regulation of gene expression of class I alcohol dehydrogenase by glucocorticoids. Proc. Natl. Acad. Sci. USA 85, 767-771.

Duester, G., Smith, M., Bilanchone, V., & Hatfield, G.W. (1986) Molecular analysis of the human class I alcohol dehydrogenase gene family and nucleotide sequence of the gene encoding the β subunit. J. Biol. Chem. 261, 2027-2033.

Duester, G., Winter, L., Stewart, M., Dong, Y., Poellinger, L., Okret, S., & Gustafsson, J.-Å. (1990) Tandem glucocorticoid responsive elements mediate induction of the human β alcohol dehydrogenase gene by glucocorticoids. Mol. Cell Biol., in press.

Duley, J.A., Harris, O., & Holmes, R. (1985) Analysis of human alcohol- and aldehyde-metabolizing isozymes by electrophoresis and isoelectric focusing. Alcohol. Clin. Exp. Res. 9, 263-271.

Eklund, H., Müller-Wille, P., Horjales, E., Futer, O., Holmquist, B., Vallee, B.L., Höög, J.-O., Kaiser, R., & Jörnvall, H. (1990) Comparison of three classes of human liver alcohol dehydrogenase. Emphasis on different substrate binding pockets. Eur. J. Biochem., in press.

Frezza, M., di Padova, C., Pozzato, G., Terpin, M., Baraona, E. & Lieber, C.S. (1990) High blood alcohol levels in women. The role of decreased gastric alcohol dehydrogenase activity and first-pass metabolism. New Engl. J. Med. 332, 95-99.

Höög, J.-O., von Bahr-Lindström, H., Hedén, L.-O., Holmquist, B., Larsson, K., Hempel, J., Vallee, B.L., & Jörnvall, H. (1987) Structure of the class II enzyme of human liver alcohol dehydrogenase: Combined cDNA and protein sequence determination of the π subunit. Biochemistry 26, 1926-1932.

Julià, P., Farrés, J., & Parés, X. (1986) Ocular alcohol dehydrogenase in the rat: Regional distribution and kinetics of the ADH-1 isoenzyme with retinol and retinal. Exp. Eye Res. 42, 305-314.

Julià, P., Farrés, J., & Parés, X. (1987) Characterization of three isoenzymes of rat alcohol dehydrogenase. Tissue distribution and physical and enzymatic properties. Eur. J. Biochem. 162, 179-189.

Jörnvall, H., Höög, J.-O., von Bahr-Lindström, H., & Vallee, B.L. (1987a) Mammalian alcohol dehydrogenases of separate classes: Intermediates between different enzymes and intraclass isozymes. Proc. Natl. Acad. Sci. USA 84, 2580-2584.

Jörnvall, H., Persson, B., & Jeffery, J. (1987b) Characteristics of alcohol/polyol dehydrogenases. The zinc-containng long-chain alchol dehydrogenases. Eur. J. Biochem. 167, 195-201.

Jörnvall, H., von Bahr-Lindström, H., & Höög, J.-O. (1989) Alcohol dehydrogenases - Structure. In: Human Metabolism of Alcohol vol. II (eds. Batt, R.D. & Crow, K.E.) CRC press, pp. 43-64.

Kaiser, R. Holmquist, B., Vallee, B.L., & Jörnvall, H. (1989) Characteristics of mammalian class III alcohol dehydrogenases, an enzyme less variable than the traditional liver enzyme of class I. Biochemistry 28, 8432-8438.

Kaiser, R., Nussrallah, B., Dam, R., Wagner, F.W., & Jörnvall, H. (1990) Avian alcohol dehydrogenase. Characterization of the quail enzyme, functional interpretations, and relationships to the different classes of mammalian alcohol dehydrogenases. Biochemistry, in press.

Koivusalo, M., Baumann, M., & Uotila, L. (1989) Evidence for the identity of glutathione-dependent formaldehyde dehydrogenase and class III alcohol dehydrogenase. FEBS Lett. 257, 105-109.

Lawn, R.M., Fritsch, E.F., Parker, R.C., Blake, G., & Maniatis, T. (1978) The isolation and characterization of linked δ- and β-globin genes from a cloned library of human DNA. Cell 15, 1157-1174.

Maniatis, T., Fritsch, E.F., & Sambrook, J. (1982) Molecular Cloning. Cold Spring Harbor Laboratory, Cold Spring Harbor, N.Y.

McPhearson, J.D., Smith, M., Wagner, C., Wasmuth, J., & Höög, J.-O. (1989) Mapping of the class II alcohol dehydrogenase gene locus to 4q22. Cytogenet. Cell Genet. 51, 1043.

Matsuo, Y., & Yokoyama, S. (1990) Cloning and sequencing of a processed pseudogene derived from a human class III alcohol dehydrogenase gene. Am. J. Hum. Genet. 46, 85-91.

Moreno, A., Boleda, M.D., Peralba, J.M., & Parés, X. (1989) Purification and characterization of a new alcohol dehydrogenase form from human stomach. Alcohol Alcoholism 24, 383.

Mount, S.M. (1982) A catalogue of splice junction signals. Nucl. Acids Res. 10, 459-472.

Mårdh, G., Dingley, A.L., Auld, D.S., & Vallee, B.L. (1986) Human class II (π) alcohol dehydrogenase has a redox-specific function in norepinephrine metabolism. Proc. Natl. Acad. Sci. USA 83, 8908-8912.

Sanger, F., Nicklen, S., & Coulson, A.R. (1977) DNA sequencing with chain termination inhibitors. Proc. Natl. Acad. Sci. USA 74, 5463-5467.

Smith, M. (1986) Genetics of human alcohol and aldehyde dehydrogenases. Adv. Hum. Genet. 15, 249-290.

Vallee, B.L. & Bazzone, T.J. (1983) Isozymes of human liver alcohol dehydrogenase. Isozymes 8, 219-244.

THE ACTIVATION OF ALCOHOLS BY LIVER ALCOHOL DEHYDROGENASE:

DEPENDENCE OF INHIBITION UPON THE pK_a LOWERING EFFECT

Y. Pocker and Joe D. Page

Department of Chemistry
University of Washington
Seattle, WA 98195

INTRODUCTION

Liver alcohol dehydrogenase (LADH) is a zinc metalloenzyme that catalyzes the reversible oxidation of alcohols to aldehydes or ketones (Pocker, 1989). The catalytic zinc ion is bound 25 Å from the protein surface at the bottom of a tunnel lined with hydrophobic amino acid residues that stabilize the binding of substrates. For most primary alcohols, including ethanol, the oxidation process follows a compulsory ordered mechanism in which NAD^+ binding precedes alcohol binding (Dalziel, 1975). Oxidation occurs within the ternary complex followed by the stepwise release of aldehyde and NADH, respectively.

Several lines of investigation indicate that oxidation within the ternary complex involves several molecular steps and protein conformations. The alcohol is initially bound with its hydroxyl group chelated to the zinc ion and its alkyl chain extended into the hydrophobic tunnel. If the metal is tetracoordinate (Eklund and Brändén, 1987), the alcohol takes the fourth ligation site by displacing the zinc bound water molecule. Alternatively, the complex may be pentacoordinate, not requiring the displacement of the zinc water molecule (Makinen et al., 1983). The electrostatic environment created by the positively charged nicotinamide ring, the Lewis acid effect of the zinc ion, and the surrounding hydrophobic residues all contribute to the ionization of the alcohol to its zinc bound alkoxide. The generation of this alkoxide complex may induce the closed conformation of the protein, as it has been shown that a negative charge at the catalytic metal stabilizes the closed conformation (Maret and Zeppezauer, 1986). In the transition to this conformation, all water molecules are believed to be expelled from the active site. The anhydrous environment and the negatively charged alkoxide facilitate the hydride transfer from the alcohol to the nicotinamide ring, completing the oxidation.

To characterize the ionization of alcohols in ternary complexes of LADH, the inhibition of ethanol oxidation by a series of electronegative alcohols was studied. Highly electronegative alcohols such as 2,2,2-trifluoroethanol and 2,2,2-trichloroethanol are not oxidized by LADH, but are potent inhibitors of enzyme action. To investigate if the acidity of an alcohol is related to its inhibitory potential, the following series of small, electronegative alcohols with pK_a values ranging from 12.37 to 15.5 were used to inhibit

Enzymology and Molecular Biology of Carbonyl Metabolism 3
Edited by H. Weiner et al., Plenum Press, New York, 1990

the oxidation of ethanol by LADH: 2,2,2-trifluoroethanol, 2,2,2-trichloroethanol, 2,2,2-tribromoethanol, 2,2-dichloroethanol, 2,2-difluoroethanol, propargyl alcohol, 3-hydroxypropionitrile, 2-chloroethanol, 2-iodoethanol, 2-methoxyethanol, ethylene glycol, and methanol. Because it was essential to the analysis that the inhibition constants reflect predominantly the electronegative nature of each alcohol, only short, non-branched alcohols with no more than three carbon atoms were chosen such that additional hydrophobic contacts within the active site would make a negligible contribution to the respective inhibition constants. The substrate competitive inhibition constants used in this analysis were determined by steady state techniques. Since several inhibition patterns were observed, the kinetic data were analyzed using four potential inhibition models and the best fit determined. In all cases, the inhibition conformed to a model containing a competitive term with respect to substrate.

These alcohol inhibitors, in most cases, form non-productive alkoxide ternary complexes that mimic substrate alcohols whose ternary alkoxide complexes are rapidly oxidized by hydride transfer. Due to the extreme reactivity of productive ternary complexes, it proved difficult to directly measure the pK_a values of these complexes. However, it has been possible to determine experimentally the relationship between competitive inhibition constants and pK_a's of the inhibitory alcohols. Using a closed cycle of reactions, it can be shown that the pK_a of an alcohol in the ternary complex is related to its pK_a as a free alcohol. This relationship has been quantified for all the alcohols used in this study.

METHODS

All solutions were prepared in a pH 7.00 buffer of KH_2PO_4 and K_2HPO_4 that was 30 mM in phosphate. The ionic strength was adjusted to 0.1 by addition of Na_2SO_4.

Enzyme Solutions. LADH (EC 1.1.1.1) was obtained from Sigma, crystallized and lyophilized with greater than 98% protein, isolated from horse livers (1-2 units/mg protein). The percent activity of the enzyme was determined by titration of the active sites with NAD^+ and pyrazole (Theorell and Yonetani, 1963). Stock solutions were prepared immediately prior to use by dissolving the protein into buffer at 0 °C. Its concentration was determined using $\epsilon_{280} = 36$ mM^{-1}cm^{-1} and corrected for its percent activity. Stock enzyme solutions were kept on ice during kinetic runs and discarded after 6-8 hours. LADH for dialysis experiments was purchased from Boehringer Mannheim as a crystalline suspension with ethanol. Prior to use, it was dialyzed for 48 hours at 4 °C at pH 7 with multiple buffer changes to remove the ethanol. Its percent activity was determined as described above. When stored below 4 °C, the stock solutions were found to retain full activity for at least two weeks.

Substrate, coenzyme and inhibitor solutions. Ethanol stock solutions were prepared by weighing absolute ethanol into a volumetric flask and diluting with buffer. NAD^+ (grade III-C, 99%) was purchased from Sigma and used without further purification. Concentrations of NAD^+ stock solutions were checked using $\epsilon_{260} = 18.0$ mM^{-1}cm^{-1}. All coenzyme solutions were prepared immediately prior to use and discarded at the end of the day. 2-Chloroethanol, 2-iodoethanol, 2-methoxyethanol, 2,2-dichloroethanol, 2,2,2-trichlorethanol, 2,2,2-tribromoethanol, 2,2,2-trifluoroethanol, 3-hydroxypropionitrile, propargyl alcohol and pyrazole were all purchased from Aldrich Chemical Co. 2,2-Difluoroethanol was purchased from PCR Inc. Methanol and ethylene glycol were purchased from J.T. Baker. All inhibitors were of the highest quality available and were used without further purification, unless they had been previously opened.   Stock inhibitor solutions were prepared immediately prior to use by weighing them into a volumetric flask and diluting with buffer. All solutions were discarded at the end of the day.

Inhibition kinetics of ethanol oxidation by electronegative alcohols were obtained by systematically varying the inhibitor and ethanol concentrations while holding NAD^+ constant. Inhibitor, ethanol and NAD^+ were first premixed in a volumetric flask at the desired concentrations. Ethanol was varied from 1 to 10 mM and NAD^+ was held constant at about 11 μM for all runs. A typical run was initiated by injecting 10-20 μl of LADH (final concentration 0.02 μM, dimer) into 3.00 ml of the above mixture and monitoring the production of NADH from its absorbance at 340 nm. Some of the less potent inhibitors: methanol, ethylene glycol, 2-methoxyethanol, 2-iodoethanol, and 2-chloroethanol were slowly oxidized by LADH. These background rates were determined under identical conditions in the absence of ethanol and their inhibition runs were corrected, accounting for the ethanol/inhibitor ratio.

Dialysis Experiments. To determine if the inhibitors had any modifying or denaturing effects on LADH, each was mixed with LADH at a concentration equal to that used in the highest inhibitory run. The inhibitors were removed by dialysis at 4 °C for 48 hours with multiple buffer changes. The resulting LADH was then used in a kinetic assay with ethanol and NAD^+ and its activity compared to a control LADH.

Instrumentation. Inhibition kinetics were measured on a Cary 210 UV-visible double beam spectrophotometer interfaced to an Apple II/e microcomputer and thermostated with a Forma Temp Jr. constant temperature bath. A Radiometer Model PHM84 research pH meter equipped with a Cole-Palmer Ag/AgCl glass electrode was used to determine pH values. Modeling of kinetic data was done on a VAX/VMS mainframe using BMDPAR (1988) of BMDP Statistical Software, Inc.

RESULTS

Initial rate measurements for the oxidation of ethanol can be described by an equation of the form (Dalziel, 1975) :

$$e/v_0 = \phi'_0 + \phi'_1/[NAD] + \phi'_2/[EtOH] + \phi'_{12}/[NAD][EtOH] \qquad (1)$$

where v_0 is the initial velocity, e is the active site concentration, and the ϕ' coefficients are kinetic constants that are related to the rate contants in Scheme I as follows: $\phi'_0 = 1/k_3$, $\phi'_1 = 1/k_1$ and $\phi'_2 = 1/k_2$. The compulsory ordered mechanism shown in Scheme I, the Theorell–Chance mechanism, is the simplest mechanism that gives an equation of this form.

Scheme I

$$E + NAD^+ \underset{k_{-1}}{\overset{k_1}{\rightleftharpoons}} E\text{-}NAD^+$$

$$E\text{-}NAD^+ + EtOH \underset{k_{-2}}{\overset{k_2}{\rightleftharpoons}} E\text{-}NADH + CH_3CHO + H^+$$

$$E\text{-}NADH \underset{k_{-3}}{\overset{k_3}{\rightleftharpoons}} E + NADH$$

Inhibition kinetics. In an effort to characterize the formation of this enzyme bound alkoxide complex, inhibition studies using a series of small, primary alcohols with pK_a's varying over three pK_a units, that act as inhibitors of ethanol oxidation, were performed. In most cases, these alcohols are competitive or mixed inhibitors of ethanol oxidation. Before modeling the inhibition data, good estimates of the ϕ'_0, ϕ'_1, ϕ'_2 and ϕ'_{12} coefficients were obtained from kinetic experiments of the oxidation of ethanol by LADH and varying concentrations of NAD^+. These values were used in the inhibition models. The data obtained from kinetic inhibition studies of the 12 different inhibitors was analyzed by four different inhibition models (*i-iv*), described below, by modification of the appropriate terms in equation 1.

(*I*) A substrate analogue, I, may compete with ethanol for the substrate binding site in E-NAD$^+$ or E-NADH. Dissociation constants K_i and K_i' for the release of I from the E-NAD$^+$-I and E-NADH-I complexes, respectively, may be calculated from the initial rate equation when these steps are incorporated into the rate equation and the mechanism is of the Theorell-Chance type. The equation takes the form (Dalziel,1975):

(i) $e/v_0 = \phi'_0 + \phi'_0(I/K_i') + \phi'_1/[NAD] + \phi'_2(1+I/K_i)/[EtOH] + \phi'_{12}/[NAD][EtOH]$

(*II*) If the inhibitor only combines with E-NAD$^+$ ($K_i' = \infty$), then the inhibition will display a competitive pattern towards ethanol and the equation takes the form:

(ii) $e/v_0 = \phi'_0 + \phi'_1/[NAD] + \phi'_2(1+I/K_i)/[EtOH] + \phi'_{12}/[NAD][EtOH]$

Figures 1 and 2 show that the inhibition data for methanol and 2,2,2-tribromoethanol fit this model.

(*III*) It has been shown (Shore and Theorell, 1966; Pocker and Raymond, 1985) that some substrate alcohols are inhibitors at high concentrations. These act by binding E-NADH to form E-NADH-I ternary complexes, which release NADH at a slower rate than E-NADH. In this type of inhibition, the turnover term, ϕ'_0 is multiplied by $(1+I/K_i'')$ where K_i'' is the dissociation constant for the release of I from the E-NADH-I complex.

(iii) $e/v_0 = \phi'_0(1+I/K_i'') + \phi'_1/[NAD] + \phi'_2/[EtOH] + \phi'_{12}/[NAD][EtOH]$

This mode of inhibition was not observed.

(*IV*) If the inhibitor competes with NAD$^+$ and forms an inactive E-I complex, with dissociation K_i''', the initial rate equation is (Dalziel,1963):

(iv) $e/v_0 = \phi'_0 + \phi'_1(1+I/K_i''')/[NAD] + \phi'_2/[EtOH] + \phi'_{12}(1+I/K_i''')/[NAD][EtOH]$

This mode of inhibition was not observed.

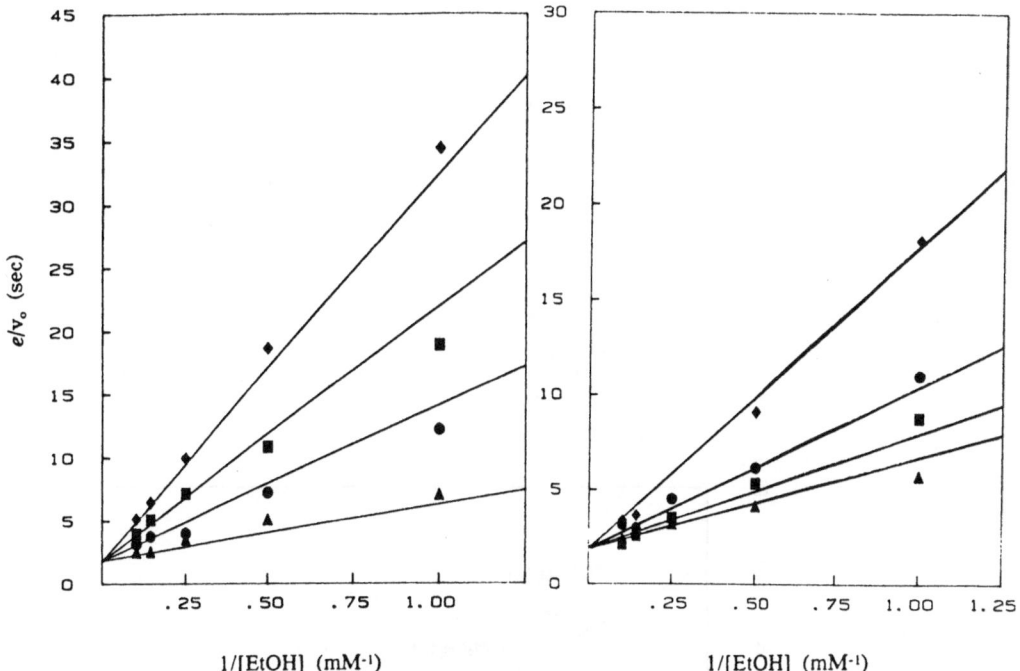

FIGURE 1. Methanol inhibition of ethanol oxidation. Lines were calculated from model (*ii*). The Active site concentration, *e* was 0.04 μM and NAD$^+$ was 12.81 μM. Methanol concentrations: (▲) 0, (●) 300, (■) 600, (♦) 1000 mM. Each symbol represents the mean of three determinations.

FIGURE 2. 2,2,2-Tribromooethanol inhibition of ethanol oxidation. Lines were calculated from model (*ii*). The Active site concentration, *e* was 0.04 μM and NAD$^+$ was 11.67 μM. 2,2,2-Tribromoethanol concentrations: (▲) 0, (■) 1.0, (●) 3.0, (♦) 9.0 mM. Each symbol represents the mean of three determinations.

The experimental data sets were tested with each model by a nonlinear regression analysis program that estimated inhibition constants and their corresponding standard deviations which were then used to calculate velocities and residuals. Comparison of the residual sum of squares and F-tests were used to clearly define the best model for each inhibitor (Mannervik,1981). Ten of the twelve inhibitors studied were best fit by the substrate competitive model (*ii*). 2-Methoxyethanol and 2-iodoethanol were best fit by the mixed model (*i*). The inhibiting potential of these alcohols is inversely proportional to their pK_a's with a slope of -0.8 as shown in Figure 3.

To assess the polar and steric contributions to the inhibition, the p$\bar{K}_i$ of each inhibitor, except 2-iodoethanol, was fit to the function p$K_i = \rho*\sigma* + \delta E_s$ according to the treatment of Pavelich and Taft (1957). This resulted in a $\rho*$ value of 1.9±0.2 and a δ value of 0.5±0.3. The larger $\rho*$ value indicates that the change in the pK_i is predominantly determined by the polar nature of the substituents. The pK_i for 2-iodoethanol deviated significantly from this trend, exhibiting a stronger inhibition not accountable for by its $\sigma*$ and E_s values. It appears that the highly polarizable iodine

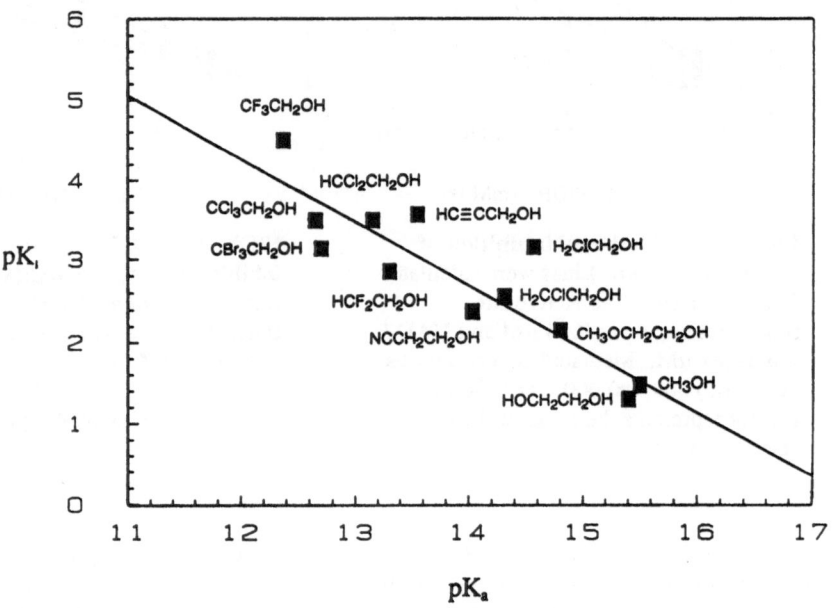

FIGURE 3. The dependence of the pK_i for ethanol oxidation on the pK_a of the free alcohol. The line was determined by linear regression for all the alcohol inhibitors shown.

substituent makes a significant hydrophobic contribution to inhibitor binding. The slope in Figure 4, $\rho*$ shows how the experimental data is fit by this function. E_s values were not available for propargyl alcohol and ethylene glycol.

Dialysis experiments. To show that the alcohols were acting by reversible inhibition and not by denaturing LADH, each inhibitor was mixed with LADH at the highest concentration used in each series. It was then removed from LADH by dialysis over the course of 48 hours and the percent activity of the LADH measured. All but three of the inhibitors returned 100% of the original activity. Methanol returned 80% of the activity and 2-chloroethanol and 2,2-dichloroethanol returned 81 and 86% of the original activity, respectively. The highest concentration of methanol used for inhibition, 1M may have been sufficient to cause changes in the tertiary structure of LADH, especially after the time course of 48 hours. In contrast, the kinetic inhibition runs are completed within 1-2 minutes and it is doubtful that any activity loss would be realized during this short period. Furthermore, studies with cryosolvents show that methanol has no significant adverse effects on the structural properties of LADH (Greeves *et al.*, 1983). In the case of the two chlorinated inhibitors, it is possible some S_N2 alkylations of the enzyme occur. However with the kinetic inhibition runs being completed within 1-2 minutes, the amount of alkylation occurring during this time course should be negligible.

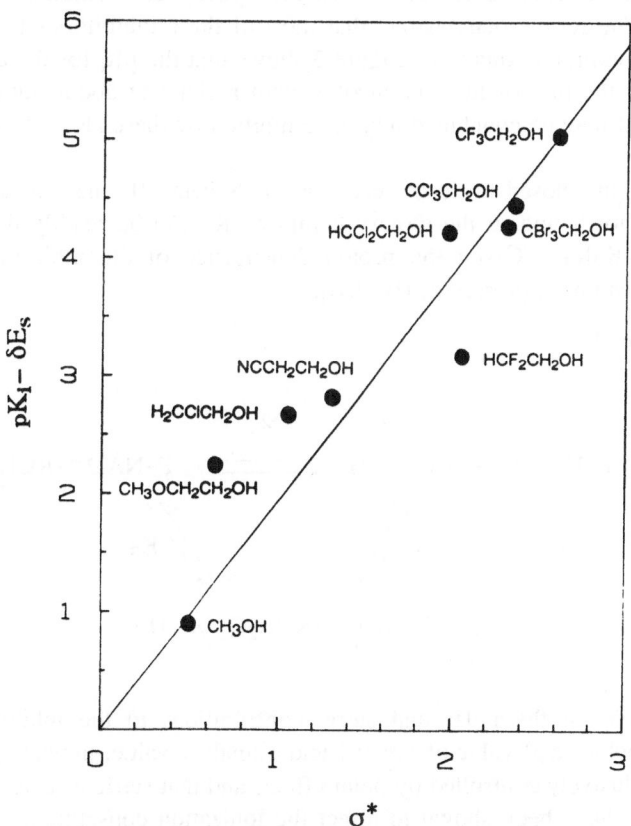

FIGURE 4. The pK_i as a function of the inhibitors polar, $\sigma*$ and steric, E_s parameters. The slope, $\rho*$ was 1.9 ± 0.2.

DISCUSSION

Several studies indicate that the oxidation of an alcohol proceeds through the formation of a zinc bound alkoxide with the release of a proton followed by hydride transfer (Kvassman and Pettersson, 1980; Pocker et al., 1987). X-ray structures with p-bromobenzyl alcohol as the inhibitor confirm that the alcohol oxygen is indeed bound to the zinc ion (Eklund et al.,1982). The inhibition of ethanol oxidation by 2,2,2-trifluoroethanol reaches a plateau at pH values higher than 7.6 suggesting that a zinc bound alkoxide is the active inhibitor (Shore et al., 1974). Studies with Co(II)LADH have shown that only charged ligands exhibit bands in the 570 nm region of the visible spectrum, as is observed in the stable Co(II)LADH-NAD$^+$-2,2,2-trifluoroethoxide ternary complex, and the transient Co(II)LADH-NAD$^+$-ethoxide complex (Sartorius et al.,1987). These results are in accord with studies which have concluded that the presence of a negative charge in the active site is necessary to induce the closed conformation in the ternary complex prior to hydride transfer (Maret and Zeppezauer, 1986).

The results presented here show that the inhibition of LADH by alcohol inhibitors is dependent upon the pK_a of the free alcohol. These experiments were performed with small primary alcohols so that the K_i value would reflect the electronegative nature of the alcohol. Large, branched alcohols that would have made additional hydrophobic contacts were avoided in order to minimize variable hydrophobic binding contributions to the inhibition. The results demonstrate that most of these inhibitors can be modeled with a competitive inhibition equation. Figure 3 shows that the pK_i for the alcohol inhibitors is dependent on the pK_a of the free alcohol with a slope of about minus one. From this dependence, it may be concluded that the inhibition by these alcohols is dependent upon a proton loss.

Using the closed cycle of reactions in Scheme II, the ionization constant for a primary alcohol bound in the ternary complex, K_3 can be readily determined from the relation $K_1 = K_2 K_3$. Given the proton dependence of the inhibition, the association constant, K_1 can be expressed as $[H^+]/K_i$.

Scheme II

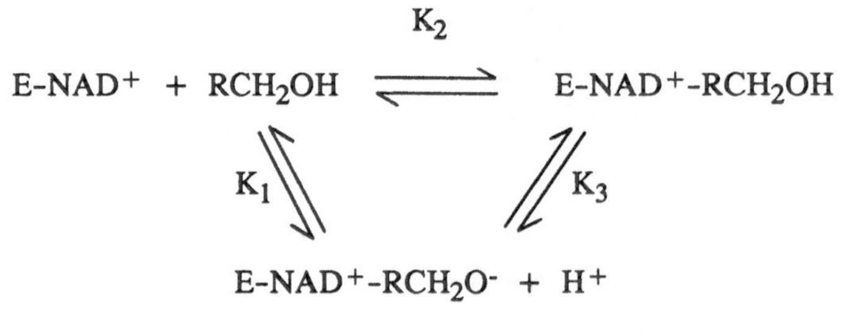

$$E\text{-NAD}^+ + RCH_2OH \underset{}{\overset{K_2}{\rightleftharpoons}} E\text{-NAD}^+\text{-}RCH_2OH$$

$$K_1 \qquad\qquad K_3$$

$$E\text{-NAD}^+\text{-}RCH_2O^- + H^+$$

Analysis of the polar and steric contributions to the inhibition (Pavelich and Taft,1957) yielded a ρ^* value of 1.9 ± 0.2 and a small δ value, indicating that the inhibition is almost exclusively controlled by polar effects and that steric effects are minor. Because polar effects have been shown to affect the ionization constants of a similar series of substituted methanols, ($\rho^* = 1.4$, Ballinger and Long,1959), it is apparent that polar effects are operative in K_3. Since steric effects are minor, it is reasonable to assume that alcohol dissociation from the E-NAD$^+$-RCH$_2$OH complex, $1/K_2$, will be essentially constant for the series of alcohols studied, except methanol which lacks a C-2 carbon. The dissociation of ethanol from the E-NAD$^+$-CH$_3$CH$_2$OH complex, $1/K_2$, was previously determined to be about 2 mM (Theorell and Yonetani, 1962), in agreement with paramagnetic NAD$^+$ analog studies of ethanol binding in the ternary complex (Mildvan and Weiner, 1969). However, a value of 11.3 mM has been recently measured by transient kinetic techniques and progress curve analysis (Sekhar and Plapp, 1990). Using these constants, two sets of ternary complex pK_a values for the alcohols used in this study have been determined (Table I).

TABLE I

Ternary Complex pK_a Values Determined from Primary Alcohol Inhibitors of Liver Alcohol Dehydrogenase.

			TERNARY COMPLEX[a]	
$\underline{pK_a}$	$\underline{\text{INHIBITOR}}$	$\underline{K_i \text{ (mM)}}$[b]	$\underline{pK_a}$[c]	$\underline{pK_a}$[d]
12.37[e]	2,2,2-Trifluoroethanol	0.032 ± 0.001	5.2	4.5
12.65[f]	2,2,2-Trichloroethanol	0.32 ± 0.01	6.2	5.5
12.70[f]	2,2,2-Tribromoethanol	0.73 ± 0.03	6.6	5.8
13.15[f]	2,2-Dichloroethanol	0.317 ± 0.004	6.2	5.5
13.30[e]	2,2-Difluoroethanol	1.4 ± 0.1	6.8	6.1
13.55[e]	Propargyl alcohol	0.27 ± 0.01	6.1	5.4
14.03[f]	3-Hydroxypropionitrile	4.2 ± 0.2	7.3	6.6
14.31[e]	2-Chloroethanol	2.8 ± 0.1	7.2	6.4
14.56[f]	2-Iodoethanol	0.70 ± 0.05	6.5	5.8
14.8[e]	2-Methoxyethanol	7.1 ± 0.5	7.6	6.8
15.4[e]	Ethylene glycol	51 ± 3[g]	8.4	7.7
15.5[e]	Methanol	34 ± 1	...	...
15.7	Water	38[h]	...	...
15.9[e]	Ethanol	50[h]	8.4	7.7[i]

[a] For the LADH-NAD$^+$-ROH complex. [b] Ethanol competitive inhibition constant ± S.D. from modeling of experimental data. [c] Calculated from Scheme II using $(1/K_2) = 2$ mM for the dissociation of ethanol from the LADH-NAD$^+$-ROH complex (Theorell and Yonetani, 1962). [d] Calculated from Scheme II using $(1/K_2) = 11.3$ mM (Sekhar and Plapp, 1990). [e] Ballinger and Long, 1960. [f] Takahashi *et al.*, 1971. [g] Statistical correction for two hydroxyl groups. [h] Extrapolated pK_i from Fig. 3. [i] A pK_a value of 7.2 as recently reported for ethanol (Sekhar and Plapp, 1990) can be estimated if one assumes $(1/K_2) = 31.6$ mM. This assumption would reduce the pK_a values of the other alcohols reported in this column by ca. 0.45 pK_a units.

Based on their similar ionization constants (Table I), the pK_a of a zinc bound water molecule in the E-NAD$^+$ complex should be comparable to that of ethanol bound in an E-NAD$^+$ complex. However it is apparent from Scheme II, that the pK_a of a ligand bound in the ternary complex is also dependent upon the ligand dissociation constant from the complex, $1/K_2$. The ethyl group provides ethanol and other C_2 alcohols additional binding contacts not present when water is the zinc ligand. Due to the lack of hydrophobic contacts, it is reasonable to assume that water will have a greater dissociation constant from the E-NAD$^+$ complex than ethanol. Thus, the pK_a of a water molecule bound in a ternary complex with E-NAD$^+$ should be less than that of ethanol bound in an equivalent complex. LADH saturated with NAD$^+$ is known to bind 2,2-bipyridine, pyrazole and decanoate with a pK_a of 7.6 (Evans and Shore, 1980; Andersson et al.,1981a; 1981b). The observation that imidazole abolishes the pH dependence of NAD$^+$ binding led to the suggestion that a zinc bound water molecule may be responsible for the pH effects (Theorell and McKinley-McKee, 1961). A pK_a of 7.6 for this zinc bound water molecule is in accord with the prediction that it be lower than that of ethanol bound in a ternary complex.

Methanol is oxidized about 30 times slower than ethanol. This difference in reactivity cannot be accounted for by electronic considerations given the closeness in pK_a of the two alcohols. The substrate binding site of LADH is composed of a highly polar region in the immediate vicinity of the zinc ion and a hydrophobic binding tunnel that extends to the solvent. Methanol, lacking a second carbon, is deficient in the additional hydrophobic binding afforded C-2 and longer alcohols as they extend into the hydrophobic tunnel. For this reason, the pK_a of methanol bound in the ternary complex could not be determined by this inhibition method. However, it is anticipated that methanol will have a larger $1/K_2$ value than ethanol and will therefore have a lower ternary complex pK_a value.

EPR studies of Co(II)LADH suggest that catalytically competent E-NAD$^+$-alcohol complexes are pentacoordinate with water making up the fifth ligation site (Makinen et al., 1983). On the other hand, x-ray studies have concluded that the free space around the zinc ion is limited by its chelating residues as well as the side chains Ser-48 and Phe-93, such that there is no room for a pentacoordinate zinc (Eklund and Brändén, 1987). If the crystalline structure corresponds to the catalytically active form in solution, then displacement of the zinc coordinated water molecule by the alcohol is a necessary step. The ionization pK_a for the zinc coordinated alcohol is dependent upon the electronic nature of the alcohol. The results show that if the alcohol has strong electron withdrawing substituents at C-2 then the ionization will occur with a low pK_a (Table I), and the subsequent hydride transfer will be inhibited or will not occur due to the loss of electron density at C-1. For alcohols with electron donating substituents at C-2 the ionization will occur with a relatively higher pK_a and the hydride transfer will be more facile.

In conclusion, the results show that the logarithm of the competitive inhibition constants are dependent upon the pK_a of the inhibiting alcohol with a slope approaching minus one, indicating that inhibition is mediated by a proton loss. The inhibiting alcohols used in this study, in most cases, were non-productive substrate analogues that competed with ethanol for the catalytic zinc ion. The ionization of an alcohol in the ternary complex with E-NAD$^+$ was shown to exhibit a pK_a dependence parallel to that of the pK_i. These substituent effects demonstrate that a key feature in the catalysis effected by LADH is the ability to lower the pK_a of the alcohol being oxidized.

REFERENCES

Andersson, P., Kvassman, J., Lindstörm, A., Olden, B. and Pettersson, G., 1981a, Effect of pH on pyrazole binding to liver alcohol dehydrogenase, *Eur. J. Biochem.* **114**: 549.

Andersson, P., Kvassman, J., Olden, B. and Pettersson, G., 1981b, Synergism between coenzyme and carboxylate binding in liver alcohol dehydrogenase, *Eur. J. Biochem.* **118**: 119.

Ballinger, P., and Long, F.A., 1960, Acid ionization constants of alcohols. II. Acidities of some substituted methanols and related compounds, *J. Am. Chem. Soc.* **82**: 795.

Dalziel, K., 1975, Kinetics and mechanism of nicotinamide-nucleotide-linked dehydrogenases, *The Enzymes* (Boyer, P.D., ed) 3rd edn., **11**: 1.

Dalziel, K., 1963, The purification of nicotinamide adenine dinucleotide and the kinetic effects of nucleotide impurities, *J. Biol. Chem.* **238**: 1538.

Eklund, H., and Brändén, C.-I., 1987, Alcohol dehydrogenase, *Biological Macromolecules and Assemblies* (Jurnak, F. and McPherson, A., eds) **3**:73.

Eklund, H., Plapp, B.V., Samama, J.-P., and Brändén, C.-I., 1982, Binding of substrate in a ternary complex of horse liver alcohol dehydrogenase, *J. Biol. Chem.* **257**: 14349.

Evans, S.A. and Shore, J.D., 1980, The role of zinc-bound water in liver alcohol dehydrogenase catalysis, *J. Biol. Chem.* **255**: 1509.

Greeves, M.A., Koerber, S.C., Dunn, M.F., and Fink, A.L., 1983, The effect of cryosolvents on the spectral and catalytic properties of liver alcohol dehydrogenase, *J. Biol. Chem.* **258**: 12184.

Kvassman, J., and Pettersson, G., 1980, Unified mechanism for proton-transfer reactions affecting the catalytic activity of liver alcohol dehydrogenase, *Eur. J. Biochem.* **103**: 565.

Makinen, M.W., Maret, W., and Yim, M.B., 1983, Neutral metal-bound water is the base catalyst in liver alcohol dehydrogenase, *Proc. Natl. Acad. Sci.* **80**: 2584.

Mannervik, B., 1981, Design and analysis of kinetic experiments for discrimination between rival models, *Kinetic Data Analysis* (Endrenyi, L., ed) p. 235.

Maret, W., and Zeppezauer, M., 1986, Influence of anions and pH on the conformational change of horse liver alcohol dehydrogenase induced by binding of oxidized nicotinamide adenine dinucleotide: binding of chloride to the catalytic metal ion, *Biochemistry* **25**: 1584.

Mildvan, A.S., and Weiner, H., 1969, Interaction of a spin-labeled analog of nicotinamide-adenine dinucleotide with alcohol dehydrogenase. II. Proton relaxation rate and electron paramagnetic resonance studies of binary and ternary complexes, *Biochemistry* **8**: 552.

Pavelich, W.A., and Taft, R.W., 1957, The evaluation of inductive and steric effects on reactivity. The methoxide ion-catalyzed rates of methanolysis of l-menthyl esters in methanol, *J. Am. Chem. Soc.* **79**: 4935.

Pocker, Y., 1989, Alcohol dehydrogenase: structure, catalysis, and site-directed mutagenesis, *Metal Ions in Biological Systems* (Sigel, H., ed) **25**: 336.

Pocker, Y., Li, H., and Page, J.D., 1987, Liver alcohol dehydrogenase: metabolic and energetic aspects, *Alcohol and Alcoholism, Suppl.* **1**: 181.

Pocker, Y., and Raymond, K.W., 1985, Liver alcohol dehydrogenase: substrate inhibition and competition between substrates, *Alcohol* **2**: 3.

Sartorius, C., Martin, G., Zeppezauer, M., and Dunn, M.F., 1987, Active-site cobalt(II)-substituted horse liver alcohol dehydrogenase: characterization of intermediates in the oxidation and reduction processes as a function of pH, *Biochemistry* **26**: 871.

Sekhar, C.V., and Plapp, B.V., 1990, Rate constants for a mechanism including intermediates in the interconversion of ternary complexes by horse liver alcohol dehydrogenase, *Biochemistry*, **29**, 4289.

Shore, J.D., and Theorell, H., 1966, Substrate inhibition effects in the liver alcohol dehydrogenase reaction, *Arch. Biochem. Biophy.* **117**, 375.

Shore J.D., Gutfreund, H., Brooks, R.L., Santiago, D., and Santiago, P., 1974, Proton equilibria and kinetics in the liver alcohol dehydrogenase reaction mechanism, *Biochemistry* **13**, 4185.

Takahashi, S., Cohen, L.A., Miller, H.K., and Peake, E., 1971, Calculation of the pK_a values of alcohols from $\sigma*$ constants and from their carbonyl frequencies of their esters, *J. Org. Chem.* **36**: 1205.

Theorell, H., and McKinely-McKee, J.S., 1961, Liver alcohol dehydrogenase. III. Kinetics in the presence of caprate, isobutyramide and imidazole, *Acta Chem. Scand.* **15**: 1811.

Theorell, H., and Yonetani, Y., 1962, Spectrophotometric demonstration of ternary liver alcohol dehydrogenase-coenzyme-substrate complexes, *Arch. Biochem. Biophys.,Suppl.* **1**: 209.

Theorell, H., and Yonetani, Y., 1963, Liver alcohol dehydrogenase-DPN-pyrazole complex: a model of a ternary intermediate in the enzyme reaction, *Biochem. Z.* **338**: 537.

GLUTATHIONE-DEPENDENT FORMALDEHYDE DEHYDROGENASE (EC 1.2.1.1):

EVIDENCE FOR THE IDENTITY WITH CLASS III ALCOHOL DEHYDROGENASE

Martti Koivusalo and Lasse Uotila

Department of Medical Chemistry
University of Helsinki
SF-00170 Helsinki, Finland

INTRODUCTION

The glutathione-dependent formaldehyde dehydrogenase (EC 1.2.1.1) is a ubiquitous cytoplasmic enzyme, which was first described in 1955 Strittmatter and Ball from beef and chicken livers. It was purified to homogeneity and characterized from human liver by Uotila and Koivusalo in 1974 (1974a) and has later also been purified from some other sources (Uotila and Koivusalo 1989, Pourmotabbed et al. 1989). Formaldehyde dehydrogenase catalyzes the following reversible reaction:

Formaldehyde + GSH + NAD$^+$ $\longleftrightarrow$ S-formylglutathione + NADH + H$^+$

The actual substrate is the hemithioacetal, S-hydroxymethylglutathione, which is formed non-enzymically from GSH and formaldehyde. The reaction product, which is a glutathione thiol ester, is irreversibly hydrolyzed to GSH and formate in the reaction catalyzed by a second specific enzyme, S-formylglutathione hydrolase (EC 3.1.2.12). This enzyme has also been purified to homogeneity and characterized from human liver (Uotila and Koivusalo 1974b) and other sources (Uotila 1989).

The mammalian alcohol dehydrogenases have been classified into three groups on the basis of their electrophoretic mobility, substrate specificity and sensitivity to inhibition by pyrazole and 4-methylpyrazole (Vallee and Bazzone 1983). The class III alcohol dehydrogenases form a distinct group, which is characterized by anodic mobility in the electrophoresis, very low affinity to ethanol and insensitivity to inhibition by pyrazole. The class III enzymes are also structurally distinct from the class I and class II enzymes (Kaiser et al. 1988) and are evolutionarily the most conserved alcohol dehydrogenase structures (Kaiser et al. 1989).

Class III alcohol dehydrogenases are found in most mammalian tissues (Juliá et al. 1987) and have been purified from human liver (Wagner et al. 1984) human brain (Beisswenger et al. 1985, Giri et al. 1989) horse liver (Kaiser et al. 1989) rat liver (Juliá et al. 1987) mouse liver (Algar et al. 1983) and from some other sources. The physiological function of class III alcohol dehydrogenases has been obscure but they have been suggested to be involved in the oxidation of long chain fatty alcohols and omega-hydroxyfatty acids (Giri et al. 1989).

In the present study we report and summarize evidence, which indicates that the glutathione-dependent formaldehyde dehydrogenase and the class III alcohol dehydrogenase are identical enzymes.

MATERIALS AND METHODS

Female rats of Wistar strain weighing around 200 g were used as experimental animals. 5´-AMP Sepharose 4 B and the materials for chromatofocusing were obtained from Pharmacia, Uppsala, Sweden. NAD(P)(H), GSH, pyrazole, 4-methylpyrazole, 12-hydroxydodecanoic acid and 16-hydroxyhexadecanoic acid were purchased from Sigma Chemical Co., St.Louis, MO, U.S.A. The alcohols, aldehydes and phenanthrolines were obtained from Aldrich-Chemie GmbH & Co., KG, Steinheim, Germany. The aldehydes were redistilled or recrystallized before use. The concentrations of the stock solutions of aldehydes and alcohols were assayed enzymically using rat liver aldehyde dehydrogenase, yeast alcohol dehydrogenase or horse liver alcohol dehydrogenase.

Formaldehyde dehydrogenase was purified from the cytosolic fraction of rat liver by a combination of 5´-AMP Sepharose affinity chromatography and chromatofocusing in a pH gradient 8-5 as described previously (Koivusalo et al. 1982, Uotila and Koivusalo 1983). The purified enzyme was homogeneous in SDS-gel electrophoresis giving only one protein band with silver staining (Merrill and Pratt 1986).

The assay mixture for formaldehyde dehydrogenase activity contained 1 mM formaldehyde, 1 mM GSH, 1.2 mM NAD^+ and enzyme in 0.1 M sodium pyrophosphate buffer pH 8.0 (Uotila and Koivusalo 1989) and that for class III alcohol dehydrogenase activity contained 1 mM n-octanol, 1.2 mM NAD^+ and enzyme in 0.1 M NaOH-glycine buffer pH 9.6 when not stated otherwise. The reduction of NAD^+ or oxidation of NADH was monitored at 340 nm and 25 $^{\circ}$C either on a Shimadzu UV-240 spectrophotometer or a nine-channel FP-9 Analyzing System (Labsystems Oy, Helsinki, Finland) connected to an Olivetti M20 computer. One unit of activity equals to 1 umol of NADH produced per min. The absorption coefficient of 6220 M^{-1} cm^{-1} was used for NADH at 340 nm. The protein concentrations were determined by the method of Lowry et al. (1951) or by a modification of the method of Bradford (1976) with bovine serum albumin as standard. Formaldehyde was assayed by the chromotropic acid method (Koivusalo 1956). The zinc content of purified formaldehyde dehydrogenase preparations was assayed after thorough dialysis of the samples against zinc-free buffer by atomic absorption spectrophotometry using a Perkin-Elmer 1100B instrument.

RESULTS AND DISCUSSION

We have determined the amino acid sequences for 5 tryptic peptides, which were obtained from homogeneous rat liver formaldehyde dehydrogenase (Koivusalo et al. 1989). The peptides contained altogether 57 amino acids. The N-terminal amino acid was blocked. All these sequences were exactly identical with those found in the recently reported total primary sequence of rat liver class III alcohol dehydrogenase (ADH-2) (Juliá et al. 1988). These results indicated that formaldehyde dehydrogenase is very closely related to alcohol dehydrogenases and especially to the class III enzymes.

We have reinvestigated the substrate specificity of rat liver formaldehyde dehydrogenase and especially the NAD-dependent oxidation of alcohols by it (Table 1). Short chain aliphatic alcohols from methanol to propanol at a concentration of 30 mM were not oxidized. There was some activity for ethanol and propanol at higher alcohol concentrations but the enzyme could not be saturated with up to 2 M ethanol. Straight chain aliphatic alcohols with 5-12 carbons were, however, active as substrates. The best alcohol substrates found were the omega-hydroxyfatty acids 12-hydroxydodecanoate and 16-hydroxyhexadecanoate. Benzyl alcohol was not used as substrate. The apparent K_m-values for n-octanol and 12-hydroxydodecanoate were 0.50 mM and 0.09 mM, respectively, at pH 9.6 with 1.2 mM NAD^+ as the coenzyme. In the oxidation of alcohols glutathione was not needed in contrast to the oxidation of formaldehyde. Adding glutathione to the reaction mixture had only a slight activating effect on the oxidation of alcohols (Table 2) probably due to an effect on thiol groups of the enzyme. The activities were not inhibited by even 10 mM pyrazole or 4-methylpyrazole (Table 2).

Formaldehyde dehydrogenase was also able to reduce aldehydes at pH 5.0 - 8.0 (Table 3). Straight chain aliphatic aldehydes with 4 - 12 carbons were used as substrates, but there was no activity with formaldehyde, acetaldehyde or propionaldehyde. Benzaldehyde was not reduced but 3-nitrobenzaldehyde and 4-nitrobenzaldehyde were used as substrates. There was no activity with 4-pyridylaldehyde or 4-carboxybenzaldehyde.

A higher pH was needed for the oxidation of alcohols (8.5 - 10.0) than for the oxidation of S-hydroxymethylglutathione (6.0 - 10.0) (Fig. 1). Thus at the physiological pH range the oxidation of alcohols was very low but the oxidation of formaldehyde was almost at maximum. The pH - activity profile for the reduction of aldehydes is presented in Fig. 2. Both NADH and NADPH could be used as coenzymes especially at the lower pH range. In the oxidation of alcohols at pH 8.5 - 10.0 only NAD^+ was used.

The substrate specificity and the pH-activity profiles of rat liver formaldehyde dehydrogenase with alcohols and aldehydes and the K_m-values determined accord well with the results reported earlier for rat liver

Table 1.   Oxidation of Alcohols by the Formaldehyde Dehydrogenase from Rat Liver

Substrate (Concentration)	Relative Activity
16-Hydroxyhexadecanoate (0.5 mM)	1.15
12-Hydroxydodecanoate (1 mM)	1.75
n-Dodecanol (1 mM)	1.10
n-Undecanol (1 mM)	1.10
n-Decanol (1 mM)	1.10
n-Nonanol (1 mM)	1.10
n-Octanol (1 mM)	1.00
n-Heptanol (1 mM)	0.67
n-Hexanol (1 mM)	0.27
n-Pentanol (1 mM)	0.05
n-Butanol (1 mM)	no activity
n-Butanol (30 mM)	0.13
n-Propanol (30 mM)	no activity
Ethanol (30 mM)	" "
Methanol (30 mM)	" "
trans-2-Hexenol (1 mM)	0.51
Benzyl Alcohol (1 mM)	no activity

Assay conditions: 0.1 M NaOH-glycine, pH 9.6, 1.2 mM NAD^+, the alcohol as given above and the enzyme. 25 °C. The relative activity with 1 mM n-octanol as the substrate has been set to 1.00. This value corresponded to 0.86 $\pm$ 0.35 U/mg protein (mean $\pm$ S.D.) for the 4 preparations used. The relative activity in the standard formaldehyde dehydrogenase assay (with formaldehyde, GSH and NAD^+ as the substrates at pH 8.0) was 3.68.

Table 2. Effects of Glutathione and Pyrazole on the Oxidation
of Alcohols by the Formaldehyde Dehydrogenase from
Rat Liver

Substrate (Concentration)	Relative Activity
n-Octanol (1 mM)	1.00
n-Octanol (1 mM) + 1 mM GSH	1.13
n-Octanol (1 mM) + 10 mM pyrazole	0.95
n-Octanol (1 mM) + 10 mM methylpyrazole	1.05

Assay conditions as in Table 1.

Table 3. Reduction of Aldehydes by the Formaldehyde
Dehydrogenase from Rat Liver

Substrate (Concentration)	Relative Activity
Formaldehyde (1 mM)	no activity
Acetaldehyde (1 mM)	" "
n-Propanal (1 mM)	" "
n-Butanal (1 mM)	0.05
n-Pentanal (1 mM)	0.20
n-Hexanal (1 mM)	0.40
n-Heptanal (1 mM)	0.42
n-Octanal (1 mM)	1.00
n-Nonanal (1 mM)	0.95
n-Decanal (0.5 mM)	0.65
n-Undecanal (0.2 mM)	0.35
n-Dodecanal (0.2 mM)	0.17
Benzaldehyde (1 mM)	no activity
3-Nitrobenzaldehyde (1 mM)	0.35
4-Nitrobenzaldehyde (1 mM)	0.20
3-Carboxybenzaldehyde (1 mM)	no activity
3-Pyridylaldehyde (1 mM)	" "
trans-2-Hexenal (1 mM)	0.35

Assay conditions: 0.1 M Na-phosphate, pH 6.0, 0.2 mM
NADH, the aldehyde as given above and the enzyme.
25 $^\circ$C. The relative activity with 1 mM n-octanal has
been set to 1.00.

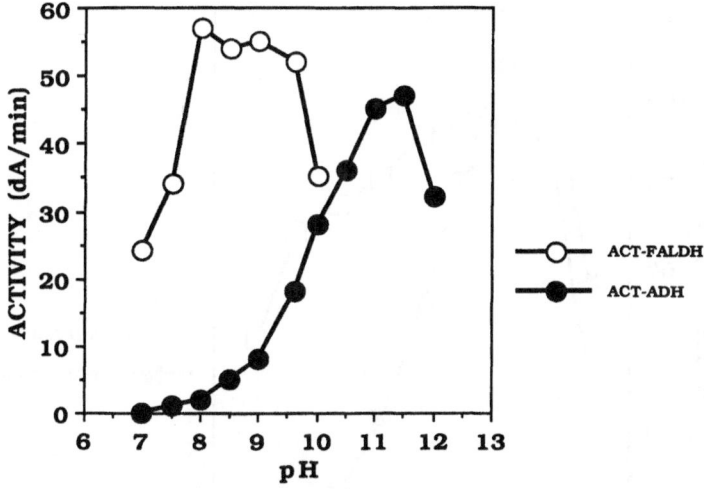

Fig. 1. pH–activity profile of the oxidation of formaldehyde (FALDH: 1 mM GSH and 1 mM formaldehyde were used as substrates) and of alcohols (ADH: 1 mM n-octanol was used as substrate) by the rat liver formaldehyde dehydrogenase. 1.2 mM NAD$^+$ was used as coenzyme in both assays. The amount of enzyme was doubled in the ADH assays.

ADH-2 (Juliá et al. 1987). The K_m-value for the oxidation of S-hydroxymethylglutathione is 0.92 uM for the rat liver formaldehyde dehydrogenase (Uotila and Koivusalo 1983). The best alcohol substrate 12-hydroxydodecanoate has about 100-fold higher K_m-value and also at least 100-fold lower specificity constant V_m/K_m than S-hydroxymethylglutathione even when measured at high pH.

We have earlier reported (Uotila and Koivusalo 1974b,1989) that formaldehyde dehydrogenase is unlikely to have a metal component because it is not very sensitive to chelating agents. We have now reinvestigated this question and we found that the rat liver formaldehyde dehydrogenase contained 3.7 g-atom zinc per mole of native dimeric enzyme. Zinc is known to be an essential component of mammalian class I-II alcohol dehydrogenases (Vallee and Bazzone 1983) and about 4 g-atom zinc per mole of native protein has been reported to be present also in most class III alcohol dehydrogenases (Vallee and Bazzone 1983, Wagner et al. 1984, Juliá et al. 1987, 1988, Kaiser et al. 1989). The only exception is the mouse liver class III alcohol dehydrogenase (B$_2$ isozyme), which has been reported (Algar et al. 1983) to contain only one zinc atom per subunit and to be insensitive to 10 mM 1,10-phenanthroline. We have found that high concentrations of 1,10-phenanthroline inhibited both the oxidation of formaldehyde and the oxidation of alcohols by the rat liver formaldehyde dehydrogenase. When 10 mM 1,10-phenanthroline was used it caused an instantaneous inhibition of 60 per cent (Fig. 3). This inhibition could be reversed by dilution but no reversion was achieved by addition of zinc under the conditions used. There was also a weak time-dependent inhibition by 1,10-phenanthroline. The non-chelating 1,7- and 4,7-phenanthrolines were better inhibitors than the chelating 1,10-phenanthroline (Fig. 3). Addition of 10-30 mM EDTA had no effect on the activity of the rat liver enzyme.

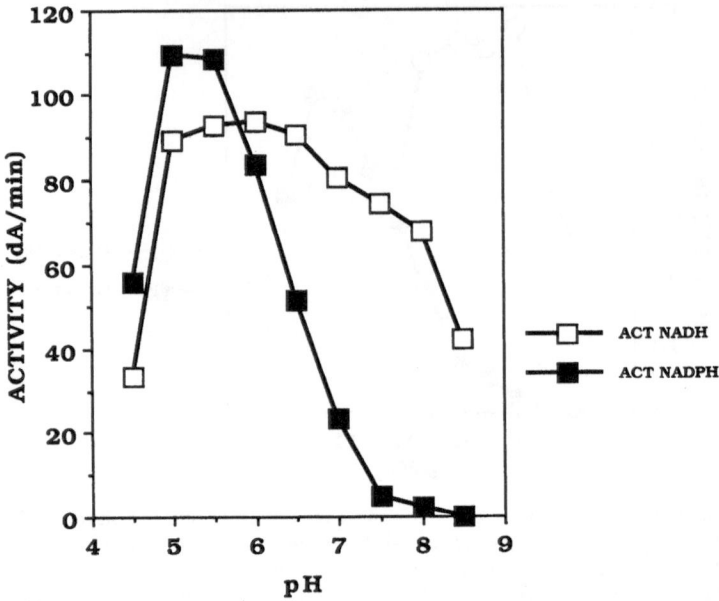

Fig. 2. pH - activity profile of the reduction aldehydes by the rat liver formaldehyde dehydrogenase. 1 mM n-octanal and 0.2 mM NADH or NADPH were used as substrates.

During the purification of rat liver formaldehyde dehydrogenase by 5'-AMP - Sepharose affinity chromatography and chromatofocusing the formaldehyde dehydrogenase activity assayed with 1 mM GSH, 1 mM formaldehyde and 1.2 mM NAD^+ in Na-pyrophosphate buffer pH 8.0 and the class III alcohol dehydrogenase activity assayed with 1 mM n-octanol, 10 mM pyrazole and 1.2 mM NAD^+ in NaOH-glycine buffer pH 9.6 were found in exactly the same fractions.

When rat liver formaldehyde dehydrogenase was studied by isoelectric focusing (Koivusalo et al. 1989) the same 1-3 main bands at pH 5.8-6.4 with some subbanding were obtained, when the gels were stained either for formaldehyde dehydrogenase activity with formaldehyde, GSH and NAD^+ as the substrates or with n-octanol and NAD^+ with or without pyrazole.

Further corroborating evidence for the identity of these two enzymes is found in earlier investigations. The molecular weights of both formaldehyde dehydrogenase (Uotila and Koivusalo 1974a,1989) and class III alcohol dehydrogenases (Algar et al. 1983, Vallee and Bazzone 1983, Wagner et al. 1984, Juliá et al. 1987, Kaiser et al. 1988) are 80 000 and the subunit molecular weights are 40 000. Both activities have been localized into cell cytosol and both have a very wide and identical tissue distribution in the mammals (Uotila and Koivusalo 1974a,1989, Juliá et al. 1987, Boleda et al. 1989). The gene for the human formaldehyde dehydrogenase has been located to chromosome 4q21-25 (Hiroshige et al. 1985, Van der Goes et al. 1985) which is the same site found for human class III (Carlock et al. 1985) and class I (Smith et al. 1985) alcohol dehydrogenases.

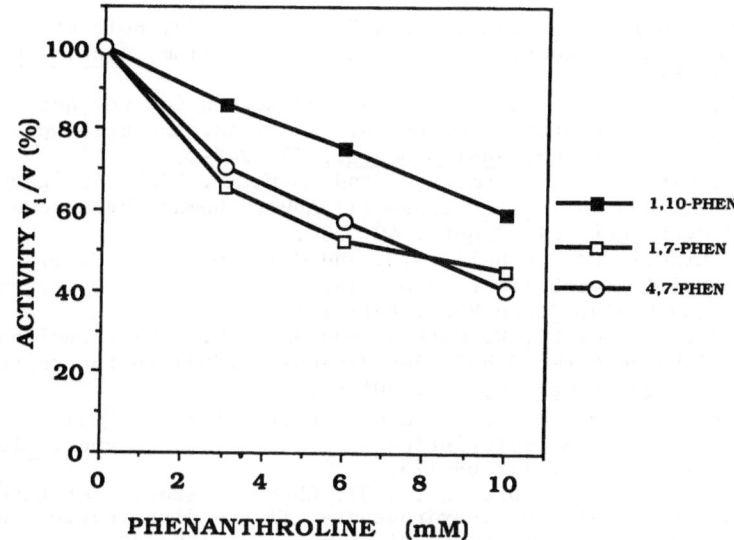

Fig. 3. Inhibition of the rat liver formaldehyde dehydrogenase by the chelating (1,10-) and non-chelating (1,7-, 4,7-) phenanthrolines. 1 mM GSH, 1 mM formaldehyde and 1.2 mM NAD$^+$ were used as substrates. NaOH-glycine buffer pH 10.

CONCLUSIONS

When the properties of formaldehyde dehydrogenase and class III alcohol dehydrogenase are compared it is seen that in addition to the sequence homology all the protein and enzyme properties of these two enzymes which have been studied are identical. We thus conclude that the glutathione-dependent formaldehyde dehydrogenase and the class III alcohol dehydrogenase are most likely identical enzymes.

ACKNOWLEDGEMENTS

These investigations have been supported by a grant from the University of Helsinki. We thank Mrs Eija Haasanen and Miss Anne Hyvönen for their skillful technical assistance and Miss Maria Saariluoma for the typing of the manuscript. We thank Georg Alfthan, M.Sc., National Public Health Institute, Helsinki, Finland for the zinc analyses.

REFERENCES

Algar,E.M., Seeley,T-L. and Holmes,R.S., 1983, Purification and molecular properties of mouse alcohol dehydrogenase isozymes, Eur.J.Biochem., 137: 139-147.
Beisswenger,T.B., Holmquist,B. and Vallee,B.L., 1985, X -ADH is the sole alcohol dehydrogenase isozyme of mammalian brains: Implications and inferences, Proc.Natl.Acad.Sci.USA, 82: 8369-8373.

Boleda,M.D., Juliá,P., Moreno,A. and Parés,X., 1989, Role of extrahepatic alcohol dehydrogenase in rat ethanol metabolism, Arch.Biochem.Biophys., 274: 74-81.

Bradford,M.M., 1976, A rapid and sensitive method for the quantitation of microgram quantities of protein utilizing the principle of protein-dye binding, Anal.Biochem., 72: 248-254.

Carlock,L., Hiroshige,S.,Wasmuth,J. and Smith,M., 1985, Assignement of the ADH5 gene coding for class III ADH to human chromosome 4:q21-4q25, Cytogenet.Cell Genet., 40: 598.

Giri,P.R., Linnoila,M., O'Neill,J.B. and Goldman,D., 1989, Distribution and possible metabolic role of class III alcohol dehydrogenase in the human brain, Brain Res., 481: 131-141.

Hiroshige,S., Carlock,L., Wasmuth,J. and Smith,M., 1985, Regional assignement of human formaldehyde dehydrogenase (FDH) to the region 4q21-4q25, Cytogenet.Cell Genet., 40: 651-652.

Juliá,P., Boleda,M.D., Farrés,J. and Parés,X., 1987, Mammalian alcohol dehydrogenase: Characteristics of class III isoenzymes, Alcohol & Alcoholism, Suppl. 1: 169-173.

Juliá,P., Farrés,J. and Parés,X., 1987, Characterization of three isoenzymes of rat alcohol dehydrogenase. Tissue distribution and enzymatic properties, Eur.J.Biochem., 162: 179-189.

Juliá,P., Parés,X. and Jörnvall,H., 1988, Rat liver alcohol dehydrogenase of class III. Primary structure, functional consequences and relationships to other alcohol dehydrogenases, Eur.J.Biochem., 172: 72-83.

Kaiser,R., Holmquist,B., Hempel,J., Vallee,B.L. and Jörnvall,H., 1988, Class III human liver alcohol dehydrogenase: A novel structural type equidistantly related to the class I and class II enzymes, Biochemistry, 27: 1132-1140.

Kaiser,R., Holmquist,B., Vallee,B.L. and Jörnvall,H., 1989, Characteristics of mammalian class III alcohol dehydrogenases, an enzyme less variable than the traditional liver enzyme of class I, Biochemistry, 28: 8432-8438.

Koivusalo,M., 1956, Studies on the metabolism of methanol and formaldehyde in the animal organism, Acta Physiol. Scand., 39: Suppl. 131: 31-34.

Koivusalo,M., Baumann,M. and Uotila,L., 1989, Evidence for the identity of glutathione-dependent formaldehyde dehydrogenase and class III alcohol dehydrogenase, FEBS Lett. 257: 105-109.

Koivusalo,M., Koivula,T. and Uotila,L., 1982, Oxidation of formaldehyde by nicotinamide nucleotide dependent enzymes, in: "Enzymology of Carbonyl Metabolism: Aldehyde Dehydrogenase and Aldo/Keto Reductase", H.Weiner and B.Wermuth, eds., pp. 155-168, Alan R. Liss, Inc., New York.

Lowry,O.H., Rosebrough,N.J., Farr,A.L. and Randall,R.J., 1951, Protein measurement with the Folin phenol reagent, J.Biol.Chem., 193: 265-275.

Merrill,C.R. and Pratt,M.E., 1986, A silver stain for the rapid quantitative detection of proteins or nucleic acids on membranes or thin layer plates, Anal.Biochem., 156: 96-110.

Pourmotabbed,T., Shih,M.J. and Creighton,D.J., 1989, Bovine liver formaldehyde dehydrogenase. Kinetic and molecular properties, J.Biol.Chem., 264: 17384-17388.

Smith,M., Duester,G., Carlock,L. and Wasmuth,J., 1985, Assignment of ADH1, ADH2 and ADH3 genes (class I ADH) to human chromosome 4q21-4q25, through use of DNA probes, Cytogenet.Cell Genet., 40: 748.

Strittmatter,P. and Ball,E.G., 1955, Formaldehyde dehydrogenase, a glutathione-dependent enzyme system, J.Biol.Chem., 213: 445-461.

Uotila,L., 1989, Glutathione thiol esterases, in: "Coenzymes and Cofactors, vol.III, Glutathione. Chemical, Biochemical and Medical Aspects, part A", D.Dolphin, R.Poulson and O.Abramovic,eds., pp. 767-804, John Wiley & Sons, Inc., New York.

Uotila,L. and Koivusalo,M., 1974a, Formaldehyde dehydrogenase from human liver. Purification, properties, and evidence for the formation of glutathione thiol esters by the enzyme, J.Biol.Chem., 249: 7653-7663.

Uotila,L. and Koivusalo,M., 1974b, Purification and properties of S-formylglutathione hydrolase from human liver, J.Biol.Chem., 249: 7664-7672.

Uotila,L. and Koivusalo,M., 1983, Formaldehyde dehydrogenase, in: "Functions of Glutathione. Biochemical, Physiological, Toxicological and Clinical Aspects", A.Larsson, S.Orrenius, A.Holmgren, B.Mannervik, eds., pp. 175-186, Raven Press, New York.

Uotila,L. and Koivusalo,M., 1989, Glutathione-dependent oxidoreductases: Formaldehyde dehydrogenase, in: "Coenzymes and Cofactors, vol. III, Glutathione. Chemical, Biochemical and Medical Aspects, part A", D.Dolphin. R.Poulson and O.Abramovic,eds., pp.517-537, John Wiley & & Sons, Inc., New York.

Vallee,B.L. and Bazzone,T.J., 1983, Isozymes of human liver alcohol dehydrogenase, in: "Isozymes: Current Topics in Biological and Medical Research, vol.8", M.C.Rattazi, J.C.Scandalios and G.S.Whitt, eds., pp. 219-244, Alan R. Liss, Inc., New York.

Van der Goes,R., Van Kessel,A.G., Hagemeijer,A., Wijnen,L.M.M. and Meera Khan,P., 1985, Localization of human FSH to 4q21-qtr, Cytogenet.Cell Genet., 40: 766.

Wagner,F.W., Parés,X., Holmquist,B. and Vallee,B.L., 1984, Physical and enzymatic properties of a class III isozyme of human liver alcohol dehydrogenase: X -ADH, Biochemistry, 23: 2193-2199.

KINETICS AND MECHANISM OF METHANOL AND FORMALDEHYDE INTERCONVERSION AND FORMALDEHYDE OXIDATION CATALYZED BY LIVER ALCOHOL DEHYDROGENASE

Y. Pocker and Hong Li

Department of Chemistry
University of Washington
Seattle, WA 98195

INTRODUCTION

Liver alcohol dehydrogenase (LADH) is a well-characterized protein. Both its primary and tertiary structures are known, as are some of the catalytic mechanisms, isozyme differences, evolutionary divergences and a number of enzymatic properties (Jörnvall, 1970; Brändén, et al., 1975; Klinman, 1981; Eklund and Brändén, 1983; Pettersson, 1987; Pocker, 1989). In contrast, relatively little is known about its interaction with methanol, although there is extensive structural homology between this compound and ethanol.

Compared with ethanol, methanol is an agent with a much more pronounced toxicity, notably, the ability to cause visual impairment (Jacobsen, et al., 1986). Moreover, there is increasing concern about the potential of formaldehyde, a metabolite of methanol, to act as a carcinogen (Clary, et al., 1983). In regard to its oxidation by LADH, methanol also shows a much slower reaction rate. There are significant differences in enzymatic behavior between methanol and other primary alcohols. Studies on the interaction of methanol with LADH will enable us to understand the biological behavior as well as the mechanism of methanol metabolism.

Due to its relatively slow reaction rate as compared to ethanol, methanol was generally believed not to be a substrate of liver alcohol dehydrogenase (Winer, 1958; Wratten and Cleland, 1965). Later studies have shown that it is a substrate of alcohol dehydrogenases from both liver and yeast (Mani, et al., 1970; Pocker, et al., 1987a & 1987b). However, prior to the current study, no detailed rationale has been put forward for the enzymatic oxidation of methanol.

The oxidation mechanism of ethanol and many other primary alcohols by LADH has been established to be an ordered bi-bi process (the upper pathway in Scheme I). That is to say, the binding of NAD^+ precedes the binding of the alcohol with the rate-limiting step being the dissociation of NADH from the enzyme-NADH complex (Wratten and Cleland, 1963). For the oxidation of methanol, however, transient and steady-state kinetic studies with deuterated analogues have shown that the rate-limiting step is the hydride transfer in the enzyme-NAD^+-methanol complex (Brooks and Shore, 1971).

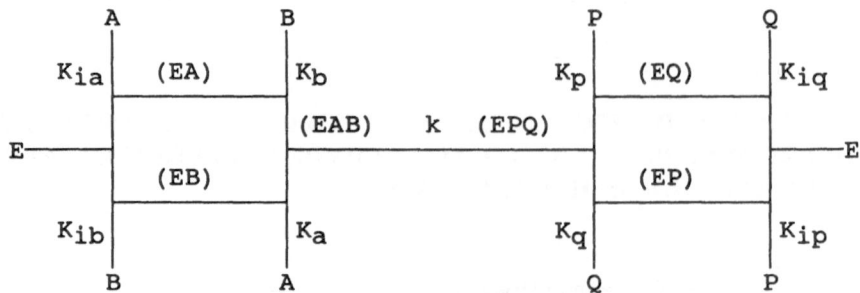

SCHEME I. Random Mechanism for Liver Alcohol Dehydrogenase. E: LADH; A: NAD^+; B: alcohol; P: aldehyde; Q: NADH.

The purpose of this study was to delineate, in detail, the kinetics and mechanism of the reactions of methanol and its metabolites with liver alcohol dehydrogenase. Our results indicate that methanol binds to a similar site in the enzyme as does ethanol, but due to the lack of an appreciable hydrophobic chain, it exhibits different kinetics and is the only simple primary alcohol found to follow a rapid equilibrium random mechanism (combined upper and lower pathways in Scheme I).

EXPERIMENTAL PROCEDURES

Horse LADH, NAD^+, and NADH were obtained from Sigma Chemical Company. Pyrazole (recrystallized twice before use) was obtained from Aldrich Chemical Company. All other chemicals used were reagent grade.

Methanol solutions were obtained by dissolving a weighed amount of absolute methanol in buffer. Formaldehyde solutions were prepared by distilling an aqueous solution of methenamine acidified with an equivalent amount of H_2SO_4. The resulting distillate was neutralized, redistilled, and its concentration checked both gravimetrically with dimethylcyclohexanedione (Weinberger, 1931), and enzymatically using an excess of NADH in the presence of LADH. LADH, NAD^+ and NADH solutions were prepared immediately before each experiment. The LADH solution was kept at 0°C to prevent loss of enzymatic activity. The concentration of LADH was checked spectrophotometrically employing an extinction coefficient, ϵ, of 3.6×10^4 $M^{-1}cm^{-1}$ at 280 nm. Enzymatic activity was measured by titration with NAD^+ in excess pyrazole (Theorell and Yonetani, 1963). NAD^+ and NADH solutions were obtained by dissolving each in buffer and assaying the concentration spectrophotometrically using extinction coefficients, ϵ, of 1.78×10^4 $M^{-1}cm^{-1}$ at 260 nm and 6.22×10^3 $M^{-1}cm^{-1}$ at 340 nm, respectively. All reactions were carried out in 30 mM phosphate buffer solution. The ionic strength was maintained at 0.1 by adding Na_2SO_4.

Absorbance measurements were performed on a Varian Cary 210 spectrophotometer interfaced to an Apple IIe computer. Fast reactions were conducted on a Durrum-Gibson Model 1300 stopped-flow spectrophotometer interfaced to a NEC Powermate desktop computer with a DAS-20 (MetraByte) A/D interface board. The kinetic software was written in Turbo Pascal by the author. Fluorescence analysis of NADH produced by formaldehyde oxidation was done on a Perkin-Elmer 650-10s fluorescence photometer.

Unless specified otherwise, enzymatic reactions were initiated by injecting a small amount of enzyme solution into a pre-mixed coenzyme-substrate solution, and the change in NADH concentration monitored at 340 nm. Fast reactions with a half-life of a few seconds or less were performed on the stopped-flow equipment. For methanol and formaldehyde oxidation, the concentration change of formaldehyde during a reaction was also monitored with chromotropic acid by employing known procedures (Frisell, et al. 1954). Initial rates were obtained by fitting the data points to polynomial curves and extrapolating to zero time.

Due to the fact that formaldehyde can be oxidized by NAD^+, caution must be taken in devising product inhibition studies. For formaldehyde inhibition on methanol oxidation, low formaldehyde or NAD^+ concentrations were used so that the rate of oxidation of formaldehyde was small compared to the rate of methanol oxidation. In a few cases, the rate of formaldehyde oxidation had to be subtracted from the observed rate in order to obtain the initial rate of methanol oxidation. NAD^+ inhibition of formaldehyde reduction was studied using low NAD^+ concentrations so that the rate of formaldehyde oxidation was negligible compared to the rate of formaldehyde reduction.

RESULTS AND DISCUSSION

1. METHANOL OXIDATION

The rate of methanol oxidation catalyzed by LADH is only about 3% of the rate of ethanol oxidation. In the case of ethanol, extensive studies have shown that the reaction follows an ordered bi-bi mechanism (Wratten and Cleland, 1963; Pettersson, 1987). The characteristic of such a reaction mechanism is that there is a burst associated with the formation of the enzyme-NADH complex, which can be observed at 325 nm, followed by a slow dissociation of this complex. The burst of enzyme-NADH formation displays a deuterium isotope effect of about 6 and is much faster than the observed rate of free NADH formation (Brooks and Shore, 1971). Thus the dissociation of the enzyme-NADH complex is the rate-limiting step.

In contrast, for the oxidation of methanol, no burst of enzyme-NADH was observed in our stopped-flow experiments. Instead, the observed result is the slow formation of free NADH. Using deuterated methanol, CD_3OH, the rate of free NADD formation is slower by a factor of about 5 compared to the rate of NADH formation from CH_3OH. These results are similar to those previously reported (Brook and Shore, 1971) and indicate that the hydride transfer step, that is, the conversion of the ternary complex, is rate-limiting for methanol oxidation.

Product inhibition studies were performed, and results are shown in Figure 1 through Figure 4. When the NAD^+ concentration is varied, the product inhibition pattern is competitive for NADH as the product inhibitor and noncompetitive for formaldehyde. When the methanol concentration is varied, the product inhibition pattern is competitive for formaldehyde and noncompetitive for NADH. These patterns fit the rapid equilibrium random mechanism (Gulbinsky and Cleland, 1968).

The kinetic parameters for the oxidation of methanol measured according to a rapid equilibrium random mechanism are listed in Table I. The k_{cat} for methanol oxidation, which corresponds to the rate constant for conversion of the LADH-NAD^+-

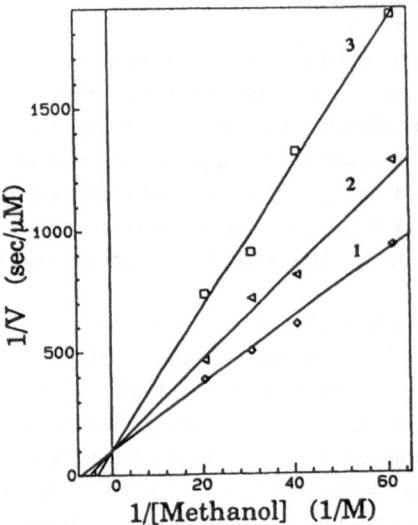

Figure 1. Product inhibition of horse liver alcohol dehydrogenase by formaldehyde with methanol as variable substrate. Enzyme, 5.0×10^{-8}M. NAD$^+$, 5.00×10^{-4}M. Formaldehyde: (1) 0, (2) 2.03×10^{-4}M, (3) 4.88×10^{-4}M.

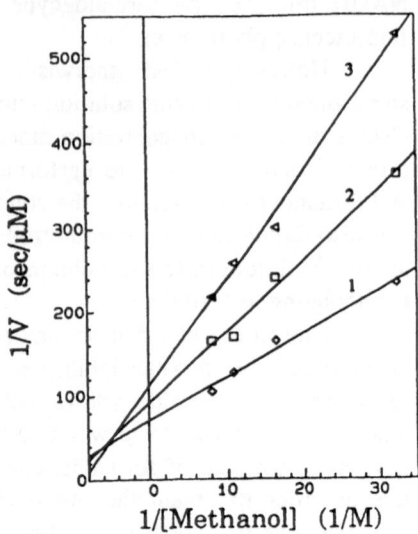

Figure 2. Product inhibition of horse liver alcohol dehydrogenase by NADH with methanol as variable substrate. Enzyme, 2.3×10^{-7}M. NAD$^+$, 5.70×10^{-4} M. NADH: (1) 0, (2) 2.94×10^{-6}M, (3) 5.88×10^{-6}M.

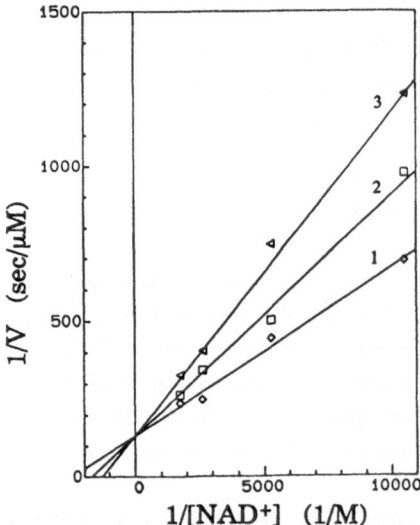

Figure 3. Product inhibition of horse liver alcohol dehydrogenase by NADH with NAD$^+$ as variable substrate. Enzyme, 2.3×10^{-7}M. Methanol, 3.10×10^{-2}M. NADH: (1) 0, (2) 2.00×10^{-6}M, (3) 5.00×10^{-6}M.

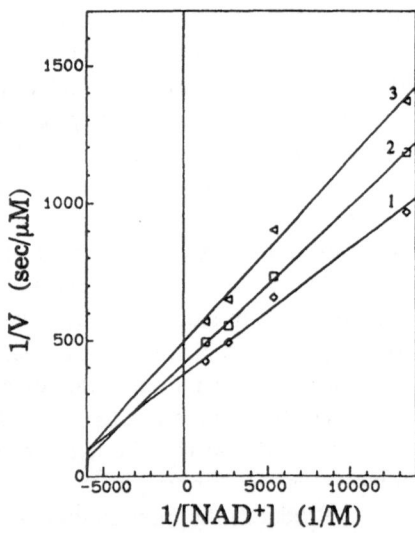

Figure 4. Product inhibition of horse liver alcohol dehydrogenase by formaldehyde with NAD$^+$ as variable substrate. Enzyme, 5.0×10^{-8}M. Methanol, 2.45×10^{-2}M. Formaldehyde: (1) 0, (2) 1.27×10^{-4}M, (3) 2.54×10^{-4}M.

318

methanol ternary complex, is about 30 times smaller than the reported value of 3.1 s^{-1} for ethanol oxidation (Theorell, et al., 1961). A k_{cat} of 0.022 s^{-1} was measured (Table I) for the oxidation of CD_3OH, which is about 5 times smaller than that of CH_3OH, supporting the idea that the hydride transfer step is rate-limiting. The dissociation constant for the LADH-NAD$^+$ complex obtained also agrees with the previously reported value (Sund and Theorell, 1963), in which a different technique was employed.

The constants for the dissociation of methanol from the LADH-methanol (K_{ib}) and the LADH-NAD$^+$-methanol complex (K_b) are very similar. This indicates that the binding of NAD$^+$ does not affect the binding of methanol greatly, and is different from the oxidation of ethanol and other primary alcohols where the binding of NAD$^+$ promotes the binding of alcohol to the enzyme resulting in a preferred order mechanism. The similarity of K_{ib} and K_b also indicates that deprotonation of methanol has not occurred during binding, because the positive charge of NAD$^+$ would have stabilized the zinc bound methoxide, thereby decreasing the value of K_b.

For CD_3OH oxidation, the dissociation constants for NAD$^+$, K_{ia} and K_a do not change significantly from those for CH_3OH. Furthermore, the dissociation constants, K_{ib} and K_b, are similar for both CD_3OH and CH_3OH. This is not surprising since the deuterated substrate should in principle only affect the hydride transfer step and not coenzyme or substrate binding to the enzyme.

Pyrazole, an effective inhibitor of ethanol oxidation (Theorell and Yonetani, 1963), was used to inhibit the oxidation of methanol. The results show that pyrazole inhibits the reaction competitively with a K_i value of 3.5×10^{-7} M. This is similar to the value of 3.9×10^{-7} M obtained for pyrazole inhibition of ethanol oxidation (Pocker and Raymond, 1980). This inhibitor binds to the enzyme-NAD$^+$ complex as an inner sphere ligand of the catalytic zinc (Shore and Gilleland, 1970) and thus blocks the substrate binding site. The close correspondence in inhibition patterns and inhibition constants noted for pyrazole in regard to the oxidation of both methanol and ethanol indicates that methanol binding to the catalytic zinc is similar to that of ethanol.

The difference in the reaction mechanism and the magnitude of rate constants between methanol and ethanol can be explained by the hydrophobic interaction in the enzyme-NAD$^+$-alcohol complex. Within the alcohol binding site of LADH there are two binding regions, a hydroxyl binding region and a hydrophobic binding region (Dalziel and Dickinson, 1967). When a substrate binds to the enzyme, the interaction between the hydrophobic binding region and the hydrophobic chain on the substrate stabilizes substrate binding. Methanol, having a chain of only one carbon atom, cannot bind as ethanol and other primary alcohols to the hydrophobic region. Because the methyl group is not held in the proper orientation for hydride transfer, methanol exhibits both a weak binding constant and a slow turnover rate.

The lack of a hydrophobic chain also accounts for the random mechanism for methanol oxidation. Early isotope exchange measurements, and analysis of substrate inhibition or activation patterns at high substrate concentration show that the LADH reaction can be generalized as a random mechanism in Scheme I (Bränden, et al., 1975; Kamlay and Shore, 1983). However, for most alcohols and aldehydes, the ordered bi-bi mechanism (upper path in Scheme I) has been established over concentration ranges where Michaelis-Menten kinetics are obeyed. The preferred ordered mechanism for most substrates is followed because NAD$^+$ binding to LADH results in a protein conformational change (Coates, et al., 1977; Hardman, 1981). This conformational change alters the space within the hydrophobic and hydroxyl binding regions, stabilizing substrate binding and greatly enhancing the upper pathway in Scheme I. In contrast, the alternation of the hydrophobic binding region, has little or no effect on methanol binding, and as a result, both the upper and the lower pathways in Scheme I are catalytically significant for methanol oxidation.

2. REDUCTION OF FORMALDEHYDE

The rate of formaldehyde reduction is much higher than that of methanol oxidation but still much slower than the reduction of acetaldehyde. The results of product inhibition studies are shown in Figures 5 through 8. When the NADH concentration is varied, the product inhibition pattern is competitive for NAD^+ and noncompetitive for methanol. When the formaldehyde concentration is varied, the product inhibition pattern is competitive for methanol and noncompetitive for NAD^+. These patterns suggest that a rapid equilibrium random mechanism is followed.

The constants calculated according to a rapid equilibrium random mechanism are listed in Table I. It is noted that the rate for the turnover of LADH-NADH-formaldehyde complex is about 500 times larger than that of LADH-NAD^+-methanol complex. The binding of formaldehyde has little or no effect on the observed binding constants of NADH. The value of the dissociation constant for the enzyme-NADH complex is in agreement with the value obtained by equilibrium studies (Sund and Theorell, 1963).

The lack of hydrophobic chain in formaldehyde also explains the random mechanism and the slower turnover rate as compared to acetaldehyde. The observation that both pathways in Scheme I are fully catalytic for aldehyde reduction has also been reported with DACA as the substrate (Andersson, et al., 1984).

Formaldehyde exhibited substrate inhibition at concentrations higher than 0.2 M. This results from the formation of LADH-NAD^+-formaldehyde complex. As will be shown in next section, this complex is also reactive.

Table I. Parameters for the Reaction of Methanol and Formaldehyde with Liver Alcohol Dehydrogenase.

Reaction	k_{cat} (s^{-1})	K_a (mM)	K_b (mM)	K_{ia} (mM)	K_{ib} (mM)
MeOH + NAD^+	0.12±0.02	0.19±0.02	25±4	0.24±0.03	32±4
CD_3OH + NAD^+	0.022±0.003	0.16±0.02	28±4	0.20±0.03	34±4
HCHO + NADH	59±5	1.4×10^{-3} $\pm 2 \times 10^{-4}$	8.3±0.8	1.1×10^{-3} $\pm 2 \times 10^{-4}$	6.5±0.5
HCHO + NAD^+	0.77±0.15	0.23±0.03	7.6±0.5	0.17±0.03	5.7±0.5

Reactions were carried out at pH 7.50 in phosphate buffer. T = 25.0 ± 0.1°C, The constants are defined in Scheme I.

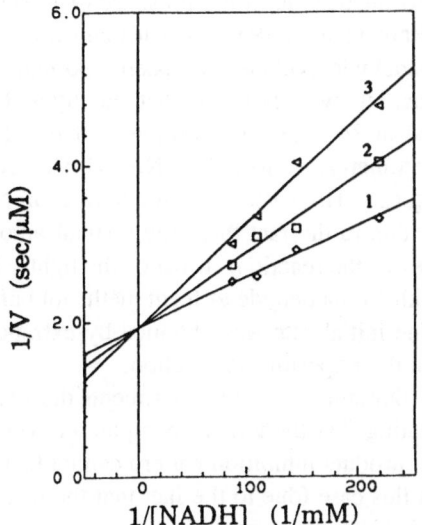

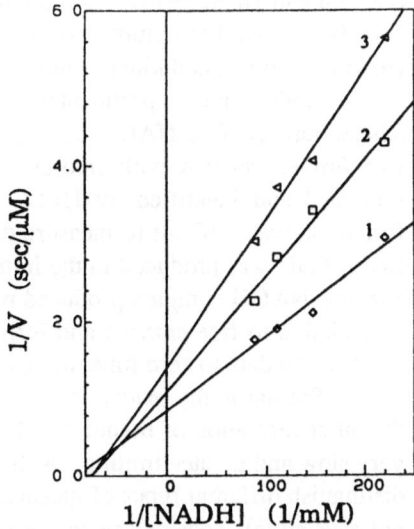

Figure 5. Product inhibition of horse liver alcohol dehydrogenase by NAD^+ with NADH as variable substrate. Enzyme, 2.0×10^{-8}M. Formaldehyde, 4.00×10^{-3}M. NAD^+: (1) 0, (2) 2.84×10^{-4}M, (3) 4.16×10^{-4}M.

Figure 6. Product inhibition of horse liver alcohol dehydrogenase by methanol with NADH as variable substrate. Enzyme, 2.0×10^{-8}M. Formaldehyde, 8.00×10^{-3}M. Methanol: (1) 0, (2) 3.00×10^{-2}M, (3) 6.00×10^{-2}M.

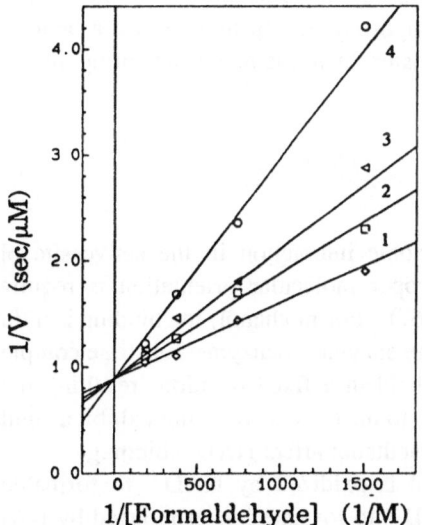

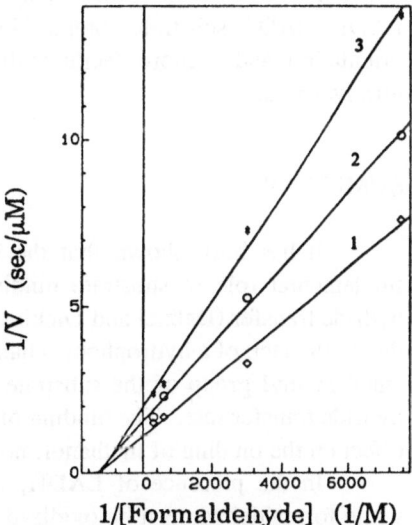

Figure 7. Product inhibition of horse liver alcohol dehydrogenase by methanol with formaldehyde as variable substrate. Enzyme, 2.0×10^{-8}M. NADH, 1.00×10^{-5} M. Methanol: (1) 0, (2) 4.13×10^{-2}M, (3) 8.25×10^{-2}M. (4) 0.165M.

Figure 8. Product inhibition of horse liver alcohol dehydrogenase by NAD^+ with formaldehyde as variable substrate. Enzyme, 2.5×10^{-8}M. NADH, 9.00×10^{-6} M. NAD^+: (1) 0, (2) 1.52×10^{-4}M, (3) 3.04×10^{-4}M.

321

3. OXIDATION OF FORMALDEHYDE

Studies on the oxidation of aldehydes catalyzed by LADH have been reported (Hinson and Neal, 1972). Dismutation of formaldehyde in the presence of NAD^+ and LADH has also been studied (Kendal and Ramanathan, 1952; Abeles and Lee, 1960). However, no detailed kinetic study on formaldehyde oxidation has been reported.

Under our experimental conditions, it was noticed that at high NAD^+ concentrations, free NADH can be observed at 340 nm. The presence of free NADH was further verified with fluorescence photometry. Also, free NADH was actually separated and identified by HPLC techniques. However, the initial rate of NADH formation was difficult to measure directly, due to the fact that only a small amount of free NADH was produced in the initial stages of the reaction. Most of the tightly bound enzyme-NADH complex produced reacts with formaldehyde to form methanol before its dissociation to free enzyme and NADH. The initial rate was obtained by extrapolation of collected data to zero time, as described in the Experimental Section.

Similar to the oxidation of methanol, our stopped-flow experiments did not show the burst formation of bound-NADH, indicating that the ternary complex conversion is very slow and is rate-limiting. Although the product inhibition pattern cannot be used to distinguish different types of mechanisms in this case (due to the fact that the reaction is not reversible), considering the principle of microscopic reversibility and the fact that formaldehyde reduction follows a rapid equilibrium random mechanism, it seems reasonable to assume that the oxidation of formaldehyde also follow a rapid equilibrium random mechanism.

The results of measurements according to a rapid equilibrium random mechanism are included in Table I. The k_{cat} of formaldehyde oxidation is about 7 times faster than that of methanol oxidation and 7 to 10 times faster than the oxidation of other aldehydes (Hinson, et al., 1972). Formaldehyde is highly hydrated in aqueous solution, and in the LADH-NAD^+-substrate complex its extra hydroxyl group may ensure a better spatial orientation and a more facile hydride transfer than methanol, resulting in a larger turnover rate.

SUMMARY

It has been shown that the hydrophobic interaction in the active-site plays a fundamental role in substrate binding. Proper molecular orientation is required for hydride transfer (Dalziel and Dickinson, 1967). For methanol, the binding is unfavored due to the lack of a hydrophobic chain. In the enzyme-coenzyme-substrate complex, the small methyl group of the substrate is not held in a fixed position, resulting in a low hydride transfer rate. The binding of NAD^+ to the enzyme does not exhibit a significant effect on the binding of methanol, nor does methanol affect NAD^+ binding.

In the presence of LADH, methanol is oxidized by NAD^+ to formaldehyde, while formaldehyde can be oxidized by NAD^+ to formate ion or reduced by NADH to methanol. These reactions follow a rapid equilibrium random mechanism. Among these three reactions, the reduction of formaldehyde is the most rapid. The rate of formaldehyde oxidation is faster than the oxidation of methanol.

Our study with these non-hydrophobic substrates provides an important bridge between the bioinorganic activation of zinc-bound water and the bioorganic oxidation of ethanol. Furthermore, it furnishes some insight into an enzymatic system that is so highly sensitive to small changes in substrate chain length that it can magnify the consequence of a modest change in substrate hydrophobicity.

REFERENCES

Abeles, R. H., and Lee, H. A., 1960, The dismutation of formaldehyde by liver alcohol dehydrogenase, J. Biol. Chem., 235:1499.

Andersson, P., Kvassman, J., Olden, B., and Pettersson, G., 1984, Catalytic significance of binary enzyme-aldehyde complexes in the liver alcohol dehydrogenase reaction, Eur. J. Biochem., 139:519.

Anderson, I., Bauer, R., and Demeter, I., 1982, Structural information concerning the catalytic metal site in horse liver alcohol dehydrogenase, obtained by perturbed angular correlation spectroscopy on ^{111}Cd, Inorg. Chim. Acta, 67:53.

Brändén, C.-I., Jörnvall, H., Eklund, H., and Furugren, B., 1975, Alcohol dehydrogenases, in: "The Enzymes", Vol. 11, 3rd ed., Boyer, P. D., ed., Academic Press, New York.

Brooks, R. L., and Shore, J. D., 1971, Effect of substrate structure on the rate of the catalytic step in the liver alcohol dehydrogenase mechanism, Biochemistry, 10:3855.

Brooks, R. L., Shore, J. D., and Gutfreund, H., 1972, The effect of pH and temperature on hydrogen transfer in the liver alcohol dehydrogenase mechanism, J. Biol. Chem., 247:2382.

Clary, J. J., Gibson, J. E., and Waritz, R. S., 1983, "Formaldehyde: Toxicology, Epidemiology, Mechanisms", Dekker: New York, N. Y.

Coates, J. H., Hardman, M. J., Shore, J. D., and Gutfreund, H., 1977, Pressure relaxation studies of isomerizations of horse liver alcohol dehydrogenase linked to NAD$^+$ binding, FEBS Lett., 18:3407.

Dalziel, K., and Dickinson, F. M., 1967, The specificities and configurations of ternary complexes of yeast and liver alcohol dehydrogenases, Biochem. J., 104:165.

Dworschack, R. T., and Plapp, B. V., 1977, pH, isotope, and substituent effect on the interconversion of aromatic substrates catalyzed by hydroxybutyrimidylated liver alcohol dehydrogenase, Biochemistry, 16(12):2716.

Eklund, H., Plapp, B. V., Samama, J.-P., and Brändén, C.-I., 1982, Binding of substrate in a ternary complex of horse liver alcohol dehydrogenase, J. Biol. Chem., 257: 14394.

Eklund, H., and Brändén, C.-I., 1983, The role of zinc in alcohol dehydrogenase, in: "Zinc Enzymes", Spiro, T. G., ed., John Wiley and Sons, New York.

Frisell, W. R., Meech, L. A., and Mackenzie, C. G., 1954, A simplified photometric analysis for serine and formaldehyde, J. Biol. Chem., 207:709.

Gulbinsky, J. S., and Cleland, W. W., 1968, Kinetic studies of escherichia coli galactokinase, Biochemistry, 7(2):566.

Hardman, M. J., 1981, The rate-determining step in the liver alcohol dehydrogenase-catalyzed reduction of acetaldehyde is an isomerization of the enzyme, Biochem. J., 195:773.

Hinson, J. A., and Neal, R. A., 1972, An examination of the oxidation of aldehydes by horse liver alcohol dehydrogenase, J. Biol. Chem., 247:7106.

Jacobsen, D., and McMartin, K. E., 1986, Methanol and ethylene glycol poisoning. Mechanism of toxicity, clinical course, diagnosis and treatment, Medical Toxicology, 1:309.

Jörnvall, H., 1970, Horse liver alcohol dehydrogenase. The primary structure of the protein chain of the ethanol-active isoenzyme, Eur. J. Biochem., 16:25.

Kamlay, M. T., and Shore, J. D., 1983, Transient kinetic studies of substrate inhibition in the horse liver alcohol dehydrogenase reaction, Arch. Biochem. Biophys., 222:59.

Kendal, L. P., and Ramanathan, A. N., 1952, Liver alcohol dehydrogenase and ester formation, Biochem. J., 52:430.

Klinman, J. P., 1981, Probes of mechanism and transition-state structure in the alcohol dehydrogenase reaction, Crit. Rev. Biochem., 10:39.

Kvassman, J., and Pettersson, G., 1978, Effect of pH on the process of ternary-complex interconversion in the liver alcohol dehydrogenase reaction, Eur. J. Biochem., 87:417.

Kvassman, J., and Pettersson, G., 1980, Unified mechanism for proton transfer reactions affecting the catalytic activity of liver alcohol dehydrogenase, Eur. J. Biochem., 103:565.

Mani, J., Pietruszko, R., and Theorell, H., 1970, Methanol activity of alcohol dehydrogenases from human liver, horse liver, and yeast, Arch. Biochem. Biophys., 140:52.

Makinen, M. W., and Yim, M. B., 1983, Neutral metal-bound water is the base catalyst in liver alcohol dehydrogenase, Proc. Natl. Acad. Sci. U. S. A., 80:2584.

Pettersson, G., 1986, Ionization properties of zinc-bound ligands in alcohol dehydrogenase, in: "Zinc Enzymes", Vol. 1., Birkhauser Boston, Inc.

Pettersson, G., 1987, Liver alcohol dehydrogenase, Crit. Rev. Biochem., 21:349.

Pocker, Y., 1989, Alcohol dehydrogenase: structure, catalysis, and site-directed mutagenesis, in: "Metal Ions in Biological Systems", Sigel, H. ed., Marcel Dekker, New York.

Pocker, Y., Li, H., and Page, J. D., 1987a, Liver alcohol dehydrogenase: metabolic and energetic aspects, Alcohol and Alcoholism, Suppl., 1:181.

Pocker, Y., Li, H., and Page, J. D., 1987b, Liver alcohol dehydrogenase: substrate orientation. Metabolic activity and energetic of enzyme catalysis, Process in Clinical and Biological Research, 232:217.

Pocker, Y., and Raymond, K. W., 1980, Kinetic and mechanistic studies of oxidation of vitamin A alcohol to vitamin A aldehyde by horse liver alcohol dehydrogenase. The inhibition by ethanol and pyrazole. in: "Alcohol and Aldehyde Metabolizing System-IV", Thurman, R. G., ed., Plenum Press, New York.

Shore, J. D., and Gilleland, M. J., 1970, Binding and kinetic studies of liver alcohol dehydrogenase-coenzyme-pyrazole complexes, J. Biol. Chem., 245:3422.

Sund, H., and Theorell, H., 1963, Alcohol dehydrogenase, in: "The Enzymes", Vol. 7, 2nd ed., Boyer, P. D., Lardy, H., Myrback, K., ed., Academic Press, New York.

Theorell, H., and McKinley-Mckee, J. S., 1961, Mechanism of action of liver alcohol dehydrogenase, Nature, 192:47.

Theorell, H., and Yonetani, T., 1963, Liver alcohol dehydrogenase-DPN-pyrazole complex: a model of a ternary intermediate in enzyme reaction, Biochemische Zeitschrift, 338:537.

Weinberger, W., 1931, A test for aldehyde using dimethylcyclohexanedione, <u>Ind. Eng. Chem. Anal. Ed.</u>, 3:365.

Winer, A. D., 1958, A note on the substrate specificity of horse liver alcohol dehydrogenase, <u>Acta Chem. Scand.</u>, 12:1695.

Wratten, C. C., and Cleland, W. W., 1963, Product inhibition studies on yeast and liver alcohol dehydrogenases, <u>Biochemistry</u>, 2:935.

Wratten, C. C., and Cleland, W. W., 1965, Kinetic studies with liver alcohol dehydrogenase, <u>Biochemistry</u>, 4:2442.

NOVEL SUBSTRATES AND INHIBITORS OF HUMAN LIVER SORBITOL

DEHYDROGENASE

Wolfgang Maret

Center for Biochemical and Biophysical Sciences and
Medicine, Harvard Medical School
Brigham and Women's Hospital, 75 Francis St.
Boston, MA 02115

INTRODUCTION

The formation of D-fructose from sorbitol, first observed
in perfused dog liver (Embden and Griesbach, 1914), was shown
to be catalyzed by sorbitol dehydrogenase (SDH)[1] from rat liver
homogenates (Blakley, 1951). SDH was later purified to homogeneity from sheep liver (Smith, 1962). Several more recent
findings prompted further investigation of structural features
of this enzyme and the role of SDH in pathological conditions.
First, the discovery of a sequence homology between sheep liver
SDH and zinc-containing alcohol dehydrogenases (ADHs) (Jeffery
et al., 1981) led to metal analyses that established SDH as a
metalloenzyme with one zinc atom per subunit of the tetramer
(Jeffery et al., 1984). In contrast, dimeric mammalian ADHs
have two zinc atoms per subunit. Second, sorbitol accumulates
in tissues affected by diabetes (Gabbay, 1973; Greene et al.,
1987) and in the liver and kidney of copper-deficient male rats
as a result of feeding fructose (Fields et al., 1989). The
elevated levels of sorbitol can upset osmoregulation and result
in cellular pathology (Burg and Kador, 1988). Also, the
increased availability of fructose associated with these
conditions causes fructosylation and crosslinking of proteins
(Walton et al., 1989).

In a recent review, Jeffery and Jörnvall (1988) pointed
out that the only primary structural data of SDH exist for the
sheep liver enzyme. Therefore, purification of SDH from human
liver was undertaken (Maret and Auld, 1988) with the aim of
further structural characterization and a study of the relationship of SDH to the three classes of ADHs found in human
liver. A brief account of the properties of human SDH follows.

---

[1]The enzyme nomenclature lists the enzyme as L-iditol:NAD$^+$
2-oxidoreductase (EC 1.1.1.14) because L-iditol is an unambi-
guous substrate having the specific configuration on both ends
of the molecule (McCorkindale and Edson, 1954).

Enzymology and Molecular Biology of Carbonyl Metabolism 3
Edited by H. Weiner *et al.*, Plenum Press, New York, 1990

In the protein sequence of human SDH, 89% of the residues are identical to the sheep liver enzyme (Karlsson et al., 1989). Roughly one quarter of the replacements are in a section of the coenzyme-binding domain. The variability between the human and the sheep enzyme is comparable to that of class I ADHs, but differs from that of class III ADHs, which are more highly conserved. Variants of human SDH had been observed by chromatographic and electrophoretic techniques (Perea et al., 1989; Nealon and Rej, 1983). The first evidence for an isozyme at the molecular level is a microheterogeneity detected at position 237 (Leu/Gln) in the coenzyme-binding domain (Karlsson et al., 1989). Since there is a single gene locus for human SDH (Donald, 1980), this polymorphism is attributed to allelic variation. Two out of the three class I ADHs also have allelic isozymes with mutations in the coenzyme-binding domain. Therefore, it appears that there is a particularly high evolutionary pressure on this part of the molecule.

SDH differs markedly from eukaryotic zinc-containing ADHs (Vallee and Auld, 1990) in zinc binding. Three of the four cysteine ligands that constitute the typical noncatalytic zinc center of ADHs are missing from the primary sequence of SDH, with the result that only one zinc atom binds per subunit (Jeffery et al., 1984; Maret and Auld, 1988). In the catalytic zinc site, computer-assisted fitting of the primary structure of sheep liver SDH to the 3D model of horse liver ADH indicated that one of the two cysteine ligands, which are present in all three classes of ADHs, is replaced by a glutamate (Eklund et al., 1985). To test the computer model and to create a spectroscopic probe for the active site, the catalytic zinc of human liver SDH was exchanged for cobalt (Maret, 1989). The electronic absorption spectra of Co^{2+}-substituted SDH support the model with one cysteine ligand. Furthermore, studies of Co^{2+}-SDH showed that the structural perturbation of the catalytic metal which is induced by coenzyme binding is different from the corresponding effect of the coenzyme in ADHs.

Human liver SDH stereospecifically oxidizes (2R,3R)-2,3-butanediol (Maret and Auld, 1988). Since the three class I ADHs also act on stereoisomers of 2,3-butanediol (Maret, unpublished results), substrates common to both class I ADHs and SDH exist. In addition, overlapping substrate specificities have been noted for the different classes of ADHs. For instance, long-chain primary alcohols are among the best substrates for ADHs from all three classes (Vallee and Bazzone, 1983). Further characterization of human liver SDH should reveal both the substrate specificity of SDH in relation to ADHs and the role of SDH in homeostasis of intracellular sorbitol and fructose concentrations.

This report describes an improved purification of human liver SDH. It also outlines the specificity of SDH toward tetritols and aromatic alcohols as substrates and dithiotetritols as inhibitors. Kinetic data on aromatic alcohols as a new class of substrates corroborate the earlier finding that SDH is a secondary alcohol dehydrogenase with relatively high stereospecificity. The inhibition of SDH by high concentrations of sorbitol could provide an explanation for the accumulation of sorbitol and fructose in complications of diabetes.

MATERIALS AND METHODS

Unless otherwise noted, chemicals were obtained from Sigma Chemical Company (St. Louis, MO) and from Fluka Chemie AG (Buchs, Switzerland). Determination of enzymatic activities, kinetic analyses, and metal-exchange experiments were described elsewhere (Maret and Auld, 1988; Maret, 1989). Substrate inhibition was analyzed as described by Cleland (1979).

Purification of Human Liver Sorbitol Dehydrogenase

Human livers with SDH activity greater than 0.05 units/mg were selected for preparations of the enzyme. All buffers were saturated with nitrogen gas. Human liver (500 g) was homogenized as described earlier (Maret and Auld, 1988) in 500 mL of 10 mM Tris/Cl⁻, pH 7.5, and dialyzed against 15 L of this buffer overnight at 4°C. In a batch procedure at room temperature, the dialysate, readjusted to pH 7.5, was layered onto 700 g of Whatman DE 52 DEAE cellulose (Clifton, NJ) equilibrated in the above buffer in a 2 L Büchner funnel. The resin was washed with the Tris buffer, the first 400 mL of eluate discarded, and a 600-800 mL fraction, containing between 70 and 90% of the activity, collected. All subsequent steps were performed at 4°C. The enzyme solution was subjected to ammonium sulfate fractionation (Maret and Auld, 1988), at which point the precipitated enzyme can be stored frozen at -70°C. The pellet was dissolved in 50 mL of 20 mM phosphate, pH 7.4, and dialyzed against 15 L of this buffer overnight. The sample was then concentrated to 15 mL, adjusted to 1 mM in NAD⁺, and applied at 1.5 mL/min to a CapGapp-sepharose column (50 x 60 mm) equilibrated with 50 mM phosphate, pH 7.4, 1 mM NAD⁺. In this affinity chromatography step all class I ADHs are bound to the resin (Lange and Vallee, 1976) whereas SDH is not bound under these conditions. However, due to weak interaction of SDH with the resin, a partial separation from the breakthrough of bulk protein can be achieved. The pooled SDH activity corresponding to 10% peak height before the peak and including all of the tailing activity was concentrated to 100 mL and dialyzed overnight against 15 L of 5 mM Hepes, pH 7.5. The sample was then loaded at 1 mL/min onto an Accell CM column (15 x 300 mm; Millipore, Waters Associates, Milford, MA), the column washed at 3 mL/min until the optical density at 280 nm was below 0.01 and then developed with a linear gradient from 0 to 0.1 M NaCl in five times the column volume. The pooled active fractions were concentrated to 10 mL and injected (2.5 mL/min) at room temperature into an HPLC system with a Protein Pak SP 5PW column (21.5 x 150 mm; Waters Associates) equilibrated with 1 mM sodium phosphate, pH 7.7 (buffer A). Buffer B contained 0.1 M sodium phosphate, pH 7.7. The gradient was programmed as follows: 10 min 100% A, 20 min 90% A, 80 min 78% A, 290 min 40% A and 300 min 0% A. The given times are end times. The enzyme can be stored for two weeks under a nitrogen blanket at 4°C without any decrease in activity.

Subcellular Fractionation

Fractionation by differential centrifugation without separating light and heavy mitochondria (Graham, 1984) was performed on unfrozen human liver, obtained 12 hours post mortem. The following marker enzymes were used: NADPH-cytochrome c oxidoreductase, EC 1.6.2.4 (endoplasmic

Table 1. Purification of Human Liver Sorbitol Dehydrogenase[a]

Step	Volume (mL)	Protein (mg/mL)	Activity (units/mL)	Specific Activity (units/mg)	Overall Purification (X-fold)	Overall Recovery (%)
Crude extract	510	25	1.5	0.06	/	/
DEAE-cellulose	590	2.9	0.8	0.28	4.7	62
Ammonium sulfate fractionation	84	9.9	4.7	0.47	7.8	52
CapGapp-sepharose	63	3.9	4.7	1.2	20	39
Accell CM	20	4	14.1	3.5	58	37

[a]From 415 g of human liver

reticulum), isocitrate dehydrogenase, EC 1.1.1.41 (mitochondrial matrix), and cytochrome c oxidase, EC 1.9.3.1 (mitochondrial membrane).

RESULTS AND DISCUSSION

Enzyme Purification

A new purification scheme is described for 500 g of human liver (Table 1). It employs batch anion exchange and column affinity chromatographic steps and is more suitable for large preparations than the tandem column technique using triazine dye chromatography (Maret and Auld, 1988). Though SDH is not bound to the support in either step, sufficient purification is achieved for a final separation by conventional and HPLC cation exchange chromatography as described (Maret and Auld, 1988). The overall recovery after the Accell CM step is 37%. In comparison, the purification protocol reported previously afforded 15%. The specific activity of SDH obtained upon re-chromatography was 11.3 units/mg, corresponding to a 188-fold purification.

The subcellular distribution of SDH activity is wider than that for class III ADH (Table 2). About 8% of SDH activity was found in the microsomal fraction whereas human liver ADHs are almost exclusively cytosolic. Though the association of SDH with "microsomes" could be an artifact, it may also imply a functional link to the pentose phosphate pathway which has been localized to the endoplasmic reticulum (Bublitz and Steavenson, 1988). SDH oxidizes xylitol to D-xylulose, which enters the pentose phosphate pathway in its phosphorylated form.

Inhibitors and Substrates

Dithiothreitol and dithioerythritol have been widely used as sulfhydryl-protecting reagents for ADHs, including SDH. During the preparation of human SDH it was observed that dithio-threitol (Cleland's Reagent) inhibits the enzymatic activity.

Table 2. Subcellular Fractionation of Human Liver Homogenate

Fraction	Class III ADH[a] Total Activity (%)	SDH Total Activity (%)
Homogenate	100	100
Cellular debris	not determined	25
Mitochondria	2	2
Mitochondria-wash	0	15
Microsomes	2	8
Cytosol	70	55
Recovery	74	105

[a]Assayed with 0.5 mM 12-hydroxydodecanoic acid, 2.5 mM NAD$^+$, 0.1 M glycine, pH 10.0, 5 mM 4-methylpyrazole

Table 3. Interaction of Human Liver Sorbitol Dehydrogenase with Tetritols and Dithiotetritols[a]

Tetritol Substrates	k_{cat} (s^{-1})	K_m (mM)	Dithiotetritol Inhibitors[b]	K_I (μM)
Erythritol	/	/	Dithioerythritol	23
L-Threitol	0.31	9.1	L-Dithiothreitol	11
D-Threitol	/	/	D-Dithiothreitol	n.d.[c]

[a]Determined in 10 mM phosphate, pH 7.4, 1 mM NAD$^+$
[b]Sorbitol as substrate
[c]not determined

Therefore, the inhibition of SDH by dithiotetritols was studied. Both dithioerythritol and L-dithiothreitol inhibit sorbitol oxidation competitively with micromolar inhibition constants (Table 3). "Oxidized" DTT, trans-4,5-dihydroxy-1,2-dithiane, acted as a mixed inhibitor (K_I = 3.7 mM, from a slope replot). Among the three stereoisomers of tetritols only L-threitol is a substrate for SDH (Table 3); D-threitol and erythritol are neither substrates nor inhibitors at a concentration of 50 mM.

Preliminary studies indicated that the substrate specificity of human liver SDH extends to secondary alcohols that are not polyhydric alcohols (Maret and Auld, 1988). Since styrene glycol (1-phenyl-1,2-ethanediol) was found as a substrate for Candida utilis SDH (Dr. Siv Sellin, personal communication), styrene glycol and other aromatic alcohols were tested as substrates or inhibitors of human liver SDH (Table 4). Indeed, (R)-styrene glycol is a substrate for human liver SDH with a k_{cat}/K_m value of 2.8 x 10^4 M^{-1}s^{-1}. To obtain a quantitative measure of the extent of stereospecificity, initial velocity data were collected for the S enantiomer under the condition [S]<<K_m. From these data a k_{cat}/K_m value of 3.1 M^{-1}s^{-1} was determined. Thus, the stereospecificity toward styrene glycols as expressed by the enantiomeric ratio (the ratio between k_{cat}/K_m values of each enantiomer) is about 1 x 10^4. Compounds bearing hydroxy or methoxy substituents on the aromatic ring, e.g., (-)-epinephrine, D,L-synephrine, D,L-metanephrine, D,L-normetanephrine, or D,L-norepinephrine, are not substrates, with the notable exception of D,L-octopamine. (S)-sec-Phenethyl alcohol is a substrate (Table 4), but phenethyl alcohol is not. Likewise, racemic 2-amino-1-phenylethanol is a substrate (Table 4) whereas (S)-2-amino-2-phenylethanol (2-phenylglycinol) is a partial noncompetitive inhibitor ($K_{I(app)}$ = 2 mM). A comparison of the structure of these compounds strongly suggests that it is the secondary alcohol group at carbon atom C-2 in (R)-styrene glycol that is being oxidized. At C-1, an amino or hydroxy function or even the absence of a functional group such as in sec-phenethyl alcohol or 2,3-butanediol is compatible with the compound being a substrate (Maret and Auld, 1988; Dills et al., 1976).

In summary, these studies establish aromatic alcohols as a new class of substrates for SDH. Furthermore, the data lend additional support to the notion that SDH is a secondary alcohol

Table 4. Substrate Specificity of Human Liver Sorbitol Dehydrogenase toward Aromatic Alcohols[a]

Compound	Stereoisomer	k_{cat} (s^{-1})	K_m (M)
1-Phenyl-1,2-ethanediol	R	4.2	1.5×10^{-4}
	S	/	/
2-Amino-1-phenylethanol	racemic	2.5	2.5×10^{-3}
sec-Phenethyl alcohol	R	/	/
	S	0.4	9.3×10^{-3}

[a]Determined in 0.1 M glycine, pH 10.0, 1 mM NAD$^+$

dehydrogenase with relatively high stereospecificity. Aromatic secondary alcohols occur in the metabolism of aromatic amino acids. An involvement of SDH in the pathways of the latter has not yet been demonstrated. However, aromatic alcohols are also formed in the metabolism of styrene and ethylbenzene. Here, the action of SDH on aromatic alcohols provides an explanation for the stereospecificity observed for the detoxication of ethylbenzene (Sullivan et al., 1976; Drummond et al., 1989). Ethylbenzene is excreted exclusively as (R)-mandelic acid. This disposition is thought to proceed according to the sequence: Ethylbenzene→sec-phenethyl alcohol→acetophenone→ω-hydroxyacetophenone→1-phenyl-1,2-ethanediol(or phenylglyoxal)→(R)-mandelic acid (Drummond et al., 1990). The data in Table 4 show that SDH is expected to control the stereochemistry of at least two steps in the transformation of ethylbenzene: The enantiospecific oxidation of only (S)-sec-phenethyl alcohol; and, by inference from the observed oxidation of (R)-1-phenyl-1,2-ethanediol, the enantioselective reduction of ω-hydroxyacetophenone.

Substrate Inhibition

SDH is completely inhibited at high concentrations of sorbitol (Fig. 1) with an inhibition constant of 0.8 M. The most plausible explanation for this inhibition is the formation of the dead-end complex SDH/NADH/sorbitol. Evidence for the existence of such a species is provided by the appearance of a new electronic transition at 486 nm in the absorption spectrum of the complex Co^{2+}-SDH/NADH upon binding of sorbitol (Fig. 2).

The accumulation of sorbitol in diabetic tissue is currently ascribed to activation of the polyol pathway. However, an inhibition of SDH at high levels of sorbitol is observed (Fig. 1). In another investigation, the dissociation constant of sorbitol from the sheep liver SDH/NADH/sorbitol complex was determined as 45 mM by product inhibition studies (Christensen et al., 1975). Taken together, substrate inhibition of SDH is expected at sorbitol concentrations in the range 45-800 mM. Recent evidence suggests that intracellular sorbitol levels can accumulate to this range. As an example, sorbitol has been reported to reach "high millimolar levels" in the diabetic ocular lens (Greene et al., 1987). Also, the intracellular sorbitol concentration in an epithelial cell line from rabbit renal papilla rose to about 0.25 M when subjected to a hyper-

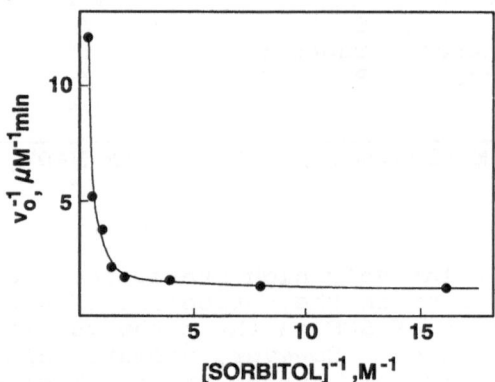

Fig. 1. Inhibition of human liver sorbitol dehydrogenase by
         sorbitol under initial velocity conditions. The graph
         shows a double reciprocal plot in the range where
         substrate inhibition is observed. Enzymatic assays
         were performed at 25°C in 10 mM phosphate, pH 7.4,
         containing 1 mM NAD^+.

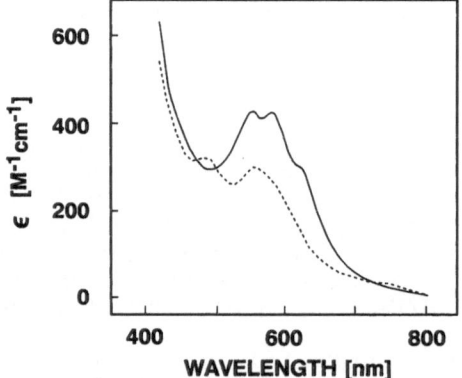

Fig. 2. Electronic absorption spectra of Co^{2+}-substituted human
         liver sorbitol dehydrogenase (0.3 mM $Co_{0.7}Zn_{0.4}$-SDH).
         Solid line: binary complex SDH/NADH (9 mM NADH); dashed
         line: ternary complex SDH/NADH/sorbitol (0.6 M
         sorbitol); buffer: 0.1 M Hepes, pH 7.0.

osmotic medium (Bagnasco et al., 1987). Therefore, the substrate inhibition of SDH could be one mechanism that contributes to accumulation of sorbitol in complications of diabetes.

ACKNOWLEDGMENT

This work was supported by a grant from the Samuel Bronfman Foundation, Inc., to the Endowment for Research in Human Biology, Inc., with funds provided by Joseph E. Seagram and Sons, Inc. I am very grateful to Dr. B. L. Vallee for his advice and encouragement. The excellent technical assistance of Douglas S. Auld and discussions with Dr. T. French are also acknowledged. I thank Dr. K. Edman for performing experiments on the subcellular distribution of alcohol dehydrogenases and sorbitol dehydrogenase.

REFERENCES

Bagnasco, S. M., Uchida, S., Balaban, R. S., Kador, P. F., and Burg, M. B., 1987, Induction of aldose reductase and sorbitol in renal medullary cells by elevated extracellular NaCl, Proc. Natl. Acad. Sci. USA, 84:1718.

Blakley, R. L., 1951, The metabolism and antiketogenic effects of sorbitol. Sorbitol dehydrogenase, Biochem. J., 49:257.

Bublitz, C., and Steavenson, S., 1988, The pentose phosphate pathway in the endoplasmic reticulum, J. Biol. Chem., 263:12849.

Burg, M. B., and Kador, P. F., 1988, Sorbitol, osmoregulation, and the complications of diabetes, J. Clin. Invest., 81:635.

Christensen, U., Tüchsen, E., and Andersen, B., 1975, Initial velocity and product inhibition studies on L-iditol:NAD oxidoreductase, Acta Chem. Scand., Ser. B, 29:81.

Cleland, W. W., 1979, Substrate inhibition, Methods Enzymol., 63:500.

Dills, W. L., and Meyer, W. L., 1976, Studies on 1-deoxy-D-fructose, 1-deoxy-D-glucitol, and 1-deoxy-D-mannitol as antimetabolites, Biochemistry, 15:4506.

Donald, L. J., Wang, H. S., and Hamerton, J. L., 1980, Assignment of sorbitol dehydrogenase locus to human chromosome 15PTER→Q21, Biochem. Genet., 18:425.

Drummond, L., Caldwell, J., and Wilson, H. K., 1989, The metabolism of ethylbenzene and styrene to mandelic acid: stereochemical considerations, Xenobiotica, 19:199.

Drummond, L., Caldwell, J., and Wilson, H. K., 1990, The stereoselectivity of 1,2-phenylethanediol and mandelic acid metabolism and disposition in the rat, Xenobiotica, 20:159.

Eklund, H., Horjales, E., Jörnvall, H., Brändén, C.-I., and Jeffery, J., 1985, Molecular aspects of functional differences between alcohol and sorbitol dehydrogenases, Biochemistry, 24:8005.

Embden, G., and Griesbach, W., 1914, Über Milchsäure- und Zuckerbildung in der isolierten Leber, Hoppe-Seyler's Z. Physiol. Chem., 91:251.

Fields, M., Lewis, C. G., and Beal, T., 1989, Accumulation of sorbitol in copper deficiency: Dependency on gender and type of dietary carbohydrate, Metab., Clin. Exp., 38:371.

Gabbay, K. H., 1973, The sorbitol pathway and the complications of diabetes, NE J. Med., 288:831.

Graham, J., 1984, Isolation of subcellular organelles and membranes, in: Centrifugation, Rickwood, D., ed., p. 161, IRL Press, Oxford, U.K.

Greene, D. L., Lattimer, S. A., and Sima, A. A. F., 1987, Sorbitol, phosphoinositides, and sodium-potassium-ATPase in the pathogenesis of diabetic complications, NE J. Med., 316:599.

Jeffery, J., Cummins, L., Carlquist, M., and Jörnvall, H., 1981, Properties of sorbitol dehydrogenase and characterization of a reactive cysteine residue reveal unexpected similarities to alcohol dehydrogenases, Eur. J. Biochem., 120:229.

Jeffery, J., Chesters, J., Mills, C., Sadler, P. J., and Jörnvall, H., 1984, Sorbitol dehydrogenase is a zinc enzyme, EMBO J., 3:357.

Jeffery, J., and Jörnvall, H., 1988, Sorbitol dehydrogenase, Adv. Enzymol., 61:47.

Karlsson, C., Maret, W., Auld, D. S., Höög, J.-O., and Jörnvall, H., 1989, Variability within mammalian sorbitol dehydrogenases. The primary structure of the human liver enzyme, Eur. J. Biochem., 186:543.

Lange, L. G., and Vallee, B. L., 1976, Double-ternary complex affinity chromatography: Preparation of alcohol dehydrogenases, Biochemistry, 15:4681.

Maret, W., and Auld, D. S., 1988, Purification and characterization of human liver sorbitol dehydrogenase, Biochemistry, 27:1622.

Maret, W., 1989, Cobalt(II)-substituted class III alcohol and sorbitol dehydrogenases from human liver, Biochemistry, 28:9944.

McCorkindale, J., and Edson, N. L., 1954, Polyol dehydrogenases. The specificity of rat-liver polyol dehydrogenase, Biochem. J., 57:518.

Nealon, D. A., and Rej, R., 1983, Human liver sorbitol dehydrogenase - Evidence for two forms, in: Selected Topics in Clinical Enzymology, Werner, M., and Goldberg, D. M., eds., Vol. 2, p. 535, de Gruyter, New York.

Perea, F. J., Vaca, G., Alvarez, C., Cantu, J. M., and Ibarra, B., 1989, Electrophoretic pattern of sorbitol dehydrogenase (EC 1.1.1.14) in human seminal plasma and spermatozoa, Ann. Genet., 32:33.

Smith, M. G., 1962, Polyol dehydrogenases. Crystallization of the L-iditol dehydrogenase of sheep liver, Biochem. J., 83:135.

Sullivan, H. R., Miller, W. M., and McMahon, R. E., 1976, Reaction pathways of in vivo stereoselective conversion of ethylbenzene to (-)mandelic acid, Xenobiotica, 6:49.

Vallee, B. L., and Bazzone, T. J., 1983, Isozymes of human liver alcohol dehydrogenase, in: Current Topics in Biological & Medical Research, Rattazzi, M. C., Scandalios, J. G., and Witt, G. S., eds., Vol. 8, p. 219, Liss, New York.

Vallee, B. L., and Auld, D. S., 1990, Zinc coordination, function, and structure of zinc enzymes and other proteins, Biochemistry, 29:5647.

Walton, D. J., McPherson, J. D., and Shilton, B. H., 1989, Fructose mediated crosslinking of proteins, in: The Maillard Reaction in Aging, Diabetes and Nutrition, Baynes, J. W., and Monnier, V. M., eds., p. 163, Liss, New York.

COMPARISON OF THE PRIMARY STRUCTURES OF

NAD(P)-DEPENDENT BACTERIAL ALCOHOL DEHYDROGENASES

Martin R. Wales and Charles A. Fewson

Department of Biochemistry
University of Glasgow
Glasgow G12 8QQ, Scotland, U.K.

INTRODUCTION

Many soluble NAD(P)-dependent alcohol dehydrogenases have been identi-fied in bacteria (MacKintosh and Fewson, 1987). They have a wide range of substrate specificities and kinetic constants and they are subject to very varied genetic regulation. Traditionally, classification of these enzymes has been according to their substrate specificities (I.U.B.-I.U.P.A.C., 1979). In order to establish the evolutionary relationships that exist amongst these enzymes, it is necessary to obtain adequate structural information (Rossman et al., 1975, Dayhoff et al., 1978). This review aims to summarise the amino acid sequence data that are available for bacterial NAD(P)-dependent alcohol dehydrogenases and to compare the structures of these enzymes, both with each other and with the eukaryotic alcohol dehydrogenases for which more information is available (Jornvall, 1986; Yokoyama et al., 1990).

There are three well established groups of NAD(P)-dependent alcohol dehydrogenases in eukaryotes and prokaryotes. The long-chain, zinc-dependent alcohol dehydrogenases, e.g. horse liver alcohol dehydrogenase (HLADH), were the first to be identified and are the best characterised group (Branden et al., 1975; Yokoyama et al., 1990). Three complete amino acid sequences for bacterial long-chain, zinc-dependent alcohol dehydrogenases have been deduced so far, and some N-terminal sequences have been determined. Short-chain, non-zinc alcohol dehydrogenases are present in various Drosophila spp (e.g. Villaroya et al., 1989) and two related polyol/sugar dehydrogenases have been identified in some bacteria (Jornvall et al., 1984). More recently, a group of iron-activated alcohol dehydrogenases has been discovered. These are mostly bacterial enzymes and several amino acid sequences are available. The only eukaryotic member of this group known so far is ADH4 from the yeast Saccharomyces cerevisiae (Williamson and Paquin, 1987). It has been proposed that these alcohol dehydrogenases and related enzymes constitute a super-family of enzymes, variations in structure being seen at different levels (Jornvall, 1986). Whether there are more groups of NAD(P)-dependent alcohol dehydrogenases remains to be seen but there is some evidence for a fourth group, at least in bacteria (Goodlove et al., 1989; M.R. Wales, J.N. Keen and C.A. Fewson, unpublished results).

The eukaryotic long-chain (approx. 370 amino acids), zinc-dependent alcohol dehydrogenases show three levels of identity of primary structure. Firstly, enzymes in one class, such as the class 1 dimeric enzymes in man and horse, have > 90% residue conservation. Secondly, enzymes of different classes, e.g. the plant and animal enzymes, show 40-60% identity. Finally, alcohol and polyol dehydrogenases, e.g. the dimeric and the tetrameric enzymes, show about 25% identity (Jornvall et al., 1987a).

Many bacterial alcohol dehydrogenases have been reported to require zinc for activity, or to contain zinc. However, not many of these enzymes have been sequenced. Complete amino acid sequences have been obtained for only the NAD-dependent, fermentative alcohol dehydrogenase from Alcaligenes eutrophus (AEADH; Jendrossek et al., 1988), the type 1 NAD-dependent alcohol dehydrogenase from Zymomonas mobilis (ZMADH1; Keshav et al., 1990) and the NADP-dependent secondary alcohol dehydrogenase from Thermoanaerobium brockii (TBADH; Peretz and Burstein, 1989) which is the first NADP-dependent enzyme in this group to have been sequenced so far. In addition, N-terminal sequences of three bacterial alcohol dehydrogenases have been found to show significant homologies with HLADH; these are Bacillus stearothermophilus alcohol dehydrogenase (BSADH; Bridgen et al., 1971; Jeck et al., 1979) and the chromosomally-encoded benzyl alcohol dehydrogenase of Acinetobacter calcoaceticus (ACBADH) and the TOL plasmid-encoded benzyl alcohol dehydrogenase of Pseudomonas putida (TOLBADH, Chalmers et al., 1990; R.M. Chalmers, J.N. Keen and C.A. Fewson, unpublished results). All six enzymes are tetramers with subunits of molecular mass about 40 kDa. The presence of zinc has been unambiguously identified only in ZMADH1 (Neale et al., 1986). The inability of chelating agents to inhibit TOLBADH or ACBADH suggests that the zinc is somehow protected or that the enzymes are not, in fact, zinc-dependent (A.J. Scott and C.A. Fewson, unpublished results; MacKintosh and Fewson, 1987).

N-Terminal Amino Acid Sequences

The N-terminal sequences of all the bacterial alcohol dehydrogenases of this group share between about 15% and 35% identity with HLADH, but there is considerably more variation than this within the bacterial sequences.

Comparison with HLADH. TBADH shares the most identity (36%) with HLADH, followed by BSADH (30%), ZMADH1 (29%), TOLBADH and ACADH (25%), then AEADH (16%). Of those residues strictly conserved amongst eukaryote sequences, Glu 49 (throughout this section numbers refer to the position in HLADH, unless stated otherwise) is completely conserved and Cys 46 probably so. Pro 31 is conserved or replaced by isoleucine or leucine. Glu 35 appears to vary the most, being conserved in two, replaced by aspartate in two, glutamine in one and phenylalanine in another. Several other residues are almost completely conserved with respect to HLADH, such as Lys 10, Ala 11 and 12 and Val 13.

Comparisons amongst the bacterial dehydrogenases. There are no strictly conserved residues in the N-terminal sequences of these enzymes other than those mentioned above. ZMADH1 and BSADH are the most similar to each other (45% identity). TOLBADH and AEADH share 36% identity, TBADH and AEADH share 32% identity, BSADH shares about 30% identity with both AEADH and TOLBADH, but the latter two are only 23% identical with each other. No other significant identities exist and some pairs of enzymes are considerably less similar to each other e.g. TBADH and TOLBADH share only 8% identity. Overall, ZMADH1 seems to be least related to the others.

Zinc Ligands

As in the eukaryotic enzymes, two of the catalytic zinc ligands are preserved in TBADH (Peretz and Burstein,1989), ZMADH1 (Keshav et al., 1990) and AEADH (Jendrossek et al.,1988), and these are Cys 46 and His 57. Unlike HLADH, there is no evidence for a structural, non-catalytic zinc atom in any of the bacterial enzymes fully sequenced. TBADH lacks a stretch of 18 amino acids, corresponding to positions 96-113 in HLADH which contain the structural zinc ligands (Cys 97, 100, 103, 111). ZMADH1 contains the structural zinc ligands but seems to lack the structural zinc itself (Neale et al.,1986). AEADH has retained only one of these possible ligands, Cys 97. In this respect, TBADH and AEADH can be considered to be analogous to sheep liver sorbitol dehydrogenase (SLSDH), a related polyol dehydrogenase which has all the catalytic zinc ligands but only one structural zinc ligand, Cys 103, and has only one zinc atom/subunit (Jeffrey and Jornvall, 1988).

Cofactor binding

NAD binding has been studied in detail in HLADH. The important residues in terms of side chain interactions with the cofactor are: Asp 223 which hydrogen bonds to the adenosine ribose, and is conserved throughout this family; Lys 228 which hydrogen bonds to the 3'-oxygen of the adenosine ribose, and is conserved or replaced by arginine; Ser 49 and His 51 which hydrogen bond nicotinamide ribose, the serine being conserved or replaced by threonine whereas the histidine is conserved or replaced by tyrosine (mostly) or serine; and Arg 369 which hydrogen bonds a phosphate oxygen, and is conserved or replaced by lysine. All these residues are conserved or conservatively replaced throughout the long-chain, zinc-dependent alcohol dehydrogenases (Eklund, 1988). The GxGxxG fingerprint region of the NAD binding domain is also conserved throughout these enzymes (Wierenga et al., 1985).

Most of the coenzyme binding regions in HLADH are conserved or conservatively replaced in AEADH, ZMADHI and TBADH.

Considering those residues whose side chains bind NAD, in ZMADH1 His 51, Asp 223, Lys 228 and Arg 369 are all conserved. Ser 48 is conservatively replaced by Thr. Also the GxGxxG fingerprint region is conserved (Keshav et al., 1990). In AEADH, His 51 and Arg 223 are conserved, Ser 48 is replaced by threonine, Lys 228 by arginine and Arg 369 by lysine, all conservative substitutions(Jendrossek et al., 1988). The fingerprint region of the NAD binding domain is altered somewhat, being AxGxxG, which has structural implications for the enzyme in that the turn between the end of a β sheet and the α helix of the βαβ fold of the NAD binding domain will be less tight.

TBADH is an NADP-dependent enzyme, therefore some variation from HLADH in this region is to be expected. Ser 48 and His 51 are both conserved and Arg 369 is replaced by lysine as in AEADH. The main changes concern those residues interacting with the adenosine moiety. Asp 223 is replaced by the much smaller, uncharged glycine (198 in TBADH); this is presumably because the extra phosphate group on the 2'-oxygen of ribose would not otherwise allow proper orientation of the NADP. Also Asn 225 is replaced by arginine, the positive charge presumably interacting with the phosphate. Arg 228 is replaced by cysteine which presumably has a different function, if any, in cofactor binding. NADP-dependent enzymes often contain a GxGxxA region (Scrutton et al., 1990); however, although TBADH uses NADP it contains a GxGxxG region (Peretz and Burstein, 1989).

The long-chain, zinc-dependent alcohol dehydrogenases have been reported to transfer a hydrogen to the proR position of NAD (Schneider-Bernlöhr et al., 1986).

Conserved Residues

In long-chain, zinc-dependent alcohol dehydrogenases there are 22 strictly conserved residues (Jornvall et al., 1987a). Most predominant among them are 11 glycines and this presumably reflects the structural importance of this residue. Two prolines are also conserved, and they may have some secondary structural role. Two cysteines and one histidine are conserved (see Zinc ligands section), as are five acidic but no basic residues and one valine (Jeffrey and Jornvall, 1988).

Over the entire primary structures, ZMADH1, TBADH and AEADH share 34%, 27% and 15% identity with HLADH, respectively. There is 35% identity between TBADH and AEADH, 25% between AEADH and ZMADH1 and about 16% between TBADH and ZMADH1.

Of the 22 residues conserved in the eukaryotic enzymes, all but the two prolines are conserved in ZMADH1. Presumably in this case they do not have an essential secondary structural role to play. Eighteen residues are conserved in TBADH i.e. 11 glycines, two aspartates, one proline, one cysteine, one histidine and one glutamate. In AEADH only 15 of these 18 are conserved, with two glycines and one aspartate being replaced by two alanines and one glutamate respectively.

No secondary structural studies have been made on the bacterial enzymes but presumably they resemble HLADH with about equal amounts of α helix and β sheet.

BACTERIAL SHORT-CHAIN, NON-ZINC ALCOHOL DEHYDROGENASES

Two bacterial polyol/sugar dehydrogenases reported to be members of the short-chain (approx. 250 amino acids), non-zinc group of NAD(P)-dependent alcohol dehydrogenases have been sequenced. They are ribitol dehydrogenase from Klebsiella aerogenes (Morris et al., 1974) and glucose dehydrogenase from Bacillus megaterium (Jornvall et al., 1984). The former enzyme has been extensively studied in terms of its evolution in the laboratory (Hartley, 1985). There appear to be two areas of homology common to the Drosophila spp. alcohol dehydrogenase and the bacterial polyol/sugar dehydrogenases. Firstly, the cofactor binding region, which contains the GxGxxG fingerprint motif for NAD binding, is situated close to the N-terminal end of these enzymes. Secondly, there are homologous sequences within a small region of the C-terminal halves of these enzymes, but the function of this region is not known (Jornvall et al., 1981).

Unlike the long-chain, zinc-dependent enzymes, these enzymes transfer hydrogen to the proS position of NAD (Schneider-Bernlohr et al., 1986).

IRON-ACTIVATED ALCOHOL DEHYDROGENASES

This is a recently discovered group of alcohol dehydrogenases with four members to date, three bacterial and one from yeast. The first sequence to be published for an enzyme in this group was that for an alcohol dehydrogenase from Zymomonas mobilis which is NAD-dependent (ZMADH2; Conway et al., 1987). Since then the amino acid sequences of ADH4 from the yeast S. cerevisiae (YADH4; Williamson and Paquin, 1987), an NADP-dependent butanol-ethanol dehydrogenase from Clostridium acetobutylicum (CADH; Youngleson et al., 1989) and an NAD-dependent propanediol oxidoreductase from Escherichia coli (EPOR; Conway and Ingram, 1989) have been deduced from their gene sequences.

All four enzymes have subunits consisting of about 380 amino acids with molecular mass values of around 40 kDa. However, two are dimers and one is a tetramer. The presence of iron has been identified unambiguously only in ZMADH2 (Neale et al., 1986). The tertiary structure of YADH4 is not known and it is expressed only when a transposable element, Ty, is inserted in the yeast genome (Williamson and Paquin, 1987).

Conserved Residues

YADH4 and ZMADH2 share the greatest identity (54%) with each other (Williamson and Paquin, 1987). ZMADH2 and YADH4 each share about 40% identity with EPOR (Conway et al., 1989) and CADH (Youngleson et al., 1989). EPOR and CADH are the most distantly related, with 32% identity between them.

Comparison of all four sequences shows that 78 residues are strictly conserved. As in the other groups, glycine is the single most conserved residue, 17 of them being conserved. Of the others, 38 branched chain residues are conserved, as are nine acidic and two basic residues, seven prolines, six residues containing hydroxyl groups, four residues containing thiol groups and one aromatic residue.

Cofactor Binding

No "fingerprint" region for cofactor binding has been identified in any of these enzymes, implying that a structure different from the classical Rossman fold is involved in cofactor binding.

ZMADH2 transfers the proR hydrogen of NADH, as do the long-chain, zinc-dependent enzymes. This is energetically less favourable than proS transfer and is presumably facilitated by destabilisation of the carbonyl group by the metal ion (Glasfield and Benner, 1989). In both cases it has been postulated that the metal ion destabilises the bound carbonyl group thus allowing proR hydrogen transfer (the conformation of NADH allowing proR hydrogen transfer is a weaker reducing agent than the conformation which permits proS transfer). This may be an example of convergent evolution in these two groups of alcohol dehydrogenases (Glasfield and Benner, 1989).

Metal Binding

Histidine and cysteine residues have been implicated as metal ion binding ligands (Jornvall et al., 1987b) and three of the four conserved histidines lie between residues 260 and 280 in these enzymes, so they may be iron binding ligands (Peretz and Burstein, 1989).

Secondary Structure Prediction

An attempt has been made to predict the secondary structure of CADH, YADH4 and ZMADH2. These enzymes show a preference for forming α helices, whereas the other two groups of NAD(P)-linked alcohol dehydrogenases are predicted to have equal amounts of α helix and β sheet (Youngleson et al., 1989).

Extended helical structures are predicted between His 196 and Cys 247, and hydropathy plots indicate a major hydrophilic portion between residues 230 and 240, implying that it is a surface region of the protein. This region lies immediately prior to the postulated metal binding region and may play a role in catalysis (Youngleson et al., 1989).

A POSSIBLE FOURTH GROUP OF NAD(P)-DEPENDENT ALCOHOL DEHYDROGENASES

The only other bacterial NAD(P)-dependent alcohol dehydrogenase sequence to have been published to date is that of a fermentative alcohol dehydrogenase in E. coli encoded by the adhE gene (Goodlove et al., 1989; Clark, 1989). The protein encoded by this gene has both NAD-linked alcohol dehydrogenase and CoA-linked acetaldehyde dehydrogenase activities and is a dimer of subunit molecular mass 96 kDa containing 891 amino acids. No homologous sequences to this enzyme have been reported in the literature. However, the N-terminal amino acid sequence of an NADP-dependent alcohol dehydrogenase from A. calcoaceticus has recently been determined (M.R. Wales, J.N. Keen and C.A. Fewson unpublished results) and it shows 32% identity with residues 31-68 of the E. coli enzyme. These two enzymes may therefore be the first to be identified of a fourth group of NAD(P)-dependent alcohol dehydrogenases.

OTHER TYPES OF BACTERIAL ALCOHOL DEHYDROGENASES

In addition to those enzymes which use NAD or NADP as cofactors, there are bacterial alcohol dehydrogenases which are linked to pyrrolo-quinoline quinones, flavins or haem centres (MacKintosh and Fewson, 1987). Gram-negative methylotrophic bacteria contain periplasmic, NAD(P)-independent, dimeric methanol dehydrogenases which contain one pyrrolo-quinoline quinone group per subunit. The structural genes for methanol dehydrogenase from Paracoccus denitrificans (Harms et al., 1987) and Methylobacterium organophilum (Machlin and Hanson, 1988) have been sequenced. Both enzymes contain 599 amino acid residues per subunit with a molecular mass of about 66 kDa. There is extensive homology between the two enzymes at the DNA level (up to a maximum of 82% in sections), and an even more pronounced homology between the amino acid sequences. Enzymes from both species are synthesized with signal peptides which are apparently cleaved between alanine and asparagine residues. There is a 32-residue hydrophobic segment near the C-terminus of the M. organophilum enzyme which may anchor it to the outer face of the inner bacterial membrane.

It is an open question as to whether any of these NAD(P)-independent enzymes is evolutionarily related to the NAD(P)-dependent alcohol dehydrogenases.

CONCLUSIONS

Sequence homologies reported amongst the various groups of NAD(P)-dependent alcohol dehydrogenases are:

i) The NAD binding domain "fingerprint" motif (Wierenga et al., 1985) which is present towards the C terminal end (about position 200) of the long-chain, zinc-dependent enzymes (Eklund, 1985) and close to the N-terminal end of the short-chain, non-zinc enzymes (Benyajati et al., 1981). This motif is also present in other dehydrogenases (Scrutton et al., 1990) and presumably indicates no closer relationship between these two groups of enzymes than amongst other dehydrogenases.

(ii) The N-terminal regions of the iron-activated ZMADH and the zinc-dependent long-chain YADH1 share 20% identity; however, there appear to be no further significant similarities between these two enzymes (Neale et al., 1986). In addition, a novel NAD-dependent methanol dehydrogenase from a methylotrophic Bacillus sp., which contains one zinc atom per subunit and requires magnesium ions for activity, shares about 25% identity in the N-terminal region with both the iron-activated enzymes ZMADH2 and yeast ADH4 (J. Vonck, N. Arfman, G.E. de Vries, J. van Beeuman, E.F.J. van Bruggen and L. Dijkhuizen, private communication). However, it is not yet known whether other parts of this enzyme are homologous with the long-chain zinc-dependent dehydrogenases.

Much more work will be necessary before we can fully understand the evolutionary relationships amongst the bacterial NAD(P)-dependent alcohol dehydrogenases. As well as the multitude of known enzymes which could be sequenced, existing knowledge about primary structures needs to be put in the context of structural and mechanistic information. In addition, because the genetic code is degenerate, evolutionary comparisons based on DNA sequences will have to be made because they may well be more informative in some ways than are comparisons based on amino acid sequences (Yokoyama et al., 1990). Only when all this has been done shall we begin to understand how and why these different groups of alcohol dehydrogenases have arisen, catalysing basically the same reaction in apparently different ways.

REFERENCES

Benyajati, C., Place, A.R., Powes, D.A. and Sofer, W., 1981, Alcohol dehydrogenase of _Drosophila melanogaster_: Relationship of intervening sequences to functional domains in the protein, Proc. Natl. Acad. Sci. U.S.A. 78:2717.

Branden, C.-I., Jornvall, H., Eklund, H. and Furugren, B., 1975, Alcohol dehydrogenases, The Enzymes, Vol. Xl:103.

Bridgen, J., Kolb, E. and Harris, J.I., 1973, Amino acid sequence homology in alcohol dehydrogenases, F.E.B.S. Lett., 33:1.

Chalmers, R.M., Scott, A.J. and Fewson, C.A., 1990, Purification of the benzyl alcohol dehydrogenase and benzaldehyde dehydrogenase encoded by the TOL plasmid pWW53 of _Pseudomonas putida_ MT53 and their preliminary comparison with benzyl alcohol dehydrogenase and benzaldehyde dehydrogenases I and II from _Acinetobacter calcoaceticus_, J. Gen. Microbiol. 136:637.

Clark, D.P., 1989, The fermentation pathways of _Escherichia coli_, F.E.M.S. Microbiol. Rev., 63:223.

Conway, T., Sewell, G.W., Osman, Y.A. and Ingram, L.O., 1987, Cloning and sequencing of the alcohol dehydrogenase II gene from _Zymomonas mobilis_, J. Bacteriol. 169:2591.

Conway, T. and Ingram, L.O., 1989, Similarity of _Escherichia coli_ propanediol oxidoreductase (_fucO_ product) and an unusual alcohol dehydrogenase from _Zymomonas mobilis_ and _Saccharomyces cerevisiae_, J. Bacteriol., 171:3754.

Dayhoff, M.O., 1978, Survey of new data and computer methods of analysis, in: "Atlas of Protein Sequence and Structure" Vol. 5 supp. 3, M.O. Dayhoff, ed., pp.1-8, Nat. Biomed. Res. Found, Washington D.C.

Eklund, H., 1988, Coenzyme binding in alcohol dehydrogenase, Biochem. Soc. Trans., 17:293.

Glasfield, A. and Benner, S.A., 1989, The stereospecificity of the ferrous-ion-dependent alcohol dehydrogenase from _Zymomonas mobilis_, Eur. J. Biochem. 180:373.

Goodlove, P.E., Cunningham, P.R., Parker, J. and Clark, D.P., 1989, Cloning and sequence analysis of the fermentative alcohol-dehydrogenase-encoding gene of _Escherichia coli_, Gene, 85:209.

Harms, N., de Vries, G.E., Maurer, K., Hoogendijk, J. and Stouthamer, A.H., 1987, Isolation and nucleotide sequence of the methanol dehydrogenase structural gene from _Paracoccus denitrificans_, J. Bacteriol. 169: 3909.

Hartley, B.S., 1985, Experimental evolution of ribitol dehydrogenase, in: "Microorganisms as Model Systems for Studying Evolution", R.P. Mortlock, ed., pp.23-54, Plenum Press, New York.

I.U.B.-I.U.P.A.C., 1979, "Enzyme Nomenclature", Academic Press, New York.

Jeck, R., Woenckaus, C., Harris, J.I. and Ruswick, M.J., 1979, Identification of the amino acid residue modified in _Bacillus_ stearothermophilus alcohol dehydrogenase by the NAD analogue 4-(3-bromoacetylpyridino)butyldiphosphoadenosine, Eur. J. Biochem., 93:57.

Jeffrey, J. and Jornvall, H., 1988, Sorbitol dehydrogenase, Adv. Enzymol., 61:47.

Jendrossek, D., Steinbüchel, A. and Schlegel, H.G., 1988, Alcohol dehydrogenase gene from Alcaligenes eutrophus: Subcloning, heterologous expression in Escherichia coli, sequencing, and location of Tn5 insertions, J. Bacteriol., 170:5248.

Jornvall, H., Persson, M. and Jeffrey, J., 1981, Alcohol and polyol dehydrogenases are both divided into two protein types and structural properties cross-relate the different enzyme activities within each type, Proc. Natl. Acad. Sci. U.S.A., 78:4226.

Jornvall, H., von Bahr-Lindstrom, H., Jany, K-D., Ulmer, W. and Froschle, M., 1984, Extended superfamily of short alcohol-polyol dehydrogenases: structural similarities between glucose and ribitol dehydrogenases, F.E.B.S. Lett., 165:190.

Jornvall, H., 1986, Evolution of isozymes and different enzymes in a protein superfamily: alcohol dehydrogenases and related proteins, Chemica Scripta, 26B:231.

Jornvall, H., Persson, B. and Jeffrey, J., 1987a, Characteristics of alcohol/polyol dehydrogenases, Eur. J. Biochem., 167:195.

Jornvall, H., Hoog, J.-O., von Bahr-Lindstrom, H., Johansson, J., Kaiser, R. and Persson, B., 1987b, Alcohol dehydrogenases and aldehyde dehydrogenases, Biochem. Soc. Trans., 16:233.

Keshav, K.F., Yomano, L.P., An, H. and Ingram, L.O., 1990, Cloning of the Zymomonas mobilis structural gene encoding alcohol dehydrogenase 1 (adhA): Sequence comparison and expression in Escherichia coli, J.Bacteriol., 172:2491.

Machlin, S.M. and Hanson, R.S., 1988, Nucleotide sequence and transcriptional start site of the Methylobacterium organophilum XX methanol dehydrogenase structural gene, J. Bacteriol. 170: 4739.

MacKintosh, R.W. and Fewson, C.A., 1987, Microbial aromatic alcohol and aldehyde dehyrogenases, in: "Enzymology and Molecular Biology of Carbonyl Metabolism", H. Weiner & T.G. Flynn, eds, pp259-273, Alan R. Liss, New York.

Morris, H.R., Williams, D.H., Midwinter, G.G. and Hartley, B.S., 1974, A mass-spectrometric sequence study of the enzyme ribitol dehydrogenase from Klebsiella aerogenes, Biochem. J., 141:701.

Neale, A.D., Scopes, R.K., Kelly, J.M. and Wettenhall, R.E., 1986, The two alcohol dehydrogenases of Zymomonas mobilis, Eur. J. Biochem., 154:119.

Peretz, M. and Burstein, Y., 1989, Amino acid sequence of alcohol dehydrogenase from the thermophilic bacterium Thermoanaerobium brockii, Biochem., 28:6549.

Rossman, M.G., Liljas, A., Branden, C.-I. and Banasazak, L.J., 1975, Evolutionary and structural relationships among dehydrogenases, The Enzymes, Vol XI:61.

Schneider-Bernlöhr, H., Adolph, W. and Zeppezaur, M., 1986, Coenzyme stereospecificity of alcohol/polyol dehydrogenases: conservation of protein types vs. functional constraints, J. Am. Chem. Soc., 108:5573.

Scrutton, N.S., Berry, A. and Perham, R.N., 1990, Redesign of the coenzyme specificity of a dehydrogenase by protein engineering, Nature, 343:38.

Villaroya, A., Juan, E., Egestad, B. and Jornvall, H., 1989, The primary structure of alcohol dehydrogenase from Drosophila lebanonensis, Eur. J. Biochem., 180:191

Wierenga, R.K., De Maeyer, M.C.K. and Hol. W.G.J., 1985, Interactions of pyrophosphate moieties with α-helices in dinucleotide binding proteins, <u>Biochem.</u>, 24:1346.

Williamson, V.M. and Paquin, C.E., 1987, Homology of <u>Saccharomyces cerevisiae</u> ADH4 to an iron-activated alcohol dehydrogenase from <u>Zymomonas</u> mobilis, <u>Mol. Gen. Genet.</u>, 209:374.

Yokoyama, S., Yokoyama, R., Kinlaw, C.S. and Harry, D.E., 1990, Molecular evolution of the zinc-containing long-chain alcohol dehydrogenase genes, <u>Mol. Biol. Evol.</u>, 7:143.

Youngleson, J.S., Jones, W.A., Jones, D.T. and Woods, D.R., 1989, Molecular analysis and nucleotide sequence of the <u>adhl</u> gene encoding an NADPH-dependent butanol dehydrogenase in the Gram-positive anaerobe <u>Clostridium acetobutylicum</u>, <u>Gene</u>, 78:355.

CONTRIBUTORS

Abedinia, Mahin
Division of Science and Technology
Griffith University, Nathan
Brisbane, Qld 4111, Australia

Abriola, Darrly P.
Center of Alcohol Studies
Rutgers University
Piscataway, NJ 08855-0969

Agarwal, Dharam P.
Institute of Human Genetics
University of Hamburg
Hamburg, F.R.G.

Akera, Tai
Department of Pediatric Pharmacology
National Children's Medical Research
Center
3-35-31 Taishido
Setagaya-Ku, Tokyo 154, Japan

Algar, Elizabeth M.
Division of Science and Technology
Griffith University, Nathan
Brisbane, Qld 4111, Australia

Barth, Patrick
Laboratoire de Chimie Organique
Biologique
URA 31
Departement de Chime
Universite Louis Pasteur
1 rue Blaise Pascal
67008 Strasbourg (France)

Bassi, A.M.
University of Genoa
Genoa, Italy

Begueret, Joel
Department of Hepatogastroenterology
Hopital Haut-Leveque, 33604 Pessac
Department of Genetics and
Biochemistry
University Bordeaux II, 33076
Bordeaux France

Biellmann, Jean-Francois
Laboratoire de Chimie Organique
Biologique
URA 31
Departement de Chime
Universite Louis Pasteur
1 rue Blaise Pascal
67008 Strasbourg (France)

Biocca, M.E.
University of Turin
Department of Experimental Medicine
and Oncology
C.so Raffaello 30
10125 Torino, Italy

Blackwell, Leonard F.
Department of Chemistry and
Biochemistry
Massey University
Palmerston North, New Zealand

Blatter, Erich
Center of Alcohol Studies
Rutgers University
Piscataway, NJ 08855-0969

Bosron, William F.
Department of Biochemistry and
Molecular Biology
Indiana University School of Medicine
Indianapolis, IN 46202

Brown, Celeste J.
Department of Biochemistry and
Molecular Biology
Indiana University School of Medicine
Indianapolis, IN 46202-5122

Buckley, Paul D.
Department of Chemistry and
Biochemistry
Massey University
Palmerston North, New Zealand

Bunai, Yasuo
Department of Legal Medicine
Gifu University School of Medicine
Gifu 500, Japan

Canuto, R. A.
University of Turin
Department of Experimental Medicine
and Oncology
C.so Raffaello 30
10125 Torino, Italy

Carper, Deborah
National Eye Institute
NIH 9000 Rockville Pike
Bethesda, MD 20892

Carper, Deborah
National Eye Institute
National Institutes of Health
Bethesda, MD

Carr, Lucinda G.
Department of Biochemistry and
Molecular Biology
Indiana University School of Medicine
Indianapolis, IN 46202-5122

Cassaigne, Andre
Department of Hepatogastroenterology
Hopital Haut-Leveque, 33604 Pessac
Department of Genetics and
Biochemistry
University Bordeaux II, 33076
Bordeaux France

Chung, Stephen
Eugene Tech International
4 Pearl Court
Allendale, NJ 07401

Coutelle, Christiane
Department of Hepatogastroenterology
Hopital Haut-Leveque, 33604 Pessac
Department of Genetics and
Biochemistry
University Bordeaux II, 33076
Bordeaux France

Couzigou, Patrice
Department of Hepatogastroenterology
Hopital Haut-Leveque, 33604 Pessac
Department of Genetics and
Biochemistry
University Bordeaux II, 33076
Bordeaux France

Crabb, David W.
Indiana University School of Medicine
and VA Medical Center
Indianapolis, IN 46202

Cunningham, Suzanne J.
Department of Biochemistry
Purdue University
W. Lafayette, IN 47907

Dasgupta, Sarmila
National Eye Institute
National Institutes of Health
Bethesda, MD

Deyashiki, Yoshihiro
Department of Biochemistry
Gifu Pharmaceutical University
Gifu 502, Japan

Dipple, Katrina M,
Departments of Biochemistry and
Molecular Biology
Indiana University School of Medicine
Indianapolis, IN 46202

Dockham, Patcia A.
Department of Pharmacology
University of Minnesota
3-249 Millard Hall
435 Delaware Street S.E.
Minneapolis, MN 55455

Dorsselaer, Alain Van
Laboratoire de Chimie Organique des
Substances Naturelles
URA 31
Departement de Chimie
Universite Louis Pasteur
1 rue Blaise Pascal
67008 Strasbourg (France)

Esterbauer, H.
University of Graz
Graz, Austria

Eckey, Rolf
Institute of Human Genetics
University of Hamburg
Hamburg, F.R.G.

Edenberg, Howard J.
Department of Biochemistry and
Molecular Biology
Indiana University School of Medicine
Indianapolis, IN 46202-5122

Ehrig, Torsten
Department of Biochemistry and
Molecular Biology
Indiana University School of Medicine
Indianapolis, IN 46202

Ellison, Aaron
Division of Pediatrics
University of Washington
Seattle, WA 98195

Farres, Jaume
Department of Biochemistry
Purdue University
West Lafayette, IN 47907

Ferro, Margherita
Instituto Di Patologia Generale
Universita Di Genova
Via L. B. Alberti 2
16132 Genova, Italy

Fewson, Charles A.
Department of Biochemistry
University of Glasgow
Glasgow G12 8QQ, Scotland, U.K.

Fleury, Benoit
Department of Hepatogastroenterology
Hopital Haut-Leveque, 33604 Pessac
Department of Genetics and
Biochemistry
University Bordeaux II, 33076
Bordeaux France

Flynn, T. Geoffery
Department of Biochemistry
Queens University
Kingston, Canada

Ganzhorn, Axel J.
Department of Biochemistry
The University of Iowa
Iowa City, IA 52242

Ghenbot, Ghiorghis
Department of Biochemistry
Purdue University
W. Lafayette, IN 47907

Goedde, H. Werner
Institute of Human Genetics
University of Hamburg
Hamburg, F.R.G.

Gould, Robert M.
Department of Biochemistry
The University of Iowa
Iowa City, IA 52242

Green, David W.
Department of Biochemistry
The University of Iowa
Iowa City, IA 52242

Grimshaw, Charles E.
Department of Molecular and
Experimental Medicine
Research Institute of Scripps Clinic
La Jolla, CA 92037

Groppi, Alexis
Department of Hepatogastroenterology
Hopital Haut-Leveque, 33604 Pessac
Department of Genetics and
Biochemistry
University Bordeaux II, 33076
Bordeaux France

Grubmeyer, Charles
Department of Biology
New York University
100 Washington Square
New York, NY 10003

Hara, Akira
Department of Biochemistry
Gifu Pharmaceutical University
Gifu 502, Japan

Hempel, John
Department of Microbiology
Biochemistry and Molecular Biology
University of Pittsburg Medical School
Pittsburg, PA 15261

Hikita, Harumi
Laboratory of Physics
Meikai University, Urayasu
Chiba, 279, Japan

Hill, Jeremy P.
Department of Chemistry and
Biochemistry
Massey University
Palmerston North, New Zealand

Hjelle, J.J.
School of Pharmacy
Alcohol and Hepatobiliary Research
Centers
University of Colorado Health
Sciences Center
Denver, CO

Ho, Wei-Hsien
Department of Biochemistry and
Molecular Biology
Indiana University School of Medicine
Indianapolis, IN 46202-5122

Hohman, Thomas C.
Wyeth-Ayerst Laboratory
CN 8000
Princeton, NJ 08540

Holmes, Roger S.
Division of Science and Technology
Griffith University, Nathan
Brisbane, Qld 4111, Australia

Hoog, Jan-Olov
Department of Chemistry I
Karolinska Institutet
S-104 01 Stockholm, Sweden

Hosomi, Saburo
Laboratory of Biochemistry
Faculty of Pharmaceutical Sciences
Osaka University
Suita, Osaka 565, Japan

Hur, Man-Wook
Department of Biochemistry and
Molecular Biology
Indiana University School of Medicine
Indianapolis, IN 46202-5122

Hurley, Thomas D.
Department of Biochemistry and
Molecular Biology
Indiana University School of Medicine
Indianapolis, IN 46202

Iron, Albert
Department of Hepatogastroenterology
Hopital Haut-Leveque, 33604 Pessac
Department of Genetics and
Biochemistry
University Bordeaux II, 33076
Bordeaux France

Itabe, Hiroyuki
National Eye Institute
National Institutes of Health
Bethesda, MD

Jacobi, Tobias
Department of Biochemistry
The University of Iowa
Iowa City, IA 52242

Kador, Peter F.
National Eye Institute
National Institutes of Health
Bethesda, MD

Kanazu, Takushi
Department of Biochemistry
Gifu Pharmaceutical University
Gifu 502, Japan

Kaneko, Masayuki
National Eye Institute
National Institutes of Health
Bethesda, MD

Katoh, Setsuko
Department of Biochemistry
Meikai University
School of Dentistry, Sakado,
Saitama 350-02, Japan

Koivusalo, Martti
Department of Medical Chemistry
University of Helsinki
SF-00170 Helsinki, Finland

Kou, Ingrid
Department of Molecular Genetics and
Biochemistry
University of Pittsburgh School of
Medicine
Pittsburgh, PA 15260

Kratzer, Darla Ann
Department of Biochemistry
The University of Iowa
Iowa City, IA 52242

Lee, Mi-Ock
Department of Pharmacology
University of Minnesota
3-249 Millard Hall
435 Delaware Street S.E.
Minneapolis, MN 55455

Li, Hong
Department of Chemistry
University of Washington
Seattle, WA 98195

Li, Ting-Kai
Indiana University School of Medicine
and VA Medical Center
Indianapolis, IN 46202

Lindahl, Ronald
Department of Biochemistry and
Molecular Biology
The University of South Dakota
School of Medicine
Vermillion, SD 57069

Little, Sally A.
Division of Pediatrics
University of Washington
Seattle, WA 98195

Lumeng, Lawrence
Indiana University School of Medicine
and VA Medical Center
Indianapolis, IN 46202

Maret, Wolfgang
Center for Biochemical and
Biophysical Sciences and Medicine
Harvard Medical School
Brigham and Women's Hospital
75 Francis Street
Boston, MA 02115

Matsuura, Kazuya
Department of Legal Medicine
Gifu University School of Medicine
Gifu 500, Japan

Matsuura, Yoshiharu
Department of Veterinary Science
National Institute of Health
Tokyo, Japan

Mellencamp, Becky
Indiana University School of Medicine
and VA Medical Center
Indianapolis, IN 46202

Mirkes, Philip E.
Division of Pediatrics
University of Washington
Seattle, WA 98195

Mitchell, D.Y.
School of Pharmacy
Alcohol and Hepatobiliary Research
Centers
University of Colorado Health
Sciences Center
Denver, CO

Mizoguchi, Tadashi
Laboratory of Biochemistry
Faculty of Pharmaceutical Sciences
Osaka University
Suita, Osaka 565, Japan

Moras, Dino
Laboratoire de Cristallographie
Biologique
I.B.M.C., C.N.R.S.
15 rue Rene Descartes
67084 Strasbourg (France)

Motion, Rosemary L.
New Zealand Dairy Research Institute
Palmerston North, New Zealand

Muzio, Giuliana
University of Turin
Department of Experimental Medicine
and Oncology
C.so Raffaello 30
10125 Torino, Italy

Nakayama, Toshihiro
Department of Biochemistry
Gifu Pharmaceutical University
Gifu 502, Japan

Nanjo, Hirofumi
Laboratory of Biochemistry
Faculty of Pharmaceutical Sciences
Osaka University
Suita, Osaka 565, Japan

Nishihara, Tsutomu
Laboratory of Biochemistry
Faculty of Pharmaceutical Sciences
Osaka University
Suita, Osaka 565, Japan

Nishimura, Chihiro
Department of Pediatric Pharmacology
National Children's Medical Research
Center
3-35-31 Taishido
Setagaya-Ku, Tokyo 154, Japan

Nishinaka, Tohru
Laboratory of Biochemistry
Faculty of Pharmaceutical Sciences
Osaka University
Suita, Osaka 565, Japan

Ohya, Isao
Department of Legal Medicine
Gifu University School of Medicine
Gifu 500, Japan

Old, Susan
National Eye Institute
NIH 9000 Rockville Pike
Bethesda, MD 20892

Page, Joe D.
Department of Chemistry
University of Washington
Seattle, WA 98195

Petersen, Dennis R.
School of Pharmacy
Alcohol and Hepatobiliary Research
Centers
University of Colorado Health
Sciences Center
Denver, CO 80262

Pietruszko, Regina
Center of Alcohol Studies
Rutgers University
Piscataway, NJ 08855-0969

Plapp, Bryce V.
Department of Biochemistry
The University of Iowa
Iowa City, IA 52242

Pocker, Yeshayau
Department of Chemistry
University of Washington
Seattle, WA 98195

Podjarny, Alberto
Laboratoire de Cristallographie
Biologique
I.B.M.C., C.N.R.S.
15 rue Rene Descartes
67084 Strasbourg (France)

Poli, G.
Universita di Totino
Dipartmento di Medicina ed Oncologia
Corso Raffaello 30-10125 Torino,
Italy

Prestwich, Glenn
Chemistry Department
SUNY
Stony Brook, NY 11794

Qulali, Mona
Departments of Biochemistry and
Molecular Biology
Indiana University School of Medicine
Indianapolis, IN 46202

Reymann, Jean-Marc
Laboratoire de Chimie Organique
Biologique
URA 31
Departement de Chime
Universite Louis Pasteur
1 rue Blaise Pascal
67008 Strasbourg (France)

Rondeau, Jean-Michel
Biostructure
8 rue Gustave Hirn
67000 Strasbourg (France)

Rose, John P.
Departments of Crystallography and
Biological Sciences
University of Pittsburgh
Pittsburgh, PA 15260

Sato, Sanai
National Eye Institute
National Institutes of Health
Bethesda, MD 20892

Sawada, Hideo
Department of Biochemistry
Gifu Pharmaceutical University
Gifu 502, Japan

Shinagawa, Kazuhiko
Laboratory of Biochemistry
Faculty of Pharmaceutical Sciences
Osaka University
Suita, Osaka 565, Japan

Shinoda, Michio
Gihoku General Hospital
Gifu 501-21, Japan

Sladek, Norman E.
Department of Pharmacology
University of Minnesota
3-249 Millard Hall
435 Delaware Street S.E.
Minneapolis, MN 55455

Sueoka, Terumi
Department of Biochemistry
Meikai University
School of Dentistry, Sakado
Saitama 350-02, Japan

Tanimoto, Tsuyoshi
Division of Biological Chemistry
National Institute of Hygienic Science
Tokyo, Japan

Terada, Tomoyuki
Laboratory of Biochemistry
Faculty of Pharmaceutical Sciences
Osaka University
Suita, Osaka 565, Japan

Tete, Frederique
Laboratoire de Cristallographie
Biologique
I.B.M.C., C.N.R.S.
15 rue Rene Descartes
67084 Strasbourg (France)

Timmann, Rudiger
Institute of Human Genetics
University of Hamburg
Hamburg, F.R.G.

Umemura, Toshifumi
Laboratory of Biochemistry
Faculty of Pharmaceutical Sciences
Osaka University
Suita, Osaka 565, Japan

Uotila, Lasse
Department of Medical Chemistry
University of Helsinki
SF-00170 Helsinki, Finland

VandeBerg, John L.
Department of Genetics
Southwest Foundation for Biomedical
Research
P.O. Box 28147
San Antonio, TX 78284

Wales, Martin R.
Department of Biochemistry
University of Glasgow
Glasgow G12 8QQ, Scotland, U.K.

Wang, Bi-Cheng
Departments of Crystallography and
Biological Sciences
University of Pittsburgh
Pittsburgh, PA 15260

Wang, Thomas T.T.Y.
Department of Biochemistry
Purdue University
West Lafayette, IN 47907

Warth, Edda
Department of Biochemistry
The University of Iowa
Iowa City, IA 52242

Weiner, Henry
Department of Biochemistry
Purdue University
West Lafayette, IN 47907

Wermuth, Bendicht
Chemisches Zentrallabor
Inselspital
CH 3010 Bern

Yamaoka, Takashi
Department of Pediatric Pharmacology
National Children's Medical Research
Center
3-35-31 Taishido
Setagaya-Ku, Tokyo 154, Japan

Zheng, Chao-Feng
Department of Biochemistry
Purdue University
W. Lafayette, IN 47907

INDEX